THE SELF-ASSEMBLING BRAIN

The Self-Assembling Brain

HOW NEURAL NETWORKS GROW SMARTER

PETER ROBIN HIESINGER

PRINCETON UNIVERSITY PRESS
PRINCETON & OXFORD

Published by Princeton University Press
41 William Street, Princeton, New Jersey 08540
99 Banbury Road, Oxford OX2 6JX

press.princeton.edu

First paperback printing, 2022
Paperback ISBN 9780691241692

The Library of Congress has cataloged the cloth edition as follows:

Names: Hiesinger, Peter Robin, author.
Title: The self-assembling brain : how neural networks grow smarter / Peter Robin Hiesinger.
Description: Princeton : Princeton University Press, [2021] | Includes bibliographical references and index.
Identifiers: LCCN 2020052826 (print) | LCCN 2020052827 (ebook) | ISBN 9780691181226 (hardback) | ISBN 9780691215518 (pdf)
Subjects: LCSH: Neural networks (Computer science) | Neural circuitry—Adaptation. | Learning—Psysiological aspects. | Artificial intelligence.
Classification: LCC QA76.87 .H53 2021 (print) | LCC QA76.87 (ebook) | DDC 006.3/2—dc23
LC record available at https://lccn.loc.gov/2020052826
LC ebook record available at https://lccn.loc.gov/2020052827

British Library Cataloging-in-Publication Data is available

Editorial: Ingrid Gnerlich, María Garcia, and Whitney Rauenhorst
Production Editorial: Karen Carter
Jacket/Cover Design: Layla Mac Rory
Production: Jacqueline Poirier
Publicity: Sara Henning-Stout and Kate Farquhar-Thomson
Copyeditor: Erin Hartshorn

Jacket/Cover Credit: Neural stem cells, fluorescence light micrograph. Photo by Daniel Schroen, Cell Applications, Inc. / Science Photo Library

This book has been composed in Arno

For Nevine and Nessim,

and in memory of Gabriele Hiesinger

CONTENTS

Acknowledgments xi

Prologue xiii

Introduction 1

The Perspective of Neurobiological Information 4

The Perspective of Algorithmic Information 5

A Shared Perspective 7

The Ten Seminars 11

On Common Ground 24

The Present and the Past 26

The First Discussion: On Communication 26

The Historical Seminar: The Deeply Engrained Worship of Tidy-Looking Dichotomies 36

1 **ALGORITHMIC GROWTH** 81

1.1 Information? What Information? 83

The Second Discussion: On Complexity 83

Seminar 2: From Algorithmic Growth to Endpoint Information 91

1.2 Noise and Relevant Information 112

The Third Discussion: On Apple Trees and the Immune System 112

Seminar 3: From Randomness to Precision 120

1.3 Autonomous Agents and Local Rules 139
The Fourth Discussion: On Filopodia and Soccer Games 139
Seminar 4: From Local Rules to Robustness 146

2 OF PLAYERS AND RULES 161

2.1 The Benzer Paradox 163
The Fifth Discussion: On the Genetic Encoding of Behavior 163
Seminar 5: From Molecular Mechanisms to Evolutionary Programming 170

2.2 The Molecules That Could 186
The Sixth Discussion: On Guidance Cues and Target Recognition 186
Seminar 6: From Chemoaffinity to the Virtues of Permissiveness 192

2.3 The Levels Problem 211
The Seventh Discussion: On Context 211
Seminar 7: From Genes to Cells to Circuits 217

3 BRAIN DEVELOPMENT AND ARTIFICIAL INTELLIGENCE 237

3.1 You Are Your History 239
The Eighth Discussion: On Development and the Long Reach of the Past 239
Seminar 8: From Development to Function 245

3.2 Self-Assembly versus "Build First, Train Later" 262
The Ninth Discussion: On the Growth of Artificial Neural Networks 262
Seminar 9: From Algorithmic Growth to Artificial Intelligence 267

3.3 Final Frontiers: Beloved Beliefs and the AI-Brain Interface 287

The Tenth Discussion: On Connecting the Brain and AI 287

Seminar 10: From Cognitive Bias to Whole Brain Emulation 294

Epilogue 312

Glossary 317

References 329

Index 351

ACKNOWLEDGMENTS

THIS BOOK is the result of many discussions. I am very grateful to all the colleagues, students and friends who endured and challenged me in these discussions over the years. I have experimented with the introduction of the key concepts in basic lectures on neurodevelopment at my home university as well as in seminars around the world. The ensuing debates have influenced my thinking and writing a great deal—so much so, that nothing in this book can possibly be entirely my own.

For reading and commenting on various parts, I am particularly grateful to Bassem Hassan, Nevine Shalaby, Randolf Menzel, Michael Buszczak, Grace Solomonoff, Dietmar Schmucker, Gerit Linneweber, Adrian Rothenfluh, Sansar Sharma, Iris Salecker, Axel Borst, Ian Meinertzhagen, Arend Hintze, Roland Fleming, Tony Hope, Uwe Drescher, QL Lim, Stuart Newman, and many students in my classes and my lab over the years.

For historic information and picture material I thank Sansar Sharma, Margaret and Gloria Minsky, Cynthia Solomon, Grace Solomonoff, Robinetta Gaze and Tony Hope, the Walt Girdner family, the Estate of W. Ross Ashby, Jessica Hill from Wolfram Companies, Nikki Stephens from The Perot Museum Dallas, Nozomi Miike from the Miraikan in Tokyo, as well as the librarians from Cold Spring Harbor Image Archives, Caltech Archives, Cornell Library and the MIT Museum.

I thank Princeton University Press, and especially my editor Ingrid Gnerlich, for expert guidance and support throughout the last three years.

PROLOGUE

ONCE THERE was an alien who found an apple seed from earth. "*This looks funny,*" it said, "*I wonder what it does?*" The alien was very smart. In fact, it was the smartest alien there ever was, and it scanned the apple seed with the most sophisticated scanning machine that ever existed or that would ever exist. The scan worked well, and the alien got all the information there was in the seed. The alien immediately saw and understood the way the molecules were arranged, every single one of them, the genetic code, the entire information content of the DNA sequence. It was all there, and it was beautiful. There were patterns in the sequence, for which the computer provided a complete analysis. The computer calculated the most advanced mathematics and showed that the code was profound, it had meaning. But what did it do? And so the alien thought: "*Now that I have all the information that could ever be extracted from the seed, I want to find out what information it can unfold in time.*" And the alien took the seed and let it grow. The alien watched the tree's growth with surprise and curiosity, and it saw the branches, the leaves and finally the apples. And the alien took an apple and said "*I would never have guessed it would look like that.*" And then it took a bite.

THE SELF-ASSEMBLING BRAIN

Introduction

THERE ARE EASIER THINGS to make than a brain. Driven by the promise and resources of biomedical research, developmental neurobiologists are trying to understand how it is done. Driven by the promise and advances of computer technology, researchers in artificial intelligence (AI) are trying to create one. Both are fields of contemporary research in search of the principles that can generate an intelligent system, a thing that can predict and decide, and maybe understand or feel something. In both developmental neurobiology and AI based on artificial neural networks (ANNs), scientists study how such abilities are encoded in networks of interconnected components. The components are nerve cells, or neurons, in biological brains. In AI, the term *neuron* has been readily adopted to describe interconnected signaling components, looking back on some 70 years of ANN research. Yet, to what extent the biological analogy is useful for AI research has been a matter of debate throughout the decades. It is a question of how much biological detail is relevant and needed, a question of the type of information necessary to make a functional network. The information problem underlies both fields. What type of information is necessary to wire a brain? What do biologists mean when they say something is "encoded by genes," and how is genetic information transformed into a brain? And finally, to what extent is the same type of information required to wire up biological brains or to create artificial intelligence?

This book is about the information problem and how information unfolds to generate functional neural networks. In the case of biological brains, prior to learning, the information for developmental growth is encoded in the genome. Yet, there are no chapters about brain regions or their connectivity to read in the genome. In fact, compared to the information necessary to describe every detail necessary to make a functioning brain, there is rather little information available in the genome. Growth requires genetic information plus time and energy. Development happens in steps that occur in space and time in an ordered fashion. The outcome is a system that would require more information to describe than was needed to start its growth. By contrast, most ANNs do not grow. Typically, an artificial network with initially random connections learns from data input in a process that is reminiscent of how biological brains learn. This process also requires time and energy. Learning also occurs in steps, and the order of these steps matters. There are important similarities and differences between these stepwise, time- and energy-consuming processes. The current hope for AI based on ANNs is that the learning process is sufficient and that a developmental process analogous to biological brains can therefore be omitted. Remarkably, there was a time in neurobiology research almost a hundred years ago when scientists felt much the same about the brain itself. It was inconceivable where the information for wiring should come from other than through learning. The idea was that, just like ANNs today, the brain must initially be wired rather randomly, and subsequent learning makes use of its plasticity.[1] But if this were so, how could, say, a monarch butterfly be born with the ability to follow thousands of miles of a migration route that it has never seen before?

As temperatures drop in the fall in North America, millions of monarch butterflies migrate for up to 3,000 miles to overwinter in Mexico. Remarkably, millions of butterflies distributed over close to 3 million square miles in the north all target only a few overwintering sites that cover less than a single square mile. Many theories have been put forth as to how a butterfly could do this.[2,3] Similarly remarkable, an individual sea turtle will return over thousands of miles to the very beach where it was born—many years later. We do not know how sea turtles do it, but it is conceivable that they had learned and can remember something about a place where they had once been before. This is where the story of the monarch butterfly turns from remarkable to downright unbelievable. The butterflies that started out in the north will overwinter in the south until temperatures rise next spring. They then start flying north again, but only a few hundred miles. At different places in the southern United States they stop, mate, lay eggs and die. A new generation of monarchs picks up the trail north, but again only for a few hundred miles. It usually takes 3–5 generations for a full round trip.[2] By the time temperatures drop again in the fall in North America, a monarch butterfly is about to embark on the 3,000-mile trip south to a precise location that was last visited by its great-great-grandfather. Where is this information coming from?

The currently almost exclusive focus of AI on ANNs is a highly successful, but recent development. It followed several decades during which AI and machine learning focused on formal, symbol-processing logic approaches, rather than the somewhat enigmatic neural networks. For most of its history, AI researchers tried to avoid the complexities and messiness of biological systems altogether.[4, 5] How does information about the role of a gene for a neuronal membrane protein help to program an intelligent system? The history of AI is a history of trying to avoid unnecessary biological detail in trying to create something that so far only exists in biology. The observation begs the question what information can safely be deemed "unnecessary." To address this question, we need to look at biological and artificial brain development from the information perspective. An assumption and hope of AI research has long been that there is a shortcut to creating intelligent systems. We may not yet know what shortcuts work best, but it seems a good idea to at least know exactly what it is we are trying to leave out in attempts to create nonbiological brains. My hope is that an understanding of the way information is encoded and transformed during the making of biological brains proves useful in the discussion what can and cannot be shortcut in the making of AI. This is the story of a neurobiologist tracking down that information.

The Perspective of Neurobiological Information

The biological brain is a complicated network of connections, wired to make intelligent predictions. Common analogies for brain wiring include circuit diagrams of modern microprocessors, the electrical wiring installations in skyscrapers or the logistics of transportation networks in big cities. How are such connections made during brain development? You can imagine yourself trying to make a connection by navigating the intricate network of city streets. Except, you won't get far, at least not if you are trying to understand brain development. There is a problem with that picture, and it is this: Where do the streets come from? Most connections in the brain are not made by navigating existing streets, but by navigating streets under construction. For the picture to make sense, you would have to navigate at the time the city is growing, adding street by street, removing and modifying old ones in the process, all the while traffic is a part of city life. The map changes just as you are changing your position in it, and you will only ever arrive if the map changes in interaction with your own movements in it. The development of brain wiring is a story of self-assembly, not a global positioning system (GPS).

When engineers design the electrical wiring in a building or a computer microchip, they have the final product in mind. We make blueprints to understand and build engineered systems with precise outcomes. A blueprint shows a picture of the final product, the endpoint. A blueprint also contains all the information needed to build that product. It largely doesn't matter in what order the pieces are put in, as long as everything is in place when you flip the on switch. But there is no blueprint for brain connectivity in the genes. There is also no such information coming from the environment. If neither the genes nor the environment contain endpoint information of connectivity, what kind of information do they contribute?

Genetic information allows brains to grow. Development progresses in time and requires energy. Step by step, the developing brain finds itself in changing configurations. Each configuration serves as a new basis for the next step in the growth process. At each step, bits of the genome are activated to produce gene products that themselves change what parts of the genome will be activated next—a continuous feedback process between the genome and its products. A specific step may not have been possible before and may not be possible ever again. As growth continues, step by step, new states of organization are reached. Rather than dealing with endpoint information, the information to build the brain unfolds with time. Remarkably, there may be no other way to read the genetic information than to run the program. This is not a trivial statement to make, and it will take some explaining. If there is no way to read the genetic code other than running it, then we are principally unable to predict exact outcomes with any analytical method of the code. We can simulate it all right, but the result would not have been predictable in any way other than actually running the whole simulation. The information is in the genes, but it cannot be read like a blueprint. It really is a very different type of information that requires time and energy to unfold.

The Perspective of Algorithmic Information

Scientists in nonbiological fields are more familiar with this type of information. There is a simple game, where you draw lines of X's or O's (or black dots versus blanks) based on simple rules that produce remarkable patterns. Imagine a single X in a row of an infinite number of O's and a simple rule that determines for each triplet of X's and O's whether there is an X or an O in the next row. To find out the next line, you read the first three characters, write the output X or O underneath the center of the triplet below, then move one character and do it again for the next partially overlapping triplet. One rule, called rule 110, looks innocently enough like this:[6]

Triplet in previous row:	XXX	XXO	XOX	XOO	OXX	OXO	OOX	OOO
..determines in next row:	O	X	X	O	X	X	X	O

For example, starting with one X: .. OOOOOXOOOOO..
will lead to the next row: .. OOOOXXOOOOO..

Repeating this process again and again, using each previous line to apply the rule and write the next one below, will create a two-dimensional pattern (you will find the result in figure 2.3 on page 96). The repeated application of defined rules is an iteration. A ruleset that uses the output of each preceding step as the input of the next step defines an algorithm. The two-dimensional pattern is the outcome of algorithmic growth based on the iterative application of simple rules. But what does this game have to do with brain development? Shockingly, for the simple rule shown above, the two-dimensional pattern turns out to be so surprisingly complicated that it was proven to contain, at some point of its pattern growth process, any conceivable computation. Mathematicians call this a universal Turing machine or "Turing-complete." This is not an intuitive concept. The information content of the underlying code is absurdly low, yet it can produce infinite complexity. What is more, there is no analytical method to tell you the pattern at iteration 1,000. If you want to know, you must play the game for 1,000 rounds, writing line by line. These systems are called cellular automata and are a beloved model for a branch of mathematics and the research field of Artificial Life (ALife). Some ALifers consider AI a subfield. Many AI researcher don't care much about ALife. And neither of them care much about developmental neurobiology.

In information theory, the cellular automaton described above highlights an important alternative to describing complete endpoint information. Instead of a precise description of every detail of the pattern after 1,000 iterations, a complete description of the system is also possible by providing the few simple rules plus the instruction "apply these rules 1,000 times." The information required to generate the complete system is also known as Kolmogorov complexity in algorithmic information theory. Data compression algorithms do exactly that. An image of a uniformly blue sky is easily compressed, because its algorithmic information content is low (paint the next 10,000 pixels blue). By contrast, a picture cannot be easily compressed if every pixel has a randomly different color and no repeating patterns. In the case of the cellular automaton, Kolmogorov complexity is very low, while endpoint information required to describe the system becomes infinite with infinite iterations. The algorithmic information content required to create the system are a few instructions plus time and energy, while the endpoint information content is enormous in the case of many iterations.

The rule 110 cellular automaton provides us with a simple example of an algorithmic growth process that can generate more information based on simple rules, and yet its output can only be determined by letting it grow. "More" information is defined here as the information needed to describe the output if there were no growth process. However, in contrast to biological systems, rule 110 can only produce one fixed outcome with every iteration based on a set of rules that never change. For these reasons alone, rule 110 cannot be a sufficient model for biological systems. Yet, rule 110 teaches us that unpredictable unfolding of information is possible even with very simple rules in a deterministic system. For rule 110 there is a proof, the proof of Turing universality. For biological growth based on the genetic code, we face many more challenges: The rules are more complicated and change with every iteration of the running algorithm, and stochastic processes are central to its run. If a simple system like rule 110 can already be unpredictable, then we should not be surprised if algorithmic growth of biological systems turns out to be unpredictable. However, the proof for biological systems seems currently out of reach. The idea that information unfolding based on genomic information cannot be mathematically calculated, but instead requires algorithmic growth or a full simulation thereof, is a core hypothesis of this book.

A Shared Perspective

Biologists like to talk about the genes that contain a certain amount of information to develop the brain, including its connectivity. But in order to appreciate the information content of genes, we must understand the differences and consequences of information encoding for a self-assembling system versus a connectivity map. The genetic code contains algorithmic information to develop the brain, not information that describes the brain. It can be misleading to search for endpoint information in the genes or the mechanisms of the proteins they encode.

Address codes, navigational cues and key-and-lock mechanisms all follow such a rationale and make intuitive sense. And they all exist as molecular mechanisms, in brain wiring as elsewhere in biology. But they are part of unfolding algorithmic information, not endpoint information of brain connectivity. As the brain grows, different genes are turned on and off in a beautiful ballet in space and time, endowing each individual neuron with a myriad of properties that play out and change in communication with its neighbors. The neuron navigates as the city map grows and changes in interaction with the neuron's own movement in it.

The study of genes in developmental neurobiology is a success story from at least two perspectives. First, in the quest for molecular mechanisms. What a gene product does at any point in time and space during brain development tells us something about a part of the growth program that is currently executed. But information about a specific molecular mechanism may only be a tiny part of the information that unfolds in the wake of a random mutation in the genome. A mutation can lead to more aggressive behavior of the animal. And yet, the mutation may well affect some metabolic enzyme that is expressed in every cell of the body. The molecular function of the gene product may tell us nothing about animal behavior. How the molecular mechanism of this gene is connected to the higher order behavior may only be understood in the context of the brain's self-assembly, its algorithmic growth.

Many mutations have been found that change predispositions for behavioral traits, yet there may be only very few cases that we could reasonably call "a gene for a trait." Most gene products contribute to develop the trait in the context of many other gene products, but do not contain information about the trait itself. A mutation, selected by evolution for behavioral changes, must change either brain development or function. If the effect is developmental, then we have to face the information problem: There may be no way to know what the altered code produces other than running the entire process in time (or simulating it on a computer). There may be no shortcut. This is the problem with the street navigation analogy: You have to navigate a changing map on a path that only works if the map changes just as you are navigating it.

The full route on the map never existed, neither at the beginning nor at the end of your trip, but instead the route was made in interaction with your own actions. This is the essence of self-assembly.

We can study self-assembly either as it happens in biology or by trying to make a self-assembling system from scratch. As of 2020, biological neural networks (i.e., brains) are still unparalleled in their intelligence. But AI is on it. And yet, self-assembly is not a major focus of AI. For many years, AI focused on formal symbol-processing logic, including enormous expert systems built on decision-making trees. As recently as the early 2000s, the victory of formal, logical symbol-processing AI was declared. Since then, just when some thought we were done with neural networks, a revolution has taken place in AI research. In the few years since 2012, practically every AI system used to predict what friends or products we allegedly want has been replaced with neural networks. "Deep learning" is the name of the game in AI today.

The ANNs we use as tools today are not grown by a genetic code to achieve their initial architecture. Instead, the initial network architecture is typically randomly connected and thus contains little or no information. Information is brought into an ANN by feeding it large amounts of data based on a few relatively simple learning rules. And yet, there is a parallel to algorithmic growth: The learning process is an iterative process that requires time and energy. Every new bit of data changes the network. And the order matters, as the output of a preceding learning step becomes the input of the next. Is this a self-assembly process? Do we ultimately need algorithmic growth or self-assembly to understand and create intelligence? One obvious problem with the question is that the definition of intelligence is unclear. But the possible role of self-assembly may need some explaining, too.

In the search for answers, I went to two highly respected conferences in late summer 2018, an Artificial Life conference themed "Beyond Artificial Intelligence" by the International Society for Artificial Life and the Cold Spring Harbor meeting "Molecular Mechanisms of Neuronal Connectivity." I knew that these are two very different fields in many respects. However, my reasoning was that the artificial life and artificial intelligence communities are trying to figure out how to make something that has an existing template in biological systems. Intelligent neural networks *do* exist; I have seen them grow under a microscope. Surely, it must be interesting to AI researchers to see what their neurobiology colleagues are currently figuring out—shouldn't it help to learn from the existing thing? Surely, the neurobiologists should be equally interested in seeing what AI researchers have come up with, if just to see what parts of the self-assembly process their genes and molecules are functioning in.

Alas, there was no overlap in attendance or topics. The differences in culture, language and approaches are remarkable. The neurobiological conference was all about the mechanisms that explain bits of brains as we see them, snapshots of the precision of development. A top-down and reverse engineering approach to glimpse the rules of life. By contrast, the ALifers were happy to run simulations that create anything that looked lifelike: swarming behavior, a simple process resembling some aspect of cognition or a complicated representation in an evolved system. They pursue a bottom-up approach to investigate what kind of code can give rise to life. What would it take to learn from each other? Have developmental biologists really learned nothing to inform artificial neural network design? Have Alifers and AI researchers really found nothing to help biologists understand what they are looking at? I wanted to do an experiment in which we try to learn from each other; an experiment that, if good for nothing else, would at least help to understand what it is that we are happy to ignore.

So I assembled a seminar series, a workshop, about the common ground of both fields. The seminars are presented from the perspective of a neurobiologist who wants to know how our findings on brain development relate to the development of ANNs and the ultimate goal of artificial general intelligence. Many neurobiologists feel that ANNs are nothing like the biological template, and many AI scientists feel that their networks should not try to resemble biology more than they currently do. The seminars are therefore presented with a broad target audience in mind: there is so little common ground that it is easily shared with any basic science-educated layperson. The average neurobiologist is a layperson when it comes to AI, and most ANN developers are laypeople when it comes to neurobiology. Developmental neurobiologists may feel they are not missing anything by not following the bottom-up approach of AI, and ANN developers may feel they are safe to ignore biological detail. But to decide what is not needed, it helps to at least know what it is we are choosing to not know.

One of the best outcomes of good seminars are good discussions. And here I didn't need to search long. Going to conferences with these ideas in mind has provided me for years with experiences for how and where such discussions can go. I started writing this book with these discussions in mind. Initially, I only used them as a guide to pertinent questions and to identify problems worth discussing. As I kept on going back to my own discussions and tried to distill their meaning in writing, it turned out all too easy to lose their natural flow of logic and the associations that come with different perspectives. So I decided to present the discussions themselves. And as any discussion is only as good as the discussants, I invented four entirely fictional scientists to do all the hard work and present all the difficult problems in ten dialogs. The participants are a developmental geneticist, a neuroscientist, a robotics engineer and an AI researcher. I think they are all equally smart, and I do hope you'll like them all equally well.

The Ten Seminars

The seminars of the series build on each other, step by step. Preceding each seminar is a discussion of the four scientists who exchange questions and viewpoints in anticipation of the next seminar. The series starts with **The Historical Seminar: The Deeply Engrained Worship of Tidy-Looking Dichotomies**, a rather unusual seminar on the history of the field. The "field" being really two fields, developmental neurobiology and AI research, this seminar provides an unusual and selective historical perspective. Their shared history puts each other's individual stories in the spotlight of shared questions and troubles. Both struggle with remarkably similar tension fields between seemingly opposing approaches and perceptions. There are those who feel that the approaches, hypotheses and analyses must be rigorously defined for any outcome to be meaningful. Then there are those who feel that, like evolution, random manipulations are okay as long as one can select the ones that work—even if that means giving up some control over hypotheses, techniques or analyses.

Both fields begin their shared history by independently asking similar questions about information. The discovery of individual nerve cells itself was a subject of divisive contention. Even before scientists were sure that separable neurons exist, concerns were already raised about the information necessary to put them all together in a meaningful network. Much easier to envision the network as a randomly preconnected entity. And when early AI researchers built their very first networks with a random architecture, they did so because they felt it had to be like that in nature—where should the information have come from to specifically connect all neurons? A randomly connected network contains little or no information; the network has to grow smart through learning. In biology, the dominance of this view was challenged already in the 1940s by studies that focused on the precision and rigidity of connectivity that is not learned. This work marked a turning point that led neurobiologists to ask questions about how network information can develop based on genetic information. By contrast, today's artificial neural networks used in typical AI applications still only grow smart by learning; there is no genetic information. Yet, years in both fields played out in similar tension fields between precision and flexibility, between rigidity and plasticity. The fields may not have talked much to each other, but they mirrored each other's troubles.

The historical background forms the basis for three sessions. The first session explores the types of information that underlie biological and artificial neural networks. The second session builds on the information-theoretical basis to discuss the approaches taken by biologists to understand how genetic information leads to network information—the missing element in most ANNs. The third session connects algorithmic growth to learning and its relevance for AI.

Each session consists of three seminars. The first session starts with **Seminar 2: From Algorithmic Growth to Endpoint Information**, which deals with the difference between information required to make a system and information required to describe a system. Genes contain information to develop neuronal connectivity in brains; they don't contain information that describes neuronal connectivity in brains. We are facing one of the hardest problems right from the start, mostly because human intelligence lacks intuition for this kind of information. The core concept is algorithmic growth. A set of simple rules is sufficient to create mindboggling complexity. But what is complexity? The journey to understand information encoding is intricately linked to this question. If a cellular automaton based on a very simple rule set can produce a Turing-complete system, including unlimited complexity of patterns, where is the information coming from? The algorithmic information content of the rules is sufficient to create the entire system. This is very little information, and there is clearly no complexity there. On the other hand, the analysis of the pattern created by such a cellular automaton reveals unlimited depth. To describe the pattern requires a lot of information, something we like to call complex. All the while, the cellular automaton is a deterministic system, meaning repeated runs with the same rules will always produce the same pattern. The information for the development of this precision is somehow in the rules, but only unfolds to our eyes if the rules are applied iteratively, step by step, in a time- and energy-consuming process. This is the idea of algorithmic growth. The brain develops through algorithmic growth. Yet, in contrast to the cellular automaton, brain development includes nondeterministic processes and the rules change during growth. How useful is the analogy of the cellular automaton in light of these constraints? This question brings us back to the information that is encoded by the genetic code. When we discuss genes, we focus on biological neural networks. In the process, we learn about the type of information and the consequences of growth and self-assembly that define the network's properties. These are the types of information that are typically left out in ANN design, and they may thus serve as a survey of what exactly is cut short in AI and why.

Seminar 3: From Randomness to Precision explores what happens when we add noise to algorithmic growth. Both an elementary set of rules for a one-dimensional cellular automaton or a genetic code will deterministically produce identical results with every run in a precise computer simulation. But nature is not a precise computer simulation, or at least so we think. (Yes, the universe could be a big deterministic cellular automaton, but let's not go there for now.) Biology is famously noisy. Noise can be annoying, and biological systems may often try to avoid it. But noise is also what creates a pool of variation for evolution to select from. From bacteria recognizing and moving towards sugar to the immune system recognizing and battling alien invaders, nature is full with beautifully robust systems that only work based on fundamental random processes that create a basis for selection. We will have some explaining to do, as we transition from the idea of simple rules that yet produce unpredictably complex outcomes on one hand to perfectly random behavior of individual components that yet produce completely predictable behavior on the other hand. Intuition may be of limited help here.

Awe and excitement about brain wiring mostly focuses on the exquisite synaptic specificity of neural circuitry that ensures function. As far as specific connectivity is absolutely required for precise circuit function, synaptic specificity has to be rigid. On the other hand, the brain develops with equally awe-inspiring plasticity and robustness based on variable neuronal choices and connections. In particular, neurons that find themselves in unexpected surroundings, be it through injury or a developmental inaccuracy or perturbation, will make unspecific synapses with the wrong partners. In fact, neurons are so driven to make synapses that scientists have yet to find a mutation that would prevent them from doing so as long as they are able to grow axons and dendrites and contact each other. Neurons really want to make synapses. If the right partner can't be found, they'll do it with a wrong partner. If a wrong partner can't be found, they'll do it with themselves (so-called autapses). This is what I call the *synaptic specificity paradox*: How can synaptic specificity be sufficiently rigid and precise to ensure function, if individual neurons are happy to make unspecific synapses?

The answer is closely linked to algorithmic growth: promiscuous synapse formation can be permissible, or even required, depending on when and where it occurs as part of the algorithm. For example, many neurons have the capacity to initially form too many synapses, which contain little information. Through subsequent steps of the growth algorithm, this pool of synapses will be pruned and refined, thereby increasing the information content in the network. Rules for the weakening or strengthening of synapses are a core functional principle of all neural networks, both biological and artificial. This reminds us of brain function, learning and memory. But remarkably, neuronal activity can be part of the growth algorithm, long before there is even an opportunity for meaningful environmental input or learning. I call this *post-specification*, the specification of synapses late in the developmental algorithm, following initially more promiscuous synapse formation. By contrast, synaptic *pre-specification* occurs when only certain neurons get to see each other in space and time during their period of synaptogenic competency, i.e., the time window when they can make synaptic connections. If the patterns of the running algorithm restrict the synaptic partners that get to see each other, the problem of identifying the partner is greatly facilitated. The more spatiotemporal positions pre-specify partnerships, the more promiscuous, random synapse formation is permissible.

Random processes therefore need not be an enemy of precision in neural networks. Instead, random processes are abundantly utilized during algorithmic growth of the brain, just as in so many other biological processes. But random developmental processes do not necessarily produce variability in the outcome; they can also lead to almost perfectly precise synaptic connectivity patterns. And random developmental processes give rise to two of the most astonishing properties of biological brain wiring: flexibility and robustness. Connections not only change with experience, but also rewire in response to injury and developmental perturbation. ANNs also have some of these properties. And yet, historically, both neurobiology and AI had a rather strained relationship with randomness. Even today, most neurobiologists and ANN developers will consider noise as something to avoid, rather than as a design principle for a network. An understanding of the roles of noise will bring us closer to appreciating how to make networks flexible and robust in addition to making them with precision.

Seminar 4: From Local Rules to Robustness brings us back to the essence of self-assembly: local interactions during algorithmic growth. In order for local interactions to flexibly react to changing environments, local agents must be able to make their own decisions, independent of, and unknowing of, the larger system they create. This is the concept of autonomous agents. If the individual players of a soccer game would not make their own decisions, the game would be boring. If the players would not follow a common set of rules, the game would fall apart. The local interactions, the players' decisions and flexibility, make the game interesting (if this kind of game happens to be interesting to you) and robust. The outcome is unpredictable at the level of the individual game, but the average outcomes over seasons are remarkably predictable. Which of the two, the individual game or the average season, is more interesting is in the eye of the beholder. For biological systems the beholder is evolutionary selection. For example, whatever local molecular and cellular interactions lead to different fingerprints may leave the outcome unpredictable at the level of the individual thumb, but perfectly predictable and robust at the level of selectable functionality.

In neural networks, both development and function vitally depend on individual neurons behaving as autonomous agents. The growing tip of a neuron employs random exploration of its environment through filopodia, tiny fingerlike protrusions. The neuron must be allowed to individually and locally decide whether it likes something it senses on the left or on the right using these protrusions. Similarly, the ability to learn in both biological and artificial neural networks relies on individual neurons, and individual synapses, to adapt their function. The concept of autonomous agents has made repeated stage appearances in AI. In all cases, the actions of autonomous agents only make sense in the context of a process that develops in time. The agents' decisions and collective actions set the stage for higher order organization that develops step by step. They are part of self-assembly in space and time. And this brings us, at the end of session 1, back to the question of types of information. It is possible to describe, in arbitrary detail, the precise angles of the movements of every soccer player or a neuron's growing protrusions. However, at what level of detail a situation must be described in order to understand a distinct step of the underlying growth algorithm is not an easy question to answer.

In the second session we approach this question by diving into the realities of players and rules during the self-assembly of the biological brain prior to learning. All three seminars in this session focus on those aspects of the neurobiological history and their outcomes that are critical from the perspective of information theory: When and where *does* the information get into the network? ANNs used in AI today do not encode much information prior to learning; they are engineered and switched on for training. The second session is therefore all about biology, but with the goal to understand what it is exactly that ANNs are leaving out.

A powerful way to study brain development is experimental perturbation through mutation of the genetic code. After all, evolution did it before: genetic changes that affect development result in changes to the brain. The evolutionary approach is based on trial and error and does not require a need to predict the outcome of a genetic change as long as it can be selected. Selection of heritable, meaningful changes are evolution's way of reprogramming the brain. But what are these meaningful changes to the genome? Are there special genes for the brain and behavior, or could any mutation in the genome help to reprogram the brain through information unfolding during development?

The second session starts with **Seminar 5: From Molecular Mechanisms to Evolutionary Programming**, in which we will explore these questions by analyzing how mutations can reprogram animal behavior. We will approach the answer through a discussion of programming by evolution: If a mutation causes heritable, meaningful and selectable change, then evolution can use it to rewire and reprogram the network. For this to work, it is not necessary that the functional mechanism of the protein encoded by the mutated gene is in any intuitive or direct way related to connection specificity. Rather, the effect of a mutation has to be such that the developmental process, and the unfolding of information that comes with it, reproducibly change the network. In this way, a behavioral predisposition can certainly be caused by a single mutation, yet there need not be a single "gene for that behavior."

The fact that single mutations in the genome can reprogram animal behavior is well established. Pioneering experiments with fruit flies have surprised and enlightened this field for more than 50 years. Examples include mutants that affect courtship behavior and the internal clock that predictively guides behavior through the daily cycle of day and night. Importantly, the way such mutations were (and still are) found is based on accelerated evolution in the lab. The first step is to dramatically increase the occurrence of random mutations without any prediction as to what this might cause. The second steps is to let thousands of the randomly mutagenized animals develop. The third step is to take those flies that survived the high mutation rate and assay them for behavioral alterations. These forward genetic screens for behavioral mutants led to the successful identification of mutants with altered behavior; over the years, several genes harboring these mutations were discovered. Some of the best studied of these genes are those where a molecular function directly relates to the behavior. There are beautiful examples, but they may be the exceptions. Most mutations that modify animal behavior affect genes that function in surprising ways during developmental growth, often at many steps or in many different cells. Such mutations can lead to heritable, meaningful and selectable behavioral change, but not through specific molecular mechanisms that are related to the behavior itself. Mutations may cause unpredictable developmental alterations that nonetheless lead to reproducibly different brains based on changes in network development or function. Those are the mutations that served evolution in the slow, trial-and-error reprogramming of brains and their astonishing behavioral innate programs. There is no single gene solely responsible for the spider's ability to weave a species-specific web or the monarch butterfly's ability to migrate a route of thousands of miles over a succession of generations. If our goal is to understand the programming of a neural network that accomplishes such feats, we must step beyond the idea of a single gene coding for a single behavior. We must learn how evolution reprograms the abilities of networks, including human intelligence.

Seminar 6: From Chemoaffinity to the Virtues of Permissiveness deals with the historical and ongoing quest of developmental neurobiologists to understand underlying molecular mechanisms. The invariable hope is to draw direct lines from mutations to genes to the gene products' molecular mechanisms in trying to decipher neural network design. Neurobiologists prefer to characterize those genes whose gene products execute molecular mechanisms that make intuitive sense with respect to neuronal connectivity, hence the terms "guidance molecules" or "chemoattractants." This is such a powerful idea and prominent concept in developmental neuroscience that we need to discuss examples of such molecules and their roles during algorithmic growth in some detail.

In search of information encoding for brain wiring, the holy grail has been the search for mechanisms that *instruct* the neuron where to make a connection. The idea of instructive mechanisms contrast with permissive mechanisms, which may be necessary to allow growth, but do not guide it actively. Oddly, the most widely used attractant for the guidance of growing neuronal protrusions in experimental culture is nerve growth factor—NGF. This is a molecule that the neuron needs to grow. By providing NGF only on the left, but not on the right, we can make neurons grow robustly to the left. This is clearly instructive. But wait, it's a growth factor! The neuron simply will not grow where it is not present. That's rather permissive. Obviously, a permissive mechanism (like a growth factor) can contribute to the neuron's choice where to grow. From an information-theoretical perspective, the information for the directionality must have previously been provided in the location of the growth factor, which may lay out an entire path. The factor itself may be permissive, but the path it marks is instructive. Which brings us to the information needed to mark the path—and that brings us back to algorithmic growth where, step by step, paths may be laid out through the interactions of many autonomous agents, including the growing neuron itself. The path may not exist either at the beginning or the end of the neuron's journey, but results from the interactions of the neuron with its surroundings as it grows. Some molecules on neuronal and non-neuronal surfaces convey local and temporally restricted attractive or repulsive signals. Yet other molecular mechanisms alter the propensity of the neuron to further grow extensions at all or gain or lose the capacity to make synapses, or alter its mechanical interactions with the surroundings. In the context of algorithmic growth, the composite of all these factors determines the rules for each step in the algorithm. A genetic instruction need not be attached to a molecular mechanism of a single gene product. Instead, *composite instructions* are fleeting states of the system defined by the molecular and cellular contexts that happen to emerge at any given time and place during algorithmic growth.

Seminar 7: From Genes to Cells to Circuits is all about levels, from molecules to neural circuits, as we move towards the function of neurons in the network. How is it that in the field today the study of neural circuit function is obviously a question to be studied at the level of cells, while the study of the same neural circuit's assembly is obviously a question to be studied at the level of molecules? This brings us back to the type of information encoded in the genome and its relation to processes at other levels. Single genes usually do not describe processes at higher levels, even though a specific mutation in a single gene can heritably and meaningfully change that process.

The levels problem is particularly pertinent when we are trying to span all the levels from the immediate effects of a mutation in the genome to a behavioral trait. Genome-wide association studies try to establish probabilities for a given genomic variation to be associated with a specific behavior. The probabilistic nature of the data and the difficulty to establish causality in such experiments is directly linked to the nature of algorithmically unfolding information.

Neuronal function is the level at which grown biological networks and engineered artificial networks meet. But in the biological template neuronal activity can in fact be part of the genetically encoded growth algorithm. Neuronal activity is part of information unfolding. As we have already discussed in the context of synaptic post-specification, activity is known to kick in before there is any environmental input. Correlated neuronal activity is one of the ingredients of algorithmic growth that require a random process to function robustly. It also provides a bridge to neural network function and AI.

The third session is all about transitions. First, there is the transition from development to function in neural networks. Next, the transition from the naïve to the trained and functional network, and with it the transition from biological to artificial networks. The transition from dull to intelligent. And then there is the transition from separate biological and artificial networks to their interactive future. In all cases, the idea of information unfolding in a time- and energy-consuming manner serves as framework to assess possibilities and limitations.

In **Seminar 8: From Development to Function** we explore in what ways developmental self-assembly is relevant for network function. We will start with the burden of evolution and development for biological brains. Evolution had to work with the outputs of previous states, no matter how inappropriate they may have been when selection pressures changed. Evolutionary change happens in steps, however small, in a necessarily sequential manner. And the process takes, of course, time and energy. As a result, brains feature some remarkable and apparently nonsensical oddities that only make sense in light of development—and the way development was slowly shaped over millennia of evolutionary modification.

These kinds of biological oddities, and messiness, led computer enthusiasts who were trying to develop AI in the '80s to take pride in ignoring what their neuroscience colleagues were doing. "We can engineer things much better," they may have thought, so why learn about the nonsensical solutions biology had to put up with?

And yet, if we avoid the burden of developmental history by starting with a randomly connected network prior to learning, the first problem we are confronted with is the time and energy it takes to train. And training, again, is a stepwise, time- and energy-consuming process. The order of input matters. And the ultimate function of the network is made possible, and burdened, by its training history. We will explore how information is stored in biological and artificial networks. How does the neural network save a four-digit PIN? The amount of bits needed to store this information is clearly defined in computational terms. Yet, neural networks save and retrieve this information flexibly and robustly, even if random neurons in the network drop out. In addition, the biological network does not have a separate training and function period. Learning is inseparable from using the network; storing is inseparable from retrieving information. And again, we meet an evolutionary principle and the power of sequences in time. Many bits of information in the biological network—memories—can only be accessed by going through a sequence in time. Try saying your phone number in reverse order. How is this information content programmed, stored and retrieved?

If self-assembly is any guide, then information has to enter by changing the sequential, auto-associative network, which means it changes algorithmic information. Maybe memories should not be understood as stored entities at all, but rather as algorithmic rules sufficient to recreate them with a certain robustness, flexibility and variability. This bring us back to the cellular automaton that does not store the memory of the pattern at iteration 1,000, but instead the information to recreate this state. We will explore to what extent this process resembles algorithmic growth, and how it transitions from development to function.

In **Seminar 9: From Algorithmic Growth to Artificial Intelligence** we focus on artificial neural networks and their relationship to self-organization and algorithmic growth. We will finally also discuss definitions of *self-assembly* and *intelligence*. Most ANNs are based on the idea of an engineered design, flipping the on switch and training the network. By contrast, in biological networks the information encoding goes hand in hand with the development of the brain. The brains of a newborn, a toddler or a 10-year-old are clearly recognizable for their developmental stages morphologically, functionally and by their behavioral output. The question is whether a tedious, years-long process of self-assembly is a desirable step to create an artificially intelligent system. More specifically, is there ever a need to grow a neural network, or is training a predesigned network like in deep learning sufficient, maybe equivalent, or even superior?

A common idea in ANN development is that the product of development is only hardware infrastructure. A lot of biological idiosyncrasies can be congealed in a single parameter, like the synaptic weight. These are shortcuts that have served ANNs well for many years and many tasks. Yet, a key question associated with this reduction is how it may limit learning. In biology, the single parameter contains none of the parameter space necessary for evolutionary programming to modify an algorithmically growing network. Based on these considerations, we dive deeper into the way engineered ANNs do, and do not, function.

Finally, in **Seminar 10: From Cognitive Bias to Whole Brain Emulation**, we will discuss the consequences of algorithmic information storage in neural network for the function and interaction of biological and artificial networks. We will start with a discussion of heuristics, the probabilistic nature of any information in the network. Neural network function is less the computing of input in conjunction with stored data based on logical operations, and more a process of probabilistic alignment and selection of patterns based on previous experiences. Both biological and artificial networks are biased by their experience. An ANN that has only been trained with numbers 0 to 9 will interpret the picture of an elephant as a number from 0 to 9.

We are all well-trained neural networks, but our brains come with a history track, as do ANNs. New information is not stored independent of other safely stored information content. Instead, any new bit of information is processed in the context of the entire history of the network. The same experience means something different for every individual. And the better the information is aligned with previous experiences, the easier it is for the network to "believe" the new arrival. This simple thought has some interesting consequences for the concept of cognitive bias: in a network built on algorithmic growth, bias is a feature, not a bug of the system, whether we like it or not.

Finally, if information is stored as an algorithmic process that requires time and energy, can it be retrieved and transferred *in toto*? That is, what does the self-assembling brain teach us about the potential to upload or download our brains? If information is not stored in any dedicated bits, but as algorithmic rules sufficient to recreate that information, then bandwidth of connection may not be the biggest challenge for data transfer. In the development of AI, we continue the debate about how similar artificial systems should be to the biological analog. But if we want to extend or copy our own brains, a clearer understanding of how information is actually stored or retrieved is needed. We encounter the levels problem again. To generate artificial human intelligence, what parts of the algorithmic growth of the human brain can be cut short? In the design of ANNs, shortcuts are useful to shorten computation time by throwing out irrelevant detail. This approach works, as long as we do not need or want to simulate, say, spatially and temporally restricted modulation of many synapses through diffusible neuromodulators. But if we want to simulate human intelligence, don't we need the neuromodulators, since circuit function requires synaptic changes that depend on the neuromodulatory context? We come to the question of the AI we want to generate. The shortcuts we choose in the development of artificially intelligent systems define what intelligence we get.

On Common Ground

This book was written with readers in mind that are interested in developmental biology or AI alike. However, those deeply immersed in either field will find much that is treated too superficially or from an unfamiliar perspective. I did not attempt to provide objective overviews over either field's history or main achievements; many great works already exist on both accounts and are referenced throughout the seminars. My goal was to identify common ground, with a focus on underlying questions of information encoding. My hope is that a reader with deeper knowledge in either topic will still find reason to smile when trying to think what it may read like for someone in the other field.

I am convinced that all concepts presented here have been part of many ideas in different contexts before. Algorithmic growth in particular is not a new concept. It is implicit in all developmental processes and any attempt to understand how the genome encodes growing things. Yet, intuitive and mechanistic thinking in either field rarely considers the implications of unpredictable information unfolding. Those familiar with self-organizing systems may find most concepts presented here oversimplified, or my definition of self-assembly (seminar 9) wanting. Similarly, developmental neurobiologists are likely to find much that could have been added from the boundless list and beauty of molecular mechanisms underlying neural network development and function. But common ground lies more in the motivation, the desire to understand how neural networks grow smart, than in the details of the individual disciplines. On this account, I hope the historical perspectives presented throughout the seminars may provide helpful parallels.

I am aware that many ALife and AI researchers may feel that reading a book written by a neurobiologist is not likely to be helpful for their work, both for reasons of perspective and the inevitable focus on unhelpful biological "messiness." Similarly, some developmental neurobiologists may currently read a book or two on the application of deep learning to analyze their data, rather than to learn from the AI community about how "real" brains come to be. I started this project wishing there were a book that both would enjoy having a look at, or at least get sufficiently upset about to trigger discussion between the fields.

The Present and the Past

The First Discussion: On Communication

ALFRED (THE NEUROSCIENTIST): Hi, I'm Alfred. I like your robots.

AKI (THE ROBOTICS ENGINEER): Thanks. Are you an engineer or a programmer?

ALFRED: Neither, I study real brains.

AKI: Yikes! Define 'real.' To me, something that actually works and that I can implement is real. Nobody could yet tell me how the human brain works. You don't even know how your real brains store memories.

ALFRED: Lovely. Does that make your robots more real than biology?

AKI: We use AI systems that can do many things already better than your biological brain. How real is that?

MINDA (THE DEVELOPMENTAL GENETICIST): We use AI to analyze large datasets. It is very powerful, but it is very specialized. Surely, you are talking about two very different things here.

PRAMESH (THE ARTIFICIAL INTELLIGENCE RESEARCHER): Hi, I just overheard this. My name is Pramesh. We work on the development of artificially intelligent systems. We have helped to develop machine learning applications like those used for large sequencing datasets. We used rather simple artificial neural nets with supervised learning. I think the differences or similarities of learning with backpropagation error correction versus learning in biological neural networks are actually an interesting and quite open discussion.[7–9]

ALFRED: No way. It's neither interesting nor open. I know this stuff: supervised learning only means that you feed new patterns with a label saying what it is supposed to be. That's not how the brain learns. And backpropagation is just a smart way to correct an output error by changing the synaptic weights in the layers that produce the output. We have pretty good evidence that this is not how the brain learns.

AKI: How can you say this is not how the brain learns if you do not know how it works?

MINDA: Aren't there at least some clear similarities? As far as I understand, what you call AI is based on neural networks and the strengthening or weakening of synaptic connections, right?

PRAMESH: Yes, that's an interesting story. For decades neural networks were not very much in fashion in the AI community, but the deep learning revolution made them the AI success story we have today.[5]

MINDA: If neural networks led to the breakthrough, then the most successful AI to date is indeed more closely modeled on real brains than AI has ever been, correct?

PRAMESH: I think so, yes.

ALFRED: So close and yet so far . . .

AKI: It doesn't matter, really. We can build cars that run faster than any animal and planes that fly further than any bird. There is always something we can learn from nature, but then we can take it to do things that nature could never achieve.

ALFRED: I would rather say that birds can fly in ways that planes could never achieve. Isn't AI about trying to make a brain that is as intelligent as brains in nature?

PRAMESH: That's a big discussion in the field. There are concepts like artificial general intelligence or artificial human intelligence, as opposed to the specialized learning algorithms for sequence analyses, for example. The discussion whether AIs should be modeled closer on biological brains has been going on for many decades. For the longest time, AI researchers felt that biology is just too messy and not helpful. But many people in AI today argue we should have a closer look at the brain again.[9–11]

MINDA: Well, I can tell you that from a developmental biologist's point of view, the idea of designing and programming networks is a problem. Biological networks are very precise, evolved structures that develop based on genetic information.

ALFRED: Sure, sure, but we are not interested in how networks develop, I want to know how they function.

AKI: Exactly. Except that you don't know how yours function . . .

ALFRED: Okay, come on. Of course we know a lot about how the brain works, we're just honest about what we don't understand yet. But we got the basics, like how synapses work, and that's exactly what you adopted in your AI approaches. We also know a lot about how neural circuits compute information, and that's very different from the networks in AI. So I would say that you may have gotten the synapses right but the circuits wrong. I guess you will have to come back and learn again from real brains to make real progress.

PRAMESH: There are some applications that use biological circuits as a template, like convolutional neural networks. I do agree that there is a lot to learn from biological networks, but attempts to make things closer to certain instances of biological networks also often failed.

MINDA: This is not my field, but is it not obvious that an artificial human intelligence will have to be more similar to a human brain? In fact, I would argue that it has to develop like a brain.

ALFRED: Naaaaaaaah . . . development was yesterday . . . We should be talking about how different biological and artificial networks function. Like, how they store information. How many neurons do you need for the memory of your grandmother?

AKI: . . . how many bits do you need to store her?

ALFRED: Yeaaa, that may be the wrong language. Although, a lot of experiments have been done in which you can see single neurons being active during memory storage and retrieval. There are neurons that are only active when a monkey sees a particular face. The memory is stored in the connectivity of the neurons. In mice we can now activate single neurons to trigger a memory![12]

PRAMESH: But you can kill that neuron and a few thousand others, and the memory is still there.

ALFRED: Well, you can't simply kill a few thousand neurons and assume there is no effect. But yes, I guess there is some truth to that. It only means there is redundancy in the storage of the information.

PRAMESH: Redundant copies mean you have two or more copies of the same information somewhere. Like a hard drive backup. But that is not how the neural network stores information. There is an old idea that information in the brain is stored like in a hologram. This is at least how I think about it. If you have a holographic picture, you can cut away a corner, but the remainder of the picture contains the information of the lost corner and recreates it. Every part of the picture is stored everywhere—and whatever you cut away will never lead to a loss of a corner, just to a loss of overall resolution. The picture just gets a bit blurrier.

ALFRED: The brain definitely has different regions and not everything is stored everywhere. But it's true that there are many ideas about the brain and memory out there that say something in this direction. The analogy to the hologram has been made many times. And you are right about redundancy. I didn't mean exact copies. Clearly information in the brain is stored somewhat decentralized.

AKI: 'Decentralized' can mean a lot of things. The question about memory is simple enough. How is memory stored and how is it retrieved? The way I think about it, if I give you a four digit number, say your credit card's PIN code, I know exactly how many bits I need to store this on a computer. How many bits do you need in a brain to store those four digits?

ALFRED: Nobody can tell you that. I don't even think this is a good question.

AKI: Why?

ALFRED: The brain doesn't work that way. Many parts of a neural network can take part in memory storage. There is no such thing as a storage address, nothing like a set of neuronal connections that exactly store your four-digit PIN. You just think like that because that's what your computers and robots do. You have some address in the computer memory where some information is stored and you can read out that information only if you have that address. I think the computer is a really bad metaphor for the brain.

PRAMESH: I very much agree. But our artificial neural networks don't do that either. We use computers, but only to simulate network architectures that work very differently from a traditional computer. We have unsupervised learning techniques and deep multilayer networks, recurrent networks, modular networks, you name it. An unsupervised learning algorithm can recognize and understand principles that we never told them to find in the first place—if only we feed them enough data.

MINDA: I do not know what 'unsupervised' and 'recurrent' mean. But I guess your 'feeding of enough data' into a big computer simulation is a form of learning. Let me ask you something. Do you simply build an 'empty' network first and then train it? I would argue that for our brains learning is a type of development. You will not get around development.

PRAMESH: Yes, training an 'empty' network is a common approach. Usually the connection weights are first randomized before training. And training is definitely the most time-consuming part, but the right architecture is also key. But what do you mean with 'learning is a type of development'?

MINDA: The brain doesn't come into being fully wired with an 'empty network,' all ready to run, just without information. As the brain grows, the wiring precision develops. A newborn has few connections. In early childhood there are more neuronal connections than in an adult, and the wrong ones get pruned, leading to a precise network. Doesn't the information content grow as wiring precision develops? Think about the monarch butterfly that knows the route for thousands of miles of migration. The development and precision of the underlying neural network is genetically encoded.

ALFRED: Yeah, that's not bad. But there is a difference between innate and learned stuff.

AKI: We can train the brains of our robots to learn exactly these things. Isn't this learning process simply replacing your brain development?

MINDA: The monarch butterfly doesn't learn these things. It's in the genes.

AKI: You just said yourself the information content grows as the newborn's connectivity develops . . . how is that in the genes?

PRAMESH: Yes, I think saying 'in the genes' can be deceiving. There is no migration route of the butterflies in the genes. The genes only contain the recipe how to build the network and that then somehow contains the information.

MINDA: That sounds like semantics to me. It's the same thing. If no other information is used to develop that brain, then it's in the genes.

AKI: I am still missing where your growing information content is coming from. Anyway, how far have you geneticists and molecular people come in understanding how to build a real brain? To me, the papers look the same as thirty years ago, just more molecules and fancier techniques that nobody understands anymore.

MINDA: We have identified hundreds of genes and the precise mechanisms by which they operate during brain development. We have a framework of how genetics and precise molecular mechanisms wire a brain. We are figuring it out bit by bit. I thought this is the type of information we could share with the AI community.

PRAMESH: I am not sure this is the information we need. I personally like to listen to biologists, but I often feel they are lost in details. Someone says one protein is required to make a neuronal connection at a certain place, someone else says it's required to keep the cell alive or cause cancer or something somewhere else. I would need to extract some principle, some algorithm to program a computer with that. But we basically try to avoid that messy stuff. We have a project right now where we use computer simulations to evolve artificial intelligence systems—this goes straight to the principles.

ALFRED: Are you not worried that you are missing something that's in the biology? And what comes out of that—anything like a real brain?

AKI: We are not interested in copying biological brains. I think we have an opportunity to build an artificial general intelligence that goes way beyond the restrictions of biological brains.

ALFRED: For that you'll need more than just backpropagation as a learning mechanism in a fixed-size network. I can see the argument that amazing innate programs are somehow wired into the network based on the genetic program. That has to be part of it, but what it is that genes really encode has never made sense to me in studying neural circuit function. I mean, obviously a gene codes for some protein and that protein then is some enzyme or binds to some membrane and so on. Scientists like Minda love such molecular mechanisms. But what does it even mean to say a gene encodes a behavior? It just doesn't add up.

MINDA: Well, you just said yourself that entire behavioral programs can be controlled by genes. Of course genes code for proteins with many precise molecular mechanisms. Protein functions control developmental processes. Many proteins that interact in a coordinated way can lead to complex behaviors. In my lab, we are deciphering the underlying mechanisms. We acknowledge that it is early days in this field. I am also aware that some people call complex behaviors an emergent property. I feel this is an empty phrase.

PRAMESH: I don't like the word 'emergent' either, because it sounds like something magical happens at some point. But there is no magic . . .

MINDA: Precisely.

PRAMESH: . . . but the details about all your genes and proteins and what they do are not helpful to me either when developing AI systems. There is little I can do when you tell me you found a mechanism whereby a protein needs to be here or there in a particular neuron in a mouse. What matters is how a neural network works—the connections are important and information is encoded by making them stronger or weaker.

ALFRED: There is more to synaptic connections than 'stronger or weaker.' They can be excitatory or inhibitory, there is modulation, all kinds of stuff. I may not be a big fan of all the molecular details either, but just think about hormones or neuromodulators. They affect maybe millions of synapses at the same time; they change thresholds and states of entire sub-networks. Some brain regions are very sensitive to certain hormones and transmitter systems and many drugs work on those.[13, 14]

PRAMESH: But in the end all these things just affect synaptic strength in one way or the other . . .

ALFRED: That's not the point. In order to change what a brain does, I can swamp it with modulators that only affect some parts in a specific way. This is part of how the network works. If you want to do this in an artificial neural network, you have to either include the simulation of such a modulator—or change all the synapses that are affected by hand, one by one.

AKI: Nothing happens by hand . . . and it's still just synaptic strengths. The AI may find different ways to do what your neuromodulators do.

PRAMESH: Hmm, it's an interesting thought, though. Yes, we can include something like neuromodulators in simulations. But when it comes to evolution, . . . I guess a small change in a neuromodulator might affect a very large number of synapses and change the entire network. This is not typically done.

MINDA: We know that developing neurons change continuously. Sets of genes turn on and off continuously. The cell runs a developmental program that is partly cell autonomous, but then there are also interactions with the surroundings. All of these parts of the developmental program run with precision every time a brain develops. Are these all details that are unnecessary in your artificial neural networks?

PRAMESH: Growing artificial neural networks is very uncommon. It is not used in current applications, like the things Aki builds into her robots. But growth simulations of course exist. The same situation though: We can't really take all the detailed stuff you tell us about gene X or protein Y—that's not useful. We need rules for growth.

ALFRED: . . . and can I just say that growth is not all just super precise? Brain development is very flexible. Your genes don't encode a precise network, they allow developing neurons to make choices during development.

MINDA: I can show you some brain structures that are genetically encoded and astonishingly precise. The genes control the choices that neurons can make. These choices lead to precise neural circuit connectivity. Flexibility is encoded with similar precision.

ALFRED: . . . bit of an oxymoron, don't you think? Flexibility also produces variability during development—and I can give you some examples of brain structures for that as well.

AKI: This is exactly how your genes and neurons are not useful to me. We build intelligent robots. We don't grow them and send them to kindergarten. And we may only be ten years away from creating human intelligence.

ALFRED: Hah! I know this one! It's always been 'just ten years.' I think people said that in the '50s already about computers we would have in the '60s, then in the '70s and on and on—and I hear the same thing today—in ten years they'll overtake us![5]

PRAMESH: Yes, there is some truth to that. The history of the field of artificial intelligence is full of overenthusiastic predictions.

But then again we always overestimate the near future and underestimate the far future. Amara's law.

AKI: Didn't developmental neurobiologists in the '50s and '60s also already make equally inadequate predictions that we should know by now how brains develop and work?

MINDA: There is a lot to study. It's early days, as I said we acknowledge in the field. We still do not understand most genes. However, I think it is fair to say that we have worked out the basic concepts.

AKI: And now you are just filling in the details?

MINDA: I do think the details are absolutely critical. Of course we still find surprises. Our continued work on how molecular mechanisms precisely work is the foundation to understand the development of precise neural circuits.

PRAMESH: Maybe there are details and details. I think results from molecular studies in developmental neurobiology have not really entered my field. Currently, I really do not see how they would help me develop intelligent systems.

ALFRED: . . . and neither do they help me to understand how the brain works.

AKI: . . . and neither do your 'real' brains help us to create AI.

MINDA: I appreciate AI as a tool to analyze data. However, I have yet to learn something from a robot about the brain.

AKI: Okeydokey, now that we have established all the ways we do *not* want to talk to each other, why again are we here anyway?

PRAMESH: Well, this is a workshop on what we might learn from each other. I guess in the end we were all curious enough to at least hear out the argument once. Although I do not see what parallels there are supposed to be between developmental neurobiology and AI research. They don't even have a common language.

ALFRED: One could try to learn the other's language. . . .

AKI: That reminds me of the Babel fish in Douglas Adams's *Hitchhiker's Guide to the Galaxy*. Anyone who put it in their ears could suddenly understand all languages. And the consequence of everybody finally understanding each other were more and bloodier wars than anything else in the history of the universe. . . . [15]

ALFRED: That's jolly. Maybe it's better to have a fight than just to ignore each other.

MINDA: I would be happy to contribute our approach and findings to the AI community.

AKI: I guess I have to listen to your problems in addition to my own now . . .

PRAMESH: Unless we have the same problems.

The Historical Seminar: The Deeply Engrained Worship of Tidy-Looking Dichotomies

Thank you for showing up to the kick-off of this somewhat experimental seminar series and workshop. We will start with the shared history of the two fields brought together in this room: brain development and artificial intelligence. All of us are interested in biological or computational approaches to understand how information is encoded in neural networks. I am talking here about an elementary understanding of what it takes to build an intelligent network—be it biological or a computer simulation. I am a neurobiologist, and my motivation has always been the human brain. Where does that information come from? What do I even mean by *information*? The biologists amongst you may say: "Look, the information is in the genes for a hard-wired neural network, and for everything that is learned the information comes from the environment—nature and nurture." But how is a hard-wired neural network actually encoded in the genes? Why can we not simply read a map for *network topology* in the genome? Whereas those of you who develop artificial neural networks may feel that this is not your problem anyway. "The information is in the training set," you may say, "and we can start simply with a randomly connected net prior to training." Does this mean you only need nurture, whereas genetically encoded network topologies are just a biological problem? Well, these questions are not new. I want to tell you a bit about those who asked them before us—and why we are here today, such as we are.

Richard Feynman is often credited with the famous quote "What I cannot create, I do not understand."[16] That's where all you machine learning, deep learning and artificial intelligence folk come in, as far as I am concerned. I know you build self-driving cars and are revolutionizing how tech giants deal with information and learn about what we apparently really want. But this seminar series is not about the application of AI; it is not about fancy new technology with the goal to make a better tool to solve complicated problems or make more money. We are here to talk about what it takes to make a neural network, well, intelligent—and what we mean by that. That's our common ground. The biologist sees it in an insect, a mouse or a human brain. How is that put together? The AI programmer wants a system that learns and understands all by itself. How is that put together? My point will be that there is something common to be learned from telling our stories to each other, from telling our story together. In doing so, I will subjectively focus on some highlights and not others. Today, I will only tell you about the early years, from the 1940s to the '90s. All the new stuff will come in the following seminars. There will be nothing comprehensive about the story I tell—enough historical accounts have been written—and it is not my intention to repeat any of them.

I'll start my story with the man who coined the term "neuron" in 1891—the German anatomist Heinrich Wilhelm Gottfried von Waldeyer-Hartz.[17, 18] At the same time the Spanish neuroscientist Santiago Ramon y Cajal described autonomous anatomical units in the brain for which he would later adopt the term neuron.[19] Cajal used a newly developed staining method by the Italian pathologist Camillo Golgi that sparsely labeled individual neurons—thereby for the first time allowing scientists to see the individual trees that a forest is made of. The neuron doctrine dates back to von Waldeyer and Cajal; it comprised, in the words of Golgi, the ideas that (1) the neurons are an embryological unit, (2) the neuron, even in its adult form, is one cell and (3) the neuron is a physiological unit.[20]

This seems like a good starting point. Cajal and Golgi received the Nobel prize for their work on the neuron doctrine. Interestingly enough, Golgi used his Nobel speech to categorically refute the three hallmarks of the neuron doctrine given above. He highlighted the fundamental information problem in his Nobel speech: "So far no one has been able to exclude the possibility of there being a connection between nerve elements already from the beginning of their development. Even if these elements were originally independent, one would not be able to show that this independence was going to be maintained. . . . no arguments, on which Waldeyer supported the theory of the individuality and independence of the neuron, will stand examination."[20]

I highly recommend reading Cajal's and Golgi's speeches for the same prize in 1906, what a story! In particular, the reticular theory was based on the idea of a gapless network of nerve fibers and is often ascribed to Joseph von Gerlach in 1871.[21] However, in the book chapter published that year, von Gerlach did not actually introduce the term reticular theory. He just described what he called the "Nervenfasernetz," or the nerve fiber network. Von Gerlach was concerned with a detailed state-of-the-art anatomical description, not an information theoretical perspective of how the structure might have come to be. Based on his staining methods he could not see synaptic contact sites, but instead observed a continuous nerve fiber network in the gray and white matter of the spinal cord (fig. 1.1). Golgi would later become the most prominent proponent of the reticular theory. If there are no individual neurons, then they also do not need to specify any connections to form a meaningful network. Unfortunately, the reticular theory left unanswered how the information for the connectivity of a preformed network was encoded. From an information theory viewpoint, the reticular theory did not even try to solve how the network came to be. Although long abandoned, the idea remains strangely relevant today when discussing information encoding in neural nets, as we will see. Be that as it may, it took electron microscopy in the 1950s to ultimately settle the debate and show that the connections between neurons are synapses that indeed separate two cells.[22, 23] Yet, already in 1910 Ross

FIGURE 1.1. Depiction of neurons from the spinal cord of an ox by Joseph von Gerlach in 1871. The figure legend reads (translated from German): *"A dividing nerve fiber, whose two branches hang together with the nerve fiber network, which is connected to two nerve cells."* In other words, von Gerlach considered the "nerve fiber network" as a structure that is separate from the nerve cells. Reproduced from: von Gerlach, J. Von dem Rückenmark. in *Handbuch der Lehre von den Geweben des Menschen und der Thiere* (ed. S. Stricker) 663–693 (Engelmann, 1871)

Harrison had developed the first neuronal cultures and showed that "the living neural plate, which looked syncytial, could readily be dissociated into separate cells."[24, 25] Today, the neuron doctrine forms the basis for everything we are talking about here—and the beginning of a shared history in neurobiology and AI. The neuron doctrine establishes the need to connect, to wire up, a functional network out of independent components. Meaningful connectivity is required for biological and artificial neural networks alike.

It is not easy to put ourselves in Golgi's shoes and understand his uneasiness with the neuron doctrine. Golgi had performed careful experiments, and he had a clear view of the literature at the time. Right here, we get our first glimpse of an interesting aspect of how the human brain works: Two smart guys given the same data may draw opposite conclusions—how can that be? We'll leave this fascinating question for the last seminar in this series. For now, we will focus on the new questions it raised at the time. Indeed, the neuron doctrine opened a door to many new problems: If neurons start out independently, how do the right cells find each other to "wire up"? How do they communicate with each other? How does the network store and compute information? In short, how can a system composed of independent components self-assemble and function as a whole brain? We have not stopped searching for answers to these questions to this day.

In 1930, a 31-year-old Austrian-born biologist moved to America in search of better research opportunities. Paul Weiss had finished his baccalaureate studies, served for three years in the First World War and witnessed the roaring '20s in Europe until the financial system came crashing down in 1929. At the time of his move to America, the Great Depression was moving towards its low point. Between 1922 and 1926, Weiss had performed limb transplantation experiments on newts and frogs that revealed remarkable plasticity: after some postsurgery recovery time, the supernumerary transplanted limb would actually function, and in fact move in coordination with a neighboring limb.[26] He called this a "homologous response."[27, 28] His regeneration experiments suggested neuronal connectivity of the extra limb to the same command centers that control the correctly developed limb. Weiss further proposed that the muscle induces a corresponding adaptation in the nerve ending.[27, 29] In his resonance theory he speculated that, just like a guitar string that resonates at a particular frequency, axons would exhibit specific activity patterns that muscles would resonate with and tune to. Weiss's idea was that a connection could be induced to become specific, rather than being rigidly pre-specified. His example was the connection between the motor neuron and the target muscle. Weiss's emphasis was to explain the plasticity of the nerve innervation in accommodating a supernumerary limb, leading him to see less specificity in the connections themselves. The resonance theory emphasized plasticity and failed to explain important aspects of developmental specificity. However, the ability of axons to innervate entirely wrong areas when presented with the opportunity as well as the idea of activity-dependent induction of synaptic properties have been observed countless times and are still studied today.

In 1934, Weiss observed that cultured neurons would follow certain mechanical features, like scratches in the surface they grew on. In his contact guidance theory he extended this idea to postulate mechanical forces as a guiding principle, without need for chemical signals.[30] Again, this conclusion did not pass the test of time. And yet, just as imprecise innervation and activity-dependent inductions were later found to be important in many contexts, so did mechanical forces as contributors to axon behavior reenter the stage later on.

In his famous 1939 book *Principles of Development*, Weiss wrote "a neuroblast, though being the source of the axon, cannot also provide its itinerary. A growing nerve fiber has no inherent tendency to move in any particular direction, nor to change its direction at certain points, nor to branch and terminate after having reached a certain length. For all these actions it is dependent on directive influences of the environment, and its own endowment in this respect consists merely of the faculty of letting itself be guided."[28] And thus was another chapter opened for a field that would later be known as "axon guidance"—the study of the mechanisms that guide the cables to connect and wire up the brain. This field had started with the work of Ross Harrison, the inventor of tissue culture, and his 1910 paper based on experiments conducted in 1907, "The outgrowth of the nerve mode of protoplasmic movement."[25] He was the first to observe filopodial movements that looked like what Cajal had predicted based on his Golgi stains. Although it is not commonly depicted as such, from its inception the field has circled around the question "Where is the information coming from?" In the 1939 quote above, Weiss voices the opinion that it cannot come from the neuron itself, but somehow must be imposed on it by its environment, thus allowing for the plasticity he observed in his transplantation experiments. But where then is the information in the axon's environment coming from? Similarly, both the electric (resonant) and mechanical ideas begged the same question. On the one hand, there was the intricate and reproducible specificity of individual neuron types and their inferred connections, as described by Cajal. On the other hand, there was the ability of a limb at the wrong place to connect to the best target it could find. What type of information encoding could explain both phenomena?

In 1941 the Second World War entered its third year, the United States declared war on Japan, and Roger Sperry (fig. 1.2) concluded his PhD thesis work in the laboratory of Paul Weiss. Sperry was keenly interested in the problem of nature and nurture. In the 1920s and '30s, brain wiring was thought to be the domain of learning and psychology—nurture. The extent to which brain wiring might be heritable was highly contentious and has remained so ever since.[31] In his writings Sperry consistently assigned precision and rigidity to nature and variability and adaptability to nurture. Plasticity was considered an opposing force to specificity. He regarded his supervisor's work as erroneously caught in the "nonspecificity" and even "antiselectivity" camp. In 1963, Sperry wrote about Weiss's theory, "The 'resonance principle' of Weiss, which had remained for nearly 20 years the favored explanation of related phenomena produced by peripheral nerve derangements, was just such a scheme in which the growth of synaptic connections was conceived to be completely nonselective, diffuse, and universal in downstream contacts." His own view and efforts Sperry described as a stark contrast: "At the height of this antiselectivity movement I was led, from evidence indirect and partly behavioral, to postulate again in 1939 a form of chemical selectivity in nerve growth even more extreme in some respects than in the earlier proposals."[29, 32–35]

Sperry repeated experiments similar to those of his mentor's amphibian studies, but this time in rats. First, he transplanted hind limb muscles and shortly thereafter experimentally crossed the neuronal fibers

FIGURE 1.2. Roger Sperry at the California Institute of Technology, 1954. Photo by Walt Girdner, courtesy of the Girdner family and Caltech Archives

innervating the hind limbs. In both cases he found none of the plasticity or adaptations described by Weiss: In his graduate student paper in 1939 he reported on "Functional results of muscle transplantation in the hind limb of the albino rat" that "reversal persisted for more than 10 months with no functional adjustment."[32] In a 1941 paper on the crossing of nerves, he briefly reviewed Weiss' work on amphibians and reported: "These reports imply a plasticity of the nerve centers that is directly at variance with the results of muscle transposition in rats."[29] Sperry drew strong conclusions with far-reaching implications: "[M]ore numerous reports implying a surprising plasticity and regulatory adaptability of the central nervous system to disarrangements of the normal peripheral relations are not to be accepted without question."[29] That's a mouthful, especially for a young graduate student. Sperry depicts a field entirely dominated by the majority opinion of a high degree of developmental plasticity, which he later calls the "antiselectivity movement."[35] And here he is, questioning his PhD supervisor's key theory and positioning himself as the leading proponent of the opposite view.

Sperry moved on to do his postdoc with Karl Lashley in 1942. Lashley was renowned at that time for his theory of equipotentiality, which suggested that all portions of the cerebral cortex are equally capable of holding a specific memory. If one cortical region is damaged, another could do the job—the very essence of plasticity. Lashley further proposed that brain function reduces proportionately to the amount of cortical damage, but not in a specific way for a specific region, something he called mass action. Both of these ideas, equipotentiality and mass action, established important concepts of plasticity that would be revisited and refined over decades. Lashley went as far as assuming that, because of the presumed equal abilities of any cortical region, there would be no discrete functional regions and connectivity would be organized in a diffuse manner. So, both Sperry's PhD supervisor and postdoc supervisor were vocal proponents of some level of nonspecificity of neuronal connectivity, which they felt was required to explain plasticity. And in both his PhD and postdoc supervisors' labs, Sperry published paper after paper in direct contradiction. Viktor Hamburger later noted, "I know of nobody else who has disposed of the cherished ideas of both his doctoral and postdoctoral sponsor, both at the time acknowledged leaders in their fields."[30] Sperry received the Karl Lashley Award of the American Philosophical Society in 1976, a year after Paul Weiss.

Well, this is all part of scientific history: Vocal, opinionated proponents claim large territories and explanatory power for their favorite theories. The extents of these territories are almost never right; in many cases the theories by themselves are not sufficient to explain the phenomena they were designed to explain. And yet, at least some parts of these theories usually prove important and correct. These bits come up again and again and are referred to in new contexts, while the incorrect bits are discarded. By being vocal, the proponents gave their ideas airtime and a name—and this is what is remembered and integrated as science progresses. Other scientists avoided claiming too much and were therefore less likely to be wrong. But without sufficient airtime, some theories, however correct, may have gotten buried in scientific history. We will meet examples for this. Weiss, Lashley and Sperry all were strong proponents of important ideas with names that spawned entire scientific fields, even though they contradicted each other.

Sperry takes center stage in our shared history because his potentially greatest achievement during the early period of the 1940s through the 1960s was to spearhead a transition: Information in neural networks was not only learned, but the result of genetically encoded development. In 1965, Sperry gave a clear account of this transition: "This tendency to favor the functional and to minimize the innate and genetic factors in the developmental patterning of the brain circuits was carried to extremes during the late 20s and 30s," and in the same article: "For a long time it was believed that the developing embryo avoided the whole problem by letting the growth process spin out a random, diffuse, unstructured, essentially equipotential transmission network."[1]

This is not far from where ANNs stand today! Similar to the view of the '20s and '30s in neurobiology, today's neural-network-based AIs come into life large randomly connected. Of course there are many successful efforts to improve how the network is exactly put together, like advanced recurrent or convolutional networks. But none of the networks contain information on what they will be artificially intelligent about after training—they don't know anything yet. The AI is what it learns. By contrast, the monarch butterfly's intelligence is grown, not learned. Again Sperry in 1965: "behavioral networks are organized in advance of their use, by growth and differentiation. . . . The center of the problem has shifted from the province of psychology to that of developmental biology."

From a neurobiologists view, today's AI is stuck in the pre-Sperry days, a functionalist view of brain wiring, where information only enters through learning.[36] If information content of networks "in advance of their use, by growth and differentiation" is what differentiates biological neural networks from ANNs, then it surely is worth tracking down the relevance of growth and differentiation. We do not know yet whether it is required for AI, or whether pure learning will suffice. From this perspective, the purpose of this seminar series is to understand what developmentally unfolded information really is, and what it can achieve, from the perspective of a neurobiologist.

For neurobiology, Sperry's achievement in making brain wiring the science of genetically encoded developmental biology cannot be overstated. It was at the time completely unclear how the genetic encoding could specify connections, but to Sperry this was something that science could figure out. The important thing was that it did, with far-reaching consequences: "The extent to which our individual motor skills, sensory capacities, talents, temperaments, mannerisms, intelligence, and other behavioral traits may be products of inheritance would seem to be much greater on these terms than many of us had formerly believed possible."[37]

Sperry continued to present the field and his views as a defense of specificity made necessary by the tension field, the dichotomy, of specificity versus plasticity. In 1942, while in Lashley's lab, Sperry published his now famous eye rotation experiment for the first time, in newts.[33] When he let the axons regenerate, would they reconnect to the same places as before, such that the animal would see the world upside down? The answer is famously "yes": Sperry could show that the regenerating fibers grew back to where they originally connected. In quick succession in the following year Sperry published a set of papers on regeneration in the motor and visual systems of amphibians, solidifying his view of highly selective and rigid specification of how connectivity is encoded.[34, 38] It was the stepping stone to what he later called "chemoaffinity"—the idea that chemoaffine molecules mark the paths and targets of the growing axons and thereby determine specific connectivity. It was Sperry's answer to the question of nature versus nurture, with chemoaffinity solidly placed in the nature camp, and the "antiselectivity movement" and their plasticity being part of nurture.

In 1941, at the time Sperry graduated in Chicago and prepared to join Lashley, Warren McCulloch prepared his move to Chicago. McCulloch had been working at a neurophysiology laboratory at Yale for some years and was at 43-year-old already an experienced experimental scientist. In Chicago, he teamed up with the logician and mathematician

Walter Pitts, a brilliant young man who had run away from home as a teenager, hid in libraries and wanted to explain the world and the brain with pure logic. His thinking was inspired by *Principia Mathematica,* the hugely influential work on the foundation of modern mathematics by Alfred North Whitehead and Bertrand Russell that popularized symbolic logic.[39] The collaboration of McCulloch and Pitts led to a paper in 1943 entitled "A Logical Calculus of the Ideas Immanent in Nervous Activity."[40] They proposed to model artificial neurons, based on basic neuronal properties and McCulloch's experience as a neurophysiologist. The idea was that a network of such simulated neurons could compute and learn—but it was 1943 and computers to implement such simulations were not available to them.

The paper by McCulloch and Pitts was heavily based on formal logic. In fact, the paper only cited three references, the three books *Principia Mathematica, Principles of Mathematical Logic* by David Hilbert and Wilhelm Ackermann, and *Logical Syntax of Language* by Rudolf Carnap. The work was highly original in every aspect. It represented the first attempt to seriously draw on analogies from neurons and their networks to build a computational, artificial version with the same concept. The *McCulloch-Pitts neuron,* or artificial neuron, became the fundamental unit of neural network analysis. And thus was the field of artificial intelligence born, even though it wouldn't be called that, yet.

The McCulloch-Pitts neurons and subsequent efforts to build neural networks benefitted greatly from another neuropsychologist, whose work was firmly grounded in experimental studies, including brain surgeries and human behavior. Donald Hebb was a Canadian who found his way to physiology and psychology around the time of his 30th birthday in 1934. At that time, he had already worked for years as a laborer and teacher and had recently lost his wife to a tragic accident. He was looking for options in graduate school and eventually conducted his thesis research with Karl Lashley on the influence of visual input on the rat brain in 1936. He had moved via Chicago back to Canada for a teaching position before returning to experimental work with Lashley in 1942, the same year that Roger Sperry started his postdoctoral work with Lashley. The three psychologists spent most of their time in

Florida at the Yerkes Laboratories of Primate Biology, yet Sperry and Hebb pursued separate projects and ideas.

During his years in Florida, Hebb wrote his groundbreaking book *The Organization of Behavior*, published in 1949, which placed behavior solidly in the realm of brain function.[41] With this work, Hebb became one of the founding fathers of neuroscience, and in particular of the field of learning and memory. The *Hebbian learning rule* remains a cornerstone to this day of how we envision learning and memory to occur at the level of connections between neurons. And here I am on solid ground with all of you: Hebb's work is also a foundation of neural networks. His learning rule, most famously formulated in his 1949 book, has been dubbed "cells that fire together, wire together," called Hebbian learning or even *Hebb's law*. Hebbian learning was what brought McCulloch and Pitts neural networks to the next level. Today's learning algorithms for artificial neural network still build on Hebb's law.

The outcome of Hebbian learning is that the network contains information that was not there before—it has learned something. The most common current approach in artificial neural network design is to start with random synaptic strengths, representing a "no information" state. The information content increases when the network learns, and the network learns by being presented with examples of what we want it to learn. By contrast, biological brains do not start out in life as networks with random synapses and no information content. Biological brains grow. A spider does not learn how to weave a web; the information is encoded in its neural network through development and prior to environmental input. How the information apparent in the genome leads to the increase of information apparent in the network is something we surely would like to understand—with an eye on both biological and artificial neural networks. Today, I dare say, it appears as unclear as ever how comparable these two really are. On the one side, a combination of genetically encoded growth and learning from new input as it develops; on the other, no growth, but learning through readjusting a previously random network. In both cases, the goal is for the network to become more intelligent through increasing its information content. So, what actually is information?

FIGURE 1.3. Norbert Wiener and Claude Shannon, undated.
Picture courtesy of MIT Museum

This question brings us back to the 1940s to meet Claude Shannon (fig. 1.3), the American mathematician and engineer who is regarded as the father of information theory.[42] He is well known, amongst other things, for coining the term *bit* for a binary digit and for his work on reliable encoding of information from a sender to a receiver. Shannon wondered about the strange relationship between entropy and information. Entropy is a measure of disorder. In thermodynamics, we find that we need to invest energy to increase order. In information theory, we find that we need information to increase order. There is a remarkable relationship between energy and information, which we will discuss in a bit more detail in the next seminar. Shannon considered that a disordered system is missing information. His calculations relied heavily on probability theory and previous work by Norbert Wiener (fig. 1.3), the founding father of cybernetics and close colleague of McCulloch and Pitts. According to John Scales Avery, when Shannon discussed his "theory of missing information," Wiener suggested, "Why don't you call it entropy?" Ever since, information entropy, also called Shannon entropy, is used to describe and calculate uncertainty based on missing information.[42]

Ordered data
Data: 00000000001111111111 **Algo: 10*0, 10*1**

Disordered data
Data: 01101000101101001110 **Algo: 01101000101101001110**

FIGURE 1.4. Algorithmic information and data compression. Energy and time are required to order data (reduced entropy). Ordered data can be described in compressed form. By contrast, disordered data (increased entropy) cannot be described in compressed form. Algorithmic information theory describes the type of information sufficient to recreate (decompress) the full information content.

Let me give you a simple example to help understand both the historical and present relevance of this concept. Take a random series of 0s and 1s. You need energy or information to order the 0s and 1s. A random, maximally disordered series has maximal Shannon entropy; it "misses" information. To sort the series means to put in this missing information (or energy). Consequently, a more ordered series has less missing information and lower Shannon entropy. This state of affairs has the important side effect that the ordered system is now easier to describe. In fact, a perfectly sorted long series of *n* 0s and *m* 1s can very simply be compressed to: *n* times 0, *m* times 1 (fig. 1.4), instead of writing it out digit by digit. In the random series, we do not have that option and are forced to write out every single digit. The compressed description of the ordered series is an algorithmic description. The algorithmic description can be short; it contains less information than the complete series when read digit by digit. And yet, it contains all the information necessary to precisely write down the complete series. Where is the missing information coming from? You guessed it—just as energy was needed to create the ordered state, so do you need to invest time and energy to decode, to unfold the algorithmic information to yield the complete series from its short, algorithmic description.

This is the idea of algorithmic information theory. It was first developed based on statistical methods by Ray Solomonoff following the now famous Dartmouth AI workshop in the late '50s and early '60s.[43–45] Together with the subsequent work of Andrey Kolmogorov and Greg Chaitin, algorithmic information theory helped to describe a type of information that is sufficient to encode an apparently larger information content. The physicist Charles Bennett later pointed out how the decompression of algorithmically encoded information comes at a cost. In order to unfold *n* times 0, *m* times 1 to a string of 0s and 1s, both energy and time must be expended. Again, just as energy and time was expended to sort the digits, that is to reduce entropy, energy and time must be expended to decode the compressed description of the ordered state. Well, this should remind us that energy and time must be expended to unfold the information in the genome to grow a neural network or to train an artificial neural network.

This may not sound exactly how developmental biologists describe what they do. Nor does it seem to be much on the minds of designers of artificial neural networks, except that they suffer the consequences of computation time needed to feed their networks with big data. But both are dealing with the same problem of energy and information. The relationship of energy, information and algorithmic information theory will be as important throughout the following seminars as it was for the history throughout the decades that followed their discovery. So let me get on with my story.

Shannon had met Alan Turing first in 1943, and by the early '50s they developed chess programs together. Their approaches were based on the algorithmic logic of game trees and designed with John von Neumann's computer architecture in mind. This was the design of the first digital computers, again all based in mathematical and logical rigor. Shannon proposed an algorithm for making decisions in a chess game. The program analyzed and evaluated all possible moves and resulting board positions for a depth of two or three moves. How exactly the evaluation was done, the weights for different board configurations assigned, was critical. Neither Shannon nor Turing had access to sufficient computer power, so they actually tried it out in games in which the two players simulated computers![46] All this work was solidly steeped in the mathematically rigorous, logical tradition of those days.

In post-war 1946, Roger Sperry moved from Florida back to Chicago for his first faculty position, and Donald Hebb left Florida a year later for a professorship back in Montreal, Canada. Turing was appointed an officer of the Order of the British Empire for his work on cryptography in England, and Shannon prepared several publications on cryptography and communication that formed the groundwork for information theory. McCulloch and Pitts thought further about designs for theoretical neural networks to recognize varying visual inputs, one of the first studies on pattern recognition.[47] The same year, a 19-year-old student with the name Marvin Minsky entered Harvard as an undergraduate student.

According to his own account, Minsky was fascinated by neurology.[46] A zoology professor gave him a lab so he could figure out how the crayfish claw worked. This involved dissecting a lot of crayfish and cutting and reconnecting the nerve fibers innervating the claw. Another professor in psychology seemed to be equally happy to provide Minsky with lab space to work on ideas on how the mind works. In Minsky's words: "I imagined that the brain was composed of little relays—the neurons—and each of them had a probability attached to it that would govern whether the neuron would conduct an electric pulse; this scheme is now known technically as a stochastic neural network. . . . It turned out that a man in Montreal named Donald Hebb had come up with a similar theory, but at the time, luckily or unluckily, I didn't know of his work."

Minsky had read the 1943 paper by McCulloch and Pitts, and he wanted to build a neural network. The opportunity to do so presented itself after his move to attend graduate school in Princeton. Together with a physics student and electronics nerd named Dean Edmonds, he drew up a plan for how to build one. They convinced his mentor George Miller at Harvard of the design, and Miller organized the money to buy the components. And thus, in the summer of 1951, was the first artificial neural network built. Mind you, we are not talking about a simulation on an early computer; they literally *built* a neural network machine. Minsky again: "It had three hundred tubes and a lot of motors. It needed some automatic electric clutches, which we machined ourselves. The memory of the machine was stored in the positions of its control knobs—forty of them—and when the machine was learning it used the clutches to adjust its own knobs. We used a surplus gyropilot from a B-24 bomber to move the clutches."[46]

The machine simulated a moving entity, a "rat" in a maze, whose movements were indicated by an arrangement of lights. A light would turn on and at first move randomly, but then, by design, certain choices would be reinforced to increase the probability of the rat making that choice. Minsky: "It turned out that because of an electronic accident in our design we could put two or three rats in the same maze and follow them all. . . . The rats actually interacted with one another. If one of them found a good path, the others would tend to follow it. We sort of quit science for a while to watch the machine."[46]

Minsky and Edmonds had actually connected many of the bits randomly, so it was impossible to predict what exactly would happen. This turned out to be critical for the function of the machine. In Minsky's words: "Because of the random wiring, it had a sort of fail-safe characteristic. If one of the neurons wasn't working, it wouldn't make much of a difference—and, with nearly three hundred tubes and the thousands of connections we had soldered, there would usually be something wrong somewhere. In those days, even a radio set with twenty tubes tended to fail a lot. I don't think we ever debugged our machine completely, but that didn't matter. By having this crazy random design, it was almost sure to work, no matter how you built it."[46]

Let's let this sink in a bit. Here we are, in 1951. The McCulloch and Pitts neuron was based on the formal propositional logic as laid out in the *Principia Mathematica* and Turing's theory of computation.[39, 48] Shannon and Turing worked on intelligent programs to play chess based on precise algorithms and von Neumann computers were being built at many universities. These computers were very precise, even though they were reliant on vacuum tubes that were notoriously unreliable. If one of thousands of tubes failed, the whole machine was very likely to fail. At the same time, Minsky and Edmonds had soldered together their machine, including some three hundred vacuum tubes, with a "crazy random design" which kind of always worked, except it wasn't quite clear what it would do. In fact, in the same year of 1951, Ray Solomonoff had already begun to develop the math of random nets.[49] Solomonoff's foundational work on algorithmic information theory, like Shannon's information entropy, was deeply steeped in probability theory. Solomonoff would later call it the "discovery of algorithmic probability."[50]

Here we have a dichotomy in a nutshell, two approaches to science: One based on formal logic, the other on probability. One based on precision, the other on imprecision. One based on total control, the other on trial and error. This dichotomy has haunted AI research until today. It has haunted developmental neurobiology as well. Sperry never grew tired of emphasizing this type of dichotomy, the "diffuse non-selectivity in nerve growth versus precise specificity, functional plasticity versus functional implasticity, mechanical and non-selective versus chemical and selective."[51] As in this quote from a book review from 1971, Sperry's categorization of nature and nurture over the years left less and less wiggle room between the two. Genetically encoded meant: precise, chemically defined, developmentally rigid and specific nature. Plasticity meant: "diffuse non-selectivity" and environmental or learned variability brought in by nurture. The dichotomy was a hallmark of its time, and it takes us to this day to witness a slow appreciation for how both precise and variable outcomes can result from both specific or stochastic developmental processes. What we associate with nature or nurture may require revision, but the thought of a tidy dichotomy carries an enduring attraction.

To be sure, formally logical, precise and controllable systems are undeniably attractive to those who seek formal understanding. Imagine you are a mathematician who wants to understand and build a problem-solving machine from first principles. You have read your *Principia Mathematica*. You understand logic and what constitutes a mathematical proof. And if you build a system step by step, the math better add up; you need to control every step during assembly.

Now imagine you are a biologist who observes the astounding specificity with which growing fibers in an animal reconnect anatomically precisely where they belong. You have read your Cajal. You understand that neuronal connectivity between a vast number of exquisitely shaped axonal and dendritic morphologies must connect highly selectively for the whole to function. And if you build a system step by step, then connectivity better add up; you need to control every step during the assembly.

I think most of us can relate to this intuition for the need of specificity and control in both cases. I am less sure what drove Minsky and Edmonds to build their "crazy random design" neural network, but it certainly demonstrated something that may be just as important as precision for any network, be it artificial or in the brain: robustness. John von Neumann first introduced the idea of "Probabilistic Logics and Synthesis of Reliable Organisms from Unreliable Components" in 1956,[52] further developed by Shmuel Winograd and Jack Cowan in 1963.[53] In 1964, Michael Arbib, then a 24-year-old mathematician who had just received his PhD from work with Wiener at MIT, published the book *Brains, Machines and Mathematics,* which would become hugely influential in the field of computational neuroscience, then in its infancy.[54] Arbib reviewed the idea of robustness up to that point and found "if a modular net, designed for a specific function, is transformed to minimize its susceptibility to errors due to malfunction or death of neurons, the resultant network has a detailed randomness which is akin to that found in the detailed anatomy of the brain."[54]

Where did the idea of "randomness in the anatomy of the brain" come from? The same question was asked by Mike Gaze (fig. 1.5), a neurobiologist working on the development of neuronal connections. Born the same year as Minsky, 1927, Gaze quoted Arbib in his 1970 book *The Formations of Nerve Connections* and wondered, I quote: "This statement to some extent begs the question, since as I have already argued, we do not know to what extent the detailed structure of the brain is ordered or random. In view of these results, however, we may now reverse the argument: redundancy and especially randomization of connections were built into these automata partly because it was thought that connections in real neural systems were redundant and somewhat random. Automata so built can be shown to work; therefore we would do well to consider whether, in certain of its parts, the nervous system

FIGURE 1.5. Michael Gaze, sometime in the 1960s.
Photo courtesy of Robinetta Gaze

might now show some 'randomness' of connection also. And so-back to biology and the investigation of whether localization of function always goes along with specificity of connection."[55]

Isn't this just beautiful? Here are some mathematicians and AI people thinking about how to make their networks more robust. "Let's make 'em a bit more random," they say, "after all, that's what real brains do." In response, there comes the neurobiologist and says, "Actually, I don't know whether that's true and whether there really is anything like this randomness in the brain, but I like that your artificial networks work, so we should indeed look out for that in our brains!" Of course brain wiring turned out not to be simply random. But, remarkably, random processes during algorithmic growth not only contribute to variability, they can actually be required to ensure precision and robustness, as we will see later.

I dare say the idea of anything random about brain wiring still does not sit well with many geneticists and developmental neurobiologists to this day. It definitely didn't sit well with Sperry. As Gaze put it, "if we contemplate a nervous system completely specified in Sperry's sense, then the whole system will be expected to be morphologically rigid."[55] Sperry was indeed comfortable with morphological rigidity, as long as it didn't imply functional rigidity. In 1958, almost twenty years into his campaign against the "antiselectivity movement," Sperry argued: "Given a morphologically rigid circuit system of sufficient complexity, it is possible, for example, to get an almost unlimited variety of different responses from the same invariant stimulus simply by shifting the distribution of central facilitation."[56] Hence, morphological rigidity was very much okay with Sperry and not a problem for functional plasticity. This is indeed true for most artificial neural networks in use today. When it comes to how the brain is wired during development, Sperry's strict view equated rigidity with specificity, leaving no role for developmentally relevant plasticity, stochasticity or variability. To Sperry, nature meant specificity and nurture meant plasticity, and he regarded these two as opposites with little room for overlap.

The study of developmental brain wiring has mostly remained a study of precision to this day. According to the "specificity view," a precise network has to develop before it starts to function and exhibit plasticity. Developmental variability may be unavoidable to some extent, but only as a limit to the precision that the system is striving for. However, plasticity turned out to be not only a feature of nurture, but part of nature as well. An important intersection of neuronal activity-dependent plasticity and genetically encoded development was highlighted in the 1990s with the work of Carla Shatz (fig. 1.6). Shatz reinvoked Hebb's law as "cells that fire together wire together"—this time for the

FIGURE 1.6. Carla Shatz and Friedrich Bonhoeffer at Cold Spring Harbor Laboratory, 1990.
Courtesy of Cold Spring Harbor Laboratory Archives, NY

developmental process of specifying the connectivity of visual projections into the brain.[57] It's the same system that Sperry and Gaze had investigated, but this time in ferrets. Shatz proposed that spontaneous, random activity waves crossed the developing ferret retina and blocking neuronal activity would lead to a failure in establishing precise connectivity. Remarkably, these waves occurred before the retina could actually see anything. That really means they are part of development as determined by "nature," part of what is genetically encoded—no environment or nurture added. This is an example where a random process during development is required to make the outcome precise.[57, 58] This is another idea that we will revisit several times, as it definitely is a centerpiece of the developmental design of both artificial and biological neural networks. For now, let's just hold the thought that a role of neuronal function and network activity does not have to be opposed to genetically encoded development. On the contrary, Shatz's activity waves and the idea of "cells that fire together wire together" can be regarded as part of the unfolding information during algorithmic growth encoded by the genome.[59]

And with this, we are back to information: clearly, there is no gene in the ferret's genome that "encodes" an electrical activity wave, let alone its randomness. Just as there is no gene in the genome that encodes any given specific synaptic connection out of a billion. Yet, we often think of genes as something that encodes a property we are interested in, even Minsky: "Children can learn to speak their parents' language without an accent, but not vice versa. I suspect there is a gene that shuts off that learning mechanism when a child reaches sexual maturity. If there weren't, parents would learn their children's language, and language itself would not have developed."[46] We are entering the debate about information that encodes the making of a network, as opposed to information that describes the output state, the connections of that network.

The recipe for brain making does not only produce networks of astonishing precision, but also flexibility and robustness. The extent to which the brain could be flexible bothered some who had described far-reaching plasticity in the brain, first and foremost Lashley. As Gaze described: "learning was dependent on the amount of functional cortical tissue and not on its anatomical specialization. So confusing did Lashley find his own results that, after more than 30 years of investigation into learning and memory, he felt tempted to come to the conclusion that the phenomena he had been studying were just not possible!"[60, 61] Gaze continues: "The apparent equipotentiality of cortical tissue revealed by Lashley's work, together with anatomical evidence which suggests that there is a continued and extensive loss of neurons from the cortex during adult life has led biomathematicians to the conclusion that the vertebrate central nervous system as a whole shows, in its performance, a system error-rate which is much lower than the error-rate of its components."[55, 62]

How do neural networks achieve precision with flexibility and robustness? A seed for an answer to this question was laid already in the 1943 paper by McCulloch and Pitts, which also laid the foundation for two branches of AI research that persist to this day: the "logical versus analogical, the symbolic versus connectionist" tradition.[48, 63] Right from its inception, there was something slightly enigmatic about the idea of neural networks. The McCulloch-Pitts neuron was based on the idea of logical operations. However, the moment such artificial neurons are put together the possibility of the new network learning was imminent, bringing with it concepts of flexibility and plasticity. There is something rigid and logically precise about a mathematical model for a neuron's function. An error in such a system is likely to render the whole thing dysfunctional. Do we have to give up control to gain robustness, or at least something that works at all? The development, interaction and tension between these two branches of AI research played out over decades.

Minsky expanded his "crazy random design" network as his PhD project. In his words: "I had the naïve idea that if one could build a big enough network, with enough memory loops, it might get lucky and acquire the ability to envision things in its head. This became a field of study later. It was called self-organizing random networks. Even today, I still get letters from young students who say, 'Why are you people trying to program intelligence? Why don't you try to find a way to build a nervous system that will just spontaneously create it?' Finally, I decided that either this was a bad idea or it would take thousands or millions of neurons to make it work, and I couldn't afford to try to build a machine like that. . . . In any case, I did my thesis on ideas about how the nervous system might learn."[46]

His environment was certainly helpful to match his ambition. John Nash (*A Beautiful Mind*) worked with him, as did von Neumann, who was on Minsky's thesis committee. In 1954, with the support of Shannon, von Neumann and Wiener, Minsky returned as a junior fellow to Harvard. Two years later, in 1956, a small workshop in Dartmouth officially started and gave the new field of artificial intelligence its name. Minsky was there, as were Shannon and Solomonoff (fig. 1.7). The field started with the tension of the dichotomy of logical, mathematically rigorous control on one side and the more flexible, connectionist approaches on the other side already in the making. The different sides were immediately connected to individuals in the field because of the famous duo of Marvin Minsky and John McCarthy. McCarthy, like Minsky born in 1927, finished high school two years early and was accepted at Caltech at the age of 17, where he first heard lectures by John von Neumann. McCarthy's first faculty job was at Dartmouth in 1955, where he promptly organized the workshop that would become generally known as the birthplace of AI. It was McCarthy who gave the field its name.[5]

Both Minsky and McCarthy moved to MIT in 1958 where they intensely collaborated and shaped the field of artificial intelligence in a world-leading effort for the next five years. But their collaboration was not to last, as McCarthy's work was strongly rooted in formal logic, whereas Minsky, not surprisingly, tended to "just get it to work somehow"—his approach has even been described as having an "antilogical outlook."[48] As Minsky would say: "Commonsense reality is too disorderly to represent in terms of universally valid axioms."[63] The different approaches were to become cornerstones of the two opposite ideological poles of AI. And yet, neither of these two poles focused on neural nets. In 1956 the name of the game was symbol-processing AI. The early successes were achieved with systems that were rooted in

FIGURE 1.7. Some participants during the Dartmouth AI workshop, 1956. In back: Oliver Selfridge, Nathaniel Rochester, Marvin Minsky, John McCarthy; in front: Ray Solomonoff, Leon Harmon, and Claude Shannon.
Courtesy of the Minsky and Solomonoff families

mathematical logic—the tradition that McCarthy would hold up for the rest of his life. A program called *Logic Theorist* that Allen Newell and Herbert Simon had brought to the Dartmouth workshop was able to prove most theorems of *Principia Mathematica*. The early programs by Newell, Simon and McCarthy were all designed to use knowledge and logic to solve problems.[48]

The early pioneers (with the exception of Norbert Wiener) were seduced by the newly emerging computers. The feeling was that thinking processes could be simulated better with the new machines than by trying to simulate the brain, i.e., with physical networks of artificial neurons, like Minsky's and Edmond's.[5] In fact, Minsky himself gave up work on neural networks and focused his efforts for years to come on symbol-processing AI, though he continued his approach of getting things done even if it meant giving up some formal logic and control. Neural networks were a parallel, less regarded field. Another early parallel development was the rather meandering work by Norbert Wiener at MIT on what he called the frontier areas between engineering and biology. He thereby created the field of cybernetics and became one of the founding fathers of Artificial Life research. The basic premise was that feedback between components or routines of a system could exhibit new system properties and become unstable. As Daniel Crevier put it in his book *AI: The Tumultuous History of the Search for Artificial Intelligence* in 1993: "Wiener speculated that all intelligent behavior is the consequence of feedback mechanisms." In their 1943 paper, McCulloch and Pitts included the idea of feedback between units.[5]

By the late '50s new approaches to create intelligent systems appeared. This period saw the birth of genetic algorithms (also called machine evolution or evolutionary algorithms) by Richard Friedberg, John Holland and others.[64–66] Such evolutionary approaches were at the opposite end of formal logic approaches and were rarely applied in AI research for many years.

Neural network research continued as a parallel, sometimes considered rival, field to symbol-processing AI. Minsky and his influential school of thought disregarded neural nets as a significant approach in AI, even though the first turning point in the history of ANNs was arguably when Minsky and Edmonds had built one in 1951. While they had turned their back to neural net research thereafter, a former high school classmate of Minsky's developed a powerful computer simulation of a simple neural network that he called the Perceptron. Frank Rosenblatt (fig. 1.8) published the first computer-simulated ANN in his seminal 1958 paper "The perceptron: a probabilistic model for information storage and organization in the brain."[67] Rosenblatt's work was well publicized, leading the magazine *Science* to feature an article entitled "Human Brains Replaced?" The *New Yorker* and other magazines ran interviews that overstated the Perceptron's abilities.[5] The Perceptron was a single layer of McCulloch-Pitts neurons that received input from a sensor layer and gave output to an activator layer (fig. 1.9). The input layer comprised

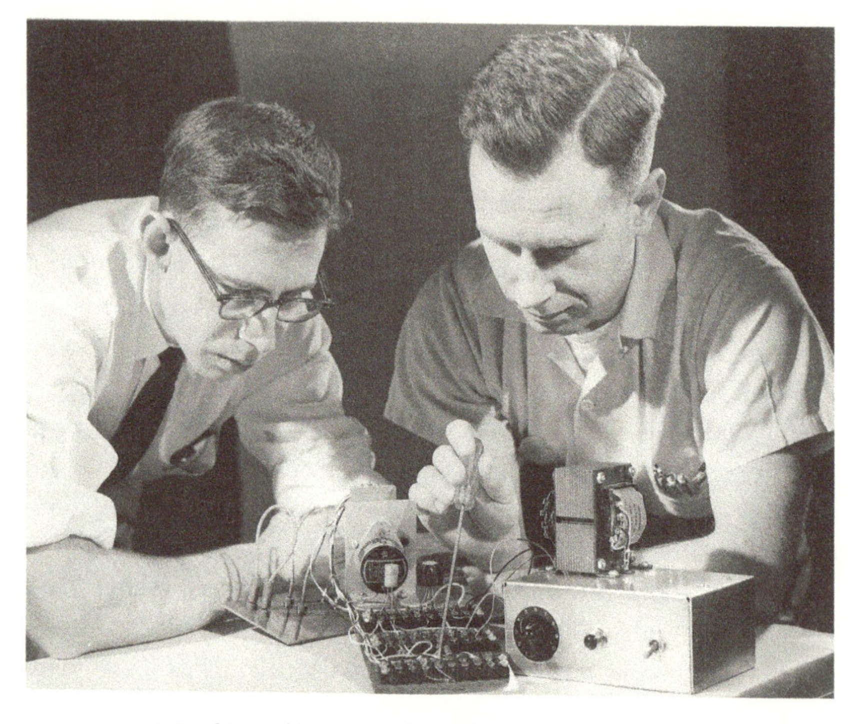

FIGURE 1.8. Frank Rosenblatt and Charles Wightman with a perceptron association unit prototype, 1958.
Courtesy of Division of Rare and Manuscript Collections, Cornell University Library

400 photocells and McCulloch-Pitts neurons that Rosenblatt called the associator unit. 512 associator units were randomly connected to the photocells. That's right, up to 40 random connections per associator unit were implemented, and their specificity was random by design (fig. 1.10). Remarkably, Rosenblatt did this probably less to mimic the haphazard "crazy random design" of the earlier Minsky-Edmonds network, but rather because he thought that this was, again, what he thought a real brain looked like. Underlying this assumption was the one question we are circling around and we will be busy with throughout this seminar series: What kind of information would allow the generation of myriads of specific contacts? Rosenblatt, like many of his contemporaries in computer engineering and mathematics, felt that there is no such information—the neurons just had to be wired more

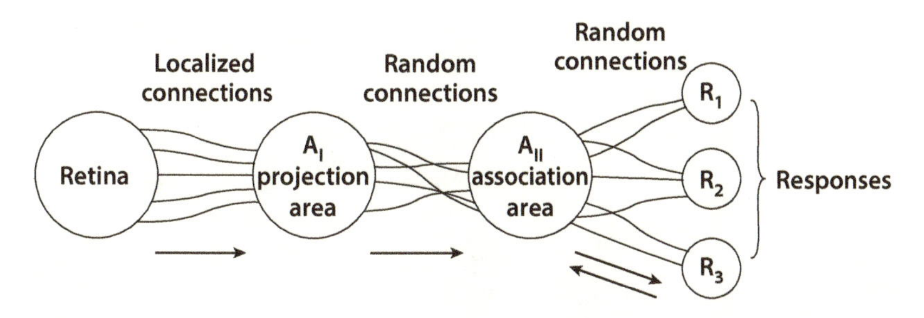

FIGURE 1.9. Reproduction of the schematic of the perceptron based on the original 1958 paper by Frank Rosenblatt. "The perceptron: a probabilistic model for information storage and organization in the brain." *Psychol Rev* 65, 386–408 (1958).

or less randomly. Relying on references to Donald Hebb, Ross Ashby and others, Rosenblatt wrote in his 1958 paper: "At birth, the construction of the most important networks is largely random, subject to a minimum number of genetic constraints."[67] These are certainly the types of words that Sperry must have felt are a reflection of what he described as the "antiselectivity movement."

Be that as it may, just like the Minsky-Edmonds network, Rosenblatt's Perceptron did work—somewhat, somehow. Its learning rule defined how it would learn the things it could, but it had fundamental limitations in the types of things it could learn at all. In 1962 Rosenblatt published a book entitled *Principles of Neurodynamics* in which he compared symbol-processing AI, to its disadvantage, with neural networks.[68] None of this sat well with Minsky. In 1969 Minsky, together with his new scientific partner Seymour Papert (fig. 1.11), spent several years working out the limitations of perceptrons and eventually published a book of their own, aptly entitled *Perceptrons*. They didn't mince their words: "Appalled at the persistent influence of perceptrons (and similar ways of thinking) on practical pattern recognition, we determined to set out our work as a book."[69] Imagine disliking a concept so much that you spend years writing a book about its limitations and then publish it under the name of the concept itself! To keep the record straight, Minsky himself as well as his students report that their interest was genuine—and had their analysis of perceptrons turned out with less obvious limitations, they might have jumped on it instead.[5] The alleged

FIGURE 1.10. Mark I Perceptron at the Cornell Aeronautical Laboratory. Courtesy of Division of Rare and Manuscript Collections, Cornell University Library

fundamental limitations were indeed overcome later, and Minsky himself would return to neural nets and work based on Rosenblatt's Perceptron years later. Today, of course, practical pattern recognition is a poster child for the success of neural networks. But starting with the 1970s, the effect of the Minsky's and Papert's book was a de facto 15-year hiatus on

FIGURE 1.11. Marvin Minsky and Seymour Papert, 1971.
Courtesy of Cynthia Solomon

work regarding neural networks of almost any kind. Rosenblatt himself drowned in a boat accident. It took until the '80s for neural networks to revive—and today, of course, ANNs have all but become synonymous with AI.

The successes of ANNs we have witnessed in recent years were not obvious at any time before 2010. While the field moved slowly following Minsky's and Papert's book in the 1970s, neural networks continued as a parallel field to mainstream AI in the 1980s. At this time, a young student who wanted to study the biological brain to understand intelligence would not necessarily receive encouragement in the AI community. Jeff Hawkins was a trained electrical engineer who wanted to make his dream of understanding intelligence a reality, but in the mid-'80s seemed to have been in the wrong place at the wrong time. In his words: "To these scientists, vision, language, robotics and mathematics were just programming problems. Computers could do anything a brain could do, and more, so why constrain your thinking by the biological messiness of nature's computer? Studying brains would limit your thinking. . . . They took an ends-justify-the means approach; they were not interested in how real brains worked. Some took a pride in ignoring neurobiology."[70] Jeff Hawkins instead went back to Silicon Valley, invented the Palm Pilot and made enough money to found his own neuroscience research institute in 2002. We will hear more about him later also.

While the field of AI looked elsewhere, research on the development of biological neural networks boomed throughout the 1970s. From Sperry's perspective, the field was mature. Already more than 30 years had passed since he started correcting the "antiselectivity movement" of the '20s and '30s. Sperry himself had continued to solidify a strong, rigid version of the chemoaffinity theory that defined individual cell-cell contacts throughout the '50s and '60s. Landmark experiments were published in 1963 by Sperry and Attardi, showing that regenerating axons, when given the chance to innervate different area of the target area, show remarkable specificity for those areas they were originally connected to.[35, 71]

Yet, at the same time as Sperry declared victory, experiments were done that seemed to require some explanation. Mike Gaze, Sansar Sharma (fig. 1.12) and others had performed regeneration experiments in which they cut half the target, the tectum, away, wondering whether now the arriving axons would simply have nowhere to go. This at least is the expected result if the target contained precise chemoaffine target definitions. However, they observed that all the axons now innervated what was left of the target, just in a compressed fashion (fig. 1.13).[55, 72] Don't worry about the details, let's just hold the thought that somehow the axons simply arrived at the target and used what was left of it, with none unable to connect. Strict chemoaffine tags of the target would not allow compression, because half of them should just be gone. Gaze, Sharma and colleagues proposed a model called "systems-matching" based on their observations of developmental flexibility. The idea is that it is not the absolute target position that is encoded (think endpoint information), but relative positions of the axons as they establish a pattern in time (think algorithmic information). Systems-matching provided an explanation for how axons from a whole or partial retina can occupy whatever whole or partial target space is available.[73]

FIGURE 1.12. Sansar Sharma in the mid 1960s at Edinburgh University. The set-up includes a perimeter to display visual stimuli and electrophysiological recording equipment to record exact visual fields (in this case of a goldfish). Electrophysiological activity recordings remain the gold standard for the mapping of functional synaptic connectivity to this day.
Courtesy of Sansar Sharma

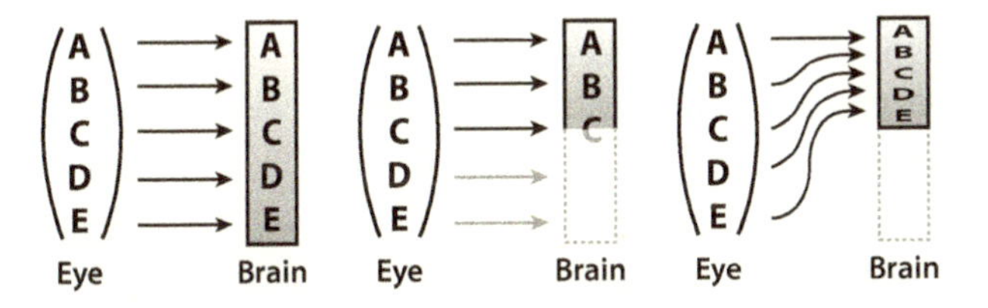

FIGURE 1.13. Classic connectivity concepts based on experiments in the visual system. On the left: Roger Sperry envisioned a precise point-to-point map from the eye to the brain based on matching *chemoaffine tags*. Middle: Removal of half the target area should remove half the target code. Right: Experiments have revealed that the projections from the eye have the ability to form a compressed map in the remaining target, arguing against a precise point-to-point map.

Sperry had an answer. He could show that, depending on the animal and the timing of the target ablation, his predicted result could still prevail. In 1976 he published a long, detailed rebuttal of those experiments that challenged his strict form of chemoaffinity with the following argument: Given the lesions were induced early enough, even a reduced target area would develop the full, albeit compressed, rigid target code. If, however, the lesion was induced too late, then the remaining target would not be able to reestablish a compressed code (fig. 1.13).[74, 75] This was a plausible explanation and Sperry recognized the importance of developmental timing. But alas, the rigid target code never squared with observations of developmental flexibility. Instead, in the same year, experiments revealed that under the conditions of the 1963 Sperry and Attardi experiments, neuronal debris left over from previous nerve crushing was sufficient to guide regenerating axons to their original targets.[76–78] Even though it provided an explanation for target-specific reinnervation results, including the eye rotations in amphibians, the debris guidance idea did not gain widespread traction. For several years the field remained in a state of some confusion caused by complications resulting from details of the experimental procedures in different organisms. For example, it remained uncommon to wait the full year it took for all axonal debris to vanish after an operation in a goldfish. Theories were peddled ranging from a pure sliding scale model with no address code to Sperry's strict chemoaffinity.[55, 79] A theory that was particularly close to the mark was put forward by Sperry's former postdoc Myong Geun Yoon based on experiments in the 1970s. Yoon argued that a systems-matching mechanism that would regard the target area purely as a "passive receiver" went too far. He instead provided evidence for a modified version that he called the "topographic regulation" hypothesis in which the tectum functions as an "active accommodator." This way, target-derived molecular functions were back on the table, albeit not as a rigid address code.[80–82]

Key target-derived molecules were finally found in the mid-'90s by Friedrich Bonhoeffer (fig. 1.6) in Germany and John Flanagan in the United States,[83, 84] as we will discuss in detail in the seminar on molecular mechanisms. The bottom line of these findings was: There are indeed molecules and they do form gradients, both as Sperry had postulated. However, these molecules do not form rigid chemoaffine target codes, but redirect incoming axons during developmental growth by repulsion. They do in fact leave the axons to sort themselves out by relative positioning. So when it comes to how information is encoded in getting the wiring done, Gaze, Sharma and colleagues were closer to the mark. Sperry is commonly cited as having predicted the molecular mechanisms based on his visionary idea of the gradient-forming molecules in the target. By contrast, those who had proposed models for relative positioning of axons, a concept Sperry disparaged until the end, are largely forgotten. The idea of a precise one-to-one mapping has proven more powerfully intuitive and simple, even when it turned out to be insufficient.

It remains unknown how convinced Sperry was by the contradictory data. According to Sansar Sharma, in 1974 Sperry acknowledged to him the validity of their work. It would take Sharma until 1981 to show that retinal axons form ordered projections even if the correct target brain regions are ablated on both sides in fish. Even in this complete absence of correct targets, retinotopic maps form—in non-visual regions of the brain.[85] However, the original work by Sharma and Gaze was previously rejected by Sperry in his capacity as editor of the journal *Experimental Neurology*. According to Sharma, Gaze decided never to send papers to American journals for the rest of his career. Sperry's forceful presentation of his scientific ideas was already obvious when he disregarded both his acclaimed PhD and postdoc supervisor's major theories based on his own first findings and a strong belief in his idea. Having spent the following more than 30 years reiterating and substantiating these views again and again, he went to great lengths defending them in 1975 and 1976. And yet, in this lengthy dealing with the new evidence, Sperry also performed the most significant departure from the concepts he had been promoting for decades. In 1976, Sperry wrote: "It was not a precise cell-to-cell connectivity that was inferred, but rather cell-to-focal tectal area with extensive terminal overlap among neighboring fibers. The competitive interaction among overlapping terminal arbors for synaptic sites and appropriate density of innervation was conceived to involve preferential, graded affinities, not all-or-none selectivity."[75] This was rather different from 1963, when he wrote "cells and fibers of the brain and cord must carry some kind of individual identification tags, presumably cytochemical in nature, by which they are distinguished one from another almost, in many regions, to the level of the single neuron; and further, that the growing fibers are extremely particular when it comes to establishing synaptic connections, each axon linking only with certain neurons to which it becomes selectively attached by specific chemical affinities."[35] Remarkably, it is the reference to "Sperry, 1963" that has become the *de facto* standard reference for chemoaffinity in molecular studies on brain wiring ever since.

I hear you, and especially the AI camp asking: Why so much about Sperry and so many of his quotes? We have reached a key moment in the historical debate that defines where we still stand today on the information question in biological and artificial neural networks alike: Sperry marked the transition to appreciating information that enters the network prior to learning, a step never fully taken for ANNs; however, with Sperry's transition came his unequivocal stance that this type of information had to be rigid and precise, a one-to-one mapping of molecules to connectivity. This is not helpful to ANN developers because it remains unclear what process is supposed to achieve the initial connectivity. We now know that the genome does not map to connectivity one-to-one, but allows neural networks to grow in a process that is not always rigid and precise. Yet again, this only begs the question where the information is coming from. This then is the core question of our workshop: *What difference does it make for the intelligence of a neural network whether connectivity is grown in an algorithmic process that is programmed by evolution, versus only learned based on a designed network with initially random connectivity?* To answer this question we need to understand what we mean with an "algorithmic process that is programmed by evolution"—and this a key focus of the following seminars.

At the time of this controversy, in 1976, remarkable papers were published by Gaze with this student Tony Hope in England, and D. J. Willshaw and Christoph von der Malsburg in Göttingen, Germany.[86, 87] Both presented computational models for how retinotectal projections could sort themselves out based on relative positioning. Both showed it could be done, at least in principle, through algorithmic growth. Willshaw and von der Malsburg concluded: "The prime value of our paper is to show that such theories of self-organization do in fact work, and they should now be taken seriously. They have the advantage of requiring only an extremely small amount of information to be specified genetically, and we have not had to make embarrassing assumptions about regulation, that is, relabelling of cells according to global observations to make mappings possible under new conditions."[86]

Well, if this sounds like a major scientific dispute, it is not commonly remembered as such in the field. When asked, most of those who witnessed the '70s in the field recall Sperry's dominance and influence. Mike Gaze has unanimously been described by his students and colleagues as a more gentle scientist. He did not promote his ideas or concerns other than by self-critical publications and ringing the bell of caution. His 1970 book is the perfect example for this. M. S. Laverack wrote in a review of Gaze's book: "It is satisfying but also disappointing to read that for almost every experiment there are several interpretations and that definitive results are difficult to obtain."[88] Gaze presented the data as he saw it—and if the data was confusing, so be it. Just because the systems-matching model was the better fit for the data as he saw it, didn't lead him to promote it in competition with chemoaffinity. A debate that could have been.

Sperry, on the other hand, did not mince his words. He also reviewed Gaze's 1970 book and found: "Occasional efforts to introduce original interpretation or to support the author's stated bias against behavioral and histological evidence in favor of the electrophysiological techniques that he himself has employed, or to assess some of the broader implications, are not the strongest features. In particular, the polar shifts in basic guideline concepts that occurred in the field in the early 1940's and which brought an about-face in the terms of interpretation (e.g., from diffuse non-selectivity in nerve growth to precise specificity, from functional plasticity to functional implasticity, from mechanical guidance to chemical guidance, from impulse specificity to connection specificity, and from neuronal specification at the organ level to specification at the individual cell level) are all left largely between the lines in this account or even missed or tossed off as minor shifts of emphasis."[51]

Sperry never lost track of the tidy dichotomy. But at the time when more and more experimental results highlighted the need to include plasticity in the developmental process itself, Sperry had shifted his focus. Already in 1976, he had been deeply immersed in his groundbreaking work on split-brain patients for more than 10 years. He received the Nobel prize for this latter work together with David Hubel and Torsten Wiesel. Hubel and Wiesel had demonstrated remarkable plasticity of cortical rewiring and had just reported in 1971 a lack of "detailed cell-to-cell specificity of geniculocortical connexions, but rather a tendency to topographic order and continuity."[89]

How different were the goings-on in AI during this time! This was partly due to the crises evoked by unmet promises, but also a different culture and an uncompromising willingness to challenge powerful ideas. Two decades since the prediction of the imminence of intelligent machines had passed, and intelligent machines seemed further than ever. In the United Kingdom, the Lighthill report led the government to withdraw most funding for any kind of AI research due to its lack of success. In the United States, a war of perspectives had broken out—and neither of the protagonists minced their words: most prominently, the Dreyfus brothers: "the relentless prophets of the omniscient computer . . . lack [a] fundamental attribute of the human brain—a few uncommitted neurons." Already in the mid-'60s, Hubert Dreyfus had published his damning book *Alchemy and Artificial Intelligence*, followed in 1972 by *What Computers Can't Do*.[90–92] Minsky simply said: "They misunderstand and should be ignored."[5] The public debate was absolutely fascinating, and even though it seemed to peter out without final endpoint, it has not truly ended to this day.

One of the hard problems in the '70s was, and still is, natural language understanding. Here as elsewhere in AI, the protagonists of the battle were not shy about expressing their opinions clearly. One key figure was Roger Schank. He had obtained his PhD in 1969 at 23 with a thesis on a model for how to represent knowledge for natural language in a computer. The big question is how meaning can be represented in an artificial intelligence. For Schank, two words with the same meaning should also have the same internal representation. His theory of language understanding and reasoning contrasted with traditional approaches.[93] The field at this point had been focused on syntax, the study of sentence structure. The American linguist Noam Chomsky was at this time a leading figure in linguistics who developed a strictly formal approach to syntax. Chomsky studied the significance of syntax in isolation, for example when a child learns grammar, stating that linguistic theory is concerned primarily with an ideal speaker-listener, "unaffected by such grammatically irrelevant conditions as memory limitations." This is a reasonable scientific approach, but to Schank this was looking with a magnifying glass into the trunk of an elephant and writing excerpts about it being an intricate tube without ever figuring out that he was studying an elephant. Schank's response to Chomsky was simply "There is no such thing as syntax." Schank would later say about himself: "I'm what is often referred to in the literature as 'the scruffy.' I'm interested in all the complicated, un-neat phenomena there are in the human mind. I believe that the mind is a hodge-podge of a range of oddball things that cause us to be intelligent, rather than the opposing view, which is that somehow there are neat, formal principles of intelligence."[94]

In fact, Roger Schank not only called himself a scruffy, he also coined the tidy dichotomy of "scruffy vs neat" in his dealings with Chomsky's formal approach that would have a lasting effect on the field of AI and beyond. Schank definitely brought fresh wind into AI efforts on natural language understanding and later said "my most important work is the attempt to get computers to be reminded the way people get reminded."[94, 95] He also noted a deep appreciation for what he observed biological brains to do and expressed this in his analysis of AI at the time. In Schank's words: "One thing that's clear to me about artificial intelligence and that, curiously, a lot of the world doesn't understand is that if you're interested in intelligent entities there are no shortcuts. . . . You can build learning machines, and the learning machine could painstakingly try to learn, but how would it learn? It would have to read the *New York Times* every day."[94, 95]

I hear some of you object vehemently—of course we can today feed hundreds of issues of the *New York Times* as data into an ANN in seconds. But feeding data has remained a stepwise process, and the time and energy it takes has turned out to be a critical bottleneck. If this doesn't sound right to you, let's hold that thought. There is a discussion to be had, and we will not evade it.

The terms "neat" and "scruffy" were quickly and readily adopted by the AI community in the '70s, albeit without clear definitions. For a while, there were a lot of neats and scruffies around. Suddenly, Minsky had always been a scruffy, and McCarthy was the archetypical neat. The McCulloch-Pitts neuron was neat, but the Rosenblatt Perceptron scruffy by design. In linguistics of course Chomsky and Schank stole the show. Symbol-processing AI had always been rather neat, while neural networks were rather more scruffy. Today, even machine learning has its neat domain, which was dominant until rather scruffy deep learning methods took hold. So let me get carried away a bit here: all of this makes Paul Weiss and Karl Lashley with their focus on plasticity sound a bit like scruffies before Roger Sperry dominated the field with a rather neat theory about morphological rigidity and molecular specificity. But who opposed the neat? Alas, I think here an outspoken scruffy may be missing, at least the label does not seem to fit the gentleman Mike Gaze. What's been going on in biology since the 1970s? Are we trapped in neat mode?

The field of AI has always been driven by progress in computer technology and has often been drawn to hyperbolic predictions. Statements about impending human-level artificial intelligence have led in the past to open war between those who considered anything done in computers as too fundamentally different from living systems. Remember the Dreyfus wars. The debate is still ongoing today, but with ever more AI systems being part of our everyday lives, the weight seems to have shifted from "if" to "when." The neat versus scruffy debate has been but a small discourse on this larger stage. Yet, it resonates with a deeper dichotomy. We recognize something as biological compared to something engineered by man immediately when we see it. Features of biological neural design that are not found in computers have been described in detail.[96] Maybe it is little wonder that it is the biological type of scruffiness that has ultimately enabled AI approaches: deep, multilayer and recurrent neural networks are more similar to brains than any other approach AI has ever taken. This is why we are looking at a combined history. AI may have veered in and out of it, but it has always come back to the brain. So, how about brain wiring? Have we embraced scruffiness? I am not sure. Sperry predicted a precise and rigid code. Many developmental biologists today are looking for that code in space and time as the brain grows. But is there such a neat solution, or what is it about scruffiness that we would need to understand? How "scruffy" is a random process really? And what is it that's supposed to be neat or scruffy anyway—the wired brain or the genetic code upon which it is built? These are just a few of the questions we'll be addressing in the following seminars. We will focus on answers that relate to the information we need to make neural networks, be it in your brain or your computer simulations.

I entitled this seminar "The deeply ingrained worship of tidy-looking dichotomies." This is a quote from the posthumously published 1962 book *Sense and Sensibilia* by the British language philosopher J. L. Austin—and it is quoted in Mike Gaze's 1970 book in the context of the tension between specificity and plasticity.[55, 97] In this history seminar, we have visited this theme time and again, across fields. So is everybody a neat or a scruffy, or is this just another tidy-looking dichotomy? My objective was not classification, but to find patterns. And yet, we must be careful about the parallels and patterns we see: There comes a point when one sees the pattern everywhere. That's how your brain works, as we will discuss in the last seminar—you may feel you have discovered the deep pattern underlying everything, how it is all connected, but beware, this is how the believer feels while digging the deepening groove of cognitive bias. So, where are we now, and how good are the parallels really?

In AI, neural networks are clearly on top, after a history of dramatic rises and falls. Many of you are protagonists in this field. Major science journals, news and conferences are occupied with the new promise of AI, finally delivering on the promises from the '50s. Only a few years ago, however, the neats had declared victory: Pamela McCorduck, in 2004: "As I write, AI enjoys a Neat hegemony, people who believe that machine intelligence, at least, is best expressed in logical, even mathematical terms."[98] Who can predict the next revolution that turns our thinking and writings of today on its head in less than 10 years?

Whether I browse through the news on AI in the public domain, new articles on deep learning in the science and business sectors, or new conferences popping up in my mailbox daily, it's almost all about applications. Concerns have been raised that there is by now some kind of alchemy to the application of ANNs, because it doesn't really matter in the end what works best, as long as it does work.[99] AI researchers have dubbed this the alchemy problem—and they are worried they do not formally understand the solutions they come up with anymore. The resurfacing of the term alchemy should evoke memories of Dreyfus' 1965 *Alchemy and Artificial Intelligence*. However, the problem today is on an entirely different footing: Dreyfus based the analogy on the idea that alchemy was a hoax and never worked, whereas worried AI researchers see their artificial neural nets working fabulously; they are just worried that they don't understand them anymore.[90, 99] Giving up neat, formal understanding does not mean giving up the hope that it works; it may in fact be required to make intelligent systems work. Evolution should make us think that this is so—again something we will need to revisit in the coming seminars several times.

We are all familiar with the classic problems and successful applications of AI: image recognition, language recognition and predictive analysis of big correlative data, like what friends or products you are most likely to have or buy. In these fields progress has been enormous and AI has become a tool, a fancy new computer program. As long as it does it ten times better than anything we had before, we don't care about the detailed inner workings of the network. That's not only scruffish, it's simply pragmatic. But this is not why we are here together. We are here to ask what we can learn from this progress about how a brain is put together, and conversely what we have learned and may yet learn from the brain about how artificial neural nets are, could or should be put together. How similar really is the function of today's ANNs to the brain? Will we make artificial general intelligence based on current ANN architectures by trusting to ever faster computers to run the simulations and feeding them ever more data? Babies are not born with a randomly connected adult network architecture that just needs to be trained with big data. Instead, their learning is part of development. Does that mean we will ultimately need to apply some of the knowledge and ideas from biological brain development to aid our making of AI?

And this brings us back to brain development. I have not talked about the busiest years in that field—the molecular age. Since the early '90s, we have found dozens of new amazing gene functions and new molecular mechanisms every month. The goal has not changed—to understand how a network is made that is capable of "intelligent" function. The concepts haven't changed much either. The name of the game is specificity. The relevance of new genes and mechanisms implicated in brain wiring is measured by their importance to contribute to wiring specificity. The search for chemoaffine Sperry-type molecules is very much alive. One approach is to dissect the underlying information gene by gene, identify those that are important and describe their function in space and time. In the end, when we put all that knowledge together in one big database or computer simulation, will we have the whole program for brain wiring? Will the biologists be able to communicate the database of molecular mechanisms to the AI community and tell them they've figured out how to build a brain?

We are here to identify parallel, common patterns. In the AI field the mantra had been for many years that biological brains are just too messy and AI developers should not feel unnecessarily constrained by what biology does. And the biological sciences have been rather suspicious of efforts to engineer something that so far only exists as the result of evolution and development. And yet, if history is any indication, we may have a lot to talk about. And it will mostly be about: information. What is the information we need to get into an AI system to make it intelligent? Are current ANNs good enough? What is the role of the exact way biological wiring diagrams are put together? And what is the role of a code at the genetic level? Do we need such a code to grow an AI? Or is feeding the information into an engineered network in the end the same as the growth of our own brains in childhood? Conversely, is brain development just a big data problem? What information could we provide for each other? Let's talk about it. Thank you.

1

Algorithmic Growth

1.1

Information? What Information?

The Second Discussion: On Complexity

AKI (THE ROBOTICS ENGINEER): Okay, this was weird. Not sure it adds up. Let's just start with one problem: the neat vs scruffy discussion was really about the culture of people working in AI in the '70s and '80s. It's not a scientific concept. These words are not properly defined, and their meaning changed with time. I know them to describe Minsky vs McCarthy, the scruffy 'hack' versus formal logic—but both worked on symbol processing AI, not neural nets. Only later did people pitch symbol processing vs neural nets using the same words—how does that make sense? It was also too much biology for me. Half of it was the history of this guy Sperry.

ALFRED (THE NEUROSCIENTIST): Well, before 'this guy Sperry' people apparently thought development only produced randomly wired networks, and all the information enters through learning. That's huge. Sperry marks a transition, a step artificial neural networks apparently never made! I liked the idea of the shared problem. It's interesting that the early computer people thought it just had to be random because of the information problem of real brains. And even Golgi thought there must be a whole network right from the start, the famous 'reticular theory.' How does a biologist or an AI programmer today think the information gets into the neural network of the butterfly?

MINDA (THE DEVELOPMENTAL GENETICIST): You need precise genetic information to build it and make it work. I think it is conceptually known how the information got into the network: it's in the butterfly's genes. There may be some loss of precision during development, but a neural circuit that works has to be sufficiently precise to do so.

PRAMESH (THE AI RESEARCHER): That sounds like genetic determinism to me. Of course there is precise information in the genes, but that's not the same as the information that describes the actual network. We need to look at information at different levels. It's similar in our AI work: We first define a precise network architecture, learning rules, etc., there we have total control. But the network has the ability to learn—we feed it a lot of information, and we may never get to know how it stored and computed that information really. Here you have less control and really don't want to have more. An unsupervised approach will even find things you never told it to find in the first place.[1]

ALFRED: I like that. I also liked that Minsky and Rosenblatt both built their networks with random connections, not precise network architecture . . . and those are the things that worked. . . .

PRAMESH: Yes, randomness and variability can make outcomes robust, but also unpredictable. In our computational evolution experiments, random processes are required to make evolution flexible and robust, but we can never predict the solution the simulation will find. It's the same with how an artificial neural network learns. It's a question of levels: you can have control at a lower level without having control over how the system operates at a higher level. This can actually be true even if a system is completely deterministic, without randomness or variability. Maybe that's how we should think about genes.

MINDA: As I said before, the genes contain the information to develop the network. In your AI work, you feed additional information for the network to learn. That's a key difference. A developing biological neural network forms precise connectivity based on nothing but the genetic program. If there is no environmental contribution, how can there be anything in the network that was not previously in the genes?

PRAMESH: Can you predict how a mutation in a gene will play out during development?

MINDA: Of course. We know hundreds, probably thousands, of developmental disease mutations where we know the outcome.

AKI: That's cheating! You only know the outcome because you've seen it. Could you predict the outcome if you had never seen it before?

MINDA: Let me see. It is true that the outcomes of *de novo* mutations are not as easy to predict, but we can usually gain insight based on previous knowledge. If we know the gene's function or other mutations, that of course helps to make predictions. We can then test those predictions.

ALFRED: . . . but then you are again predicting based on comparison with previous outcomes! How much could you really predict if you had nothing but the genome and zero knowledge about outcomes from before? It's a cool question.

MINDA: No previous data is a bit hypothetical. It would of course be more difficult to predict the effect of a given mutation. The more information I gather, the better I can predict the outcome.

PRAMESH: So, once you have seen an outcome for a given mutation, you can predict it—sometimes with 100% probability. But the point is: a mutation may play out unpredictably during development even if it always leads to the same outcome—with 100% probability.

AKI: Not clear.

PRAMESH: Well, if you run an algorithm or a machine with the same input twice and it produces the exact same output twice, then this is what we call deterministic. But even a deterministic system can function unpredictably, that's the story of deterministic chaos. It just means there is no other way to find out what a given code produces at some later point in time other than running the full simulation; there is no shortcut to prediction.[2]

ALFRED: I remember the deterministic chaos stuff from the '80s and '90s, . . . did that ever go anywhere? It always sounded like giving a catchy, fancy name to something that nobody really understood.

AKI: Yeah, as Wonko the Sane said: if we scientists find something we can't understand we like to call it something nobody else can understand. . . .[3]

MINDA: I don't see where this is supposed to go. It seems to me that if a code always produces precisely the same outcome, then it is also predictable.

PRAMESH: Well, it's not. Deterministic systems can be what is called mathematically 'undecidable.' Do you know cellular automata? People still study them a lot in artificial life research. Conway's Game of Life is quite famous. Think of a very simple rule set to create a pattern. It turns out that after a few hundred iterations, new patterns emerge that could not have been predicted based on the simple rules. There is no math on earth, no analytical method, that would have allowed one to calculate or predict these patterns. The only way to find out is to run the simulation.[4–6]

AKI: This is not very intuitive, is it? You can't generate information out of thin air.

MINDA: I agree. I don't know about deterministic chaos, but I'd say you are just describing something simple with complicated words. It's like the growth of an apple tree. There are genetically encoded rules for branching that apply over and over again as the tree grows and that creates a complex structure.

ALFRED: But when the tree grows, not every branch angle and length is defined; the environment changes that.

MINDA: Of course, everything in biology is noisy.

PRAMESH: That's true, but just forget about noise for a moment. Imagine an apple tree that grows deterministically, so if you would grow the same tree again, it would be completely identical in every detail. Just for the sake of the argument. Actually that exists, algorithmically speaking: L-systems, or Lindenmayer systems, after Aristid Lindenmayer. Anyway, you still have to decide at what level you want to describe the tree—if you want to describe the final thing, with all its branches, then it's complicated. If you only describe the rules to grow it through iteration of the same simple rules, then it's simple.[7]

AKI: I like that. It's the 'what is complexity' discussion. Each precise individual branching point is irrelevant, only the overall pattern is important. The position and speed of every air molecule in the room is irrelevant, you are probably more interested in temperature. Minda said the apple tree is complex, but what do you mean with that? If you want to describe the tree precisely, you would have to describe every single branch, the angles, lengths, thickness—it's crazy complex. But it's all irrelevant. Growing the branched structure is much simpler, that's how I think about genetics and developmental biology.

PRAMESH: Exactly. If you describe every detail of the adult tree, you'll have to describe every single branch individually, even if they are all the same. But you wouldn't do that, because you can encode and remake it by knowing the rules. So I could argue the tree is not complex, because you do not need much information to make it. That's the crux of algorithmic information theory.[8, 9]

MINDA: I agree that there is no point in describing every precise angle, because they don't matter. The tree just tries to spread leaves to increase sun exposure. The genes do not 'encode' the angles, I guess you'd say. Genes only encode outcomes that have been selected by evolution.[10]

AKI: So why did you say the apple tree is complex then?

MINDA: Okay, I see where this is going. I can tell you where the complexity is. Let's talk about neural networks and how connections are specified during development. Of course everybody understands that the brain grows, just like an apple tree. But the brain is more complex than branches of an apple tree. The precision of the tree branches is irrelevant, but the connectivity in the brain is not. The information is in the connectivity.

ALFRED: But if the brain grows like an apple tree, based on a genetic code, then maybe it just seems more complex because it is like, I don't know, billions of apple trees growing in interaction with each other, but in the end it's just growth.

AKI: In the end it's just growth! What does that mean? Of course it's way more complicated—where is the information coming from? You may not have to describe every single branch to make sense of a tree, but you have to describe every single connection to make sense of a brain.

ALFRED: Actually, individual connections by and large are irrelevant. Brains are built to continue to work just fine, even if a couple neurons randomly die.

MINDA: That's not a good argument. The wiring is still highly specific. It's just that redundancy is built in, and that again happens with precision during development.

PRAMESH: I like Alfred's 'billions of apple trees growing in interaction with each other.' The alternative is that we are missing something fundamentally different that has to happen in brain development that does not happen in trees. Fact is, if you want to describe a full brain wiring pattern, be it fly or human, it really is complex. And the information that describes the final wiring diagram is neither in the genes nor in the environment.

MINDA: What do you mean—if it is genetically encoded, then of course the information is in the genes. Of course there are billions of trees interacting, but their interactions are very complicated and defined by guidance molecules, as we call them. We don't know every detail, but that is only because we haven't collected enough data, as you might put it. But we are on the right path to understand the developmental code, and it's certainly more complicated than that of a single tree.

PRAMESH: Well, the problem is what you mean with 'genetically encoded.' People draw blueprints of electrical wiring diagrams to contain all the information you need to reproduce all important connections. But electrical wiring diagrams don't grow. You can draw a vastly complicated 'blueprint' for a brain wiring diagram as well, I guess, but it will look nothing like the information that is 'genetically encoded.' The endpoint doesn't tell you how it came to be, how it grew or what code it is based on. So, if you look at the genome as a code, then 'blueprint' is not a good metaphor.

ALFRED: . . . especially if you can't predict anything by looking at it, that's a pretty poor blueprint indeed . . .

AKI: So far so good—the distinction between endpoint information to describe a structure and information that is sufficient to grow the structure seems simple enough. Where's the problem?

MINDA: I don't see the problem either. All developmental neurobiologists agree that the brain grows and that genetic information is decoded over time. It does not make sense to relate 'what genes encode' to the endpoint without understanding the developmental process. I think all serious developmental neuroscientists will be ok with the statement that gene information does not contain a blueprint for the endpoint, but a recipe for its growth, as Richard Dawkins put it. The genes do not provide a blueprint for all synaptic connections, but instructions how to develop them. That's what 'genetically encoded' means.

ALFRED: And yet, even senior and famous colleagues in the field enjoy starting their lectures with the comparison of the number of genes available and the billions of synapses that need to be specified . . . it's all about the missing information . . .

MINDA: Yes, but as I said: The code is of course not a one-to-one genes-to-synapses code, but a developmental code: as development progresses, instructions on where to grow and where to make synapses are provided. It is still a genetic code, but a code in time, if you will. Individual genes are expressed many times at different times and places during development, often in different versions of the RNA or protein, to serve a multitude of functions. Hence, the numbers issue is solved. Trying to understand this code in space and time is the bread and butter of molecular developmental neuroscience. That's what we study every day. In a given system, say the spinal cord of the mouse or a certain brain region in a fly, for a given point in time, we can dissect the precise molecular functions of those gene products that are part of the code then and there.

ALFRED: . . . and in order to be part of a code, the assumption is generally that the signal consists of guidance cues, I know. They are either attractive or repulsive for a given growth direction or synaptic partner choice. I sat through enough of those seminars. If we think about writing down the entire code, as it unfolds throughout development, I am worried that you will find an ever growing set of instructions the deeper you dig, even for seemingly simple processes. I've always wondered how much detail you need in the end, where does it stop?

MINDA: You need as much of the code as is necessary to explain what we can test in experiments.

PRAMESH: Maybe the problem is the 'code' metaphor. The metaphor implies that a finite set of instructions exists that describes, if not the endpoint, then development as a series of many intermediate endpoints. But how many are there? As Alfred suggested, the smaller you make the step size, the more detail you need. I would use the word 'code' only for the genetic code, not the countless numbers of states and details that are produced along the way. And in the genetic code you will not find an instruction or mechanism for where a synaptic connection is made. This is the type of information that unfolds during growth. And maybe this type of unfolding information is not predictable based on the genome, even if an outcome is 100% deterministic.

AKI: Sounds like it can always be early days in your field, Minda: you can always publish another paper because there is still another level of detail to dig deeper into!

ALFRED: But the whole guidance cue idea is based on the concept that there is information in the genome that puts a protein at the right time and place to wire a brain. Are you saying that's all wrong?

PRAMESH: I don't know anything about this field and have never heard of 'guidance cues.' From my perspective, I would say, as information unfolds during development, you will find the right proteins at the right time and place for the next step of growth. Of course a protein at the right time and place can be critical for network connectivity. Based on genomic information, growth unfolds new information, new states of the system, one by one. At the point a specific protein does what it does, it is part of a state of the system that contains a lot of information that was not in the genome and that literally had to grow into that moment in time and space.

MINDA: Are you sure this is not semantics? I am worried about your genomic information versus 'new' information.

AKI: Well, we now have some information in the code that can generate a network and that's much simpler than the information you'd need to describe the final network. Then there is information in all kinds of details that happen, and some relevant information and some irrelevant information . . .

ALFRED: Sounds like all that information will be more of a problem for the neats than the scruffies! Give up control!

AKI: Sounds like something I could have said . . .

MINDA: Is this really helpful? We surely all agree that genetic information encodes the development of complex networks.

AKI: Not for AI, we don't . . .

Seminar 2: From Algorithmic Growth to Endpoint Information

In the 1940s, the Hungarian-American mathematician John von Neumann proposed a computer architecture that would hold a program and the data it operates on in the same memory. He considered this to be at least somewhat analogous to the human brain.[11] At the same time, von Neumann was fascinated by the concept of *self-replication*. He was thus one of the founding fathers of both digital computers and the field of artificial life. Together with Stanislaw Ulam at the Los Alamos National Laboratory he discovered, or really invented, cellular automata.[12] The idea behind cellular automata is simple. Start by taking a sheet of graph paper and a pencil. Every white square on the paper may have one of two states: you can leave it white or blacken it with your pencil. Now invent a rule for what should happen to a blackened square depending on its neighbors. For example, having less than two neighbors means it must die and having three neighbors means it replicates, so another square becomes black. If you apply the rule you can draw a new pattern of black and white squares—and apply the rule again. Von Neumann and Ulam recognized quickly that, depending on the input pattern and rules, surprisingly complicated pattern grow with every iteration of the process. The process is one of algorithmic growth: the application of a set of rules such that the output of the preceding step becomes the input of the next step. In the case of cellular automata, the same rules are applied for every iteration of the algorithm.

The idea was picked up and publicized in 1970 when the British mathematician John Conway invented the Game of Life (fig. 2.1).[13] He proposed four simple rules: (1) zero or one neighbor and alive: the cell dies, (2) two or three neighbors and alive: the cell persists, (3) three neighbors and dead: the cell comes to life and (4) four or more neighbors and alive: the cell dies. As an interactive game, Conway's Game of Life is boring: you set the rules once at the beginning, not knowing what they will do, and then you just sit there, investing time and energy to run your computer, hope for the best and observe. The reason for its immense popularity, and its foundation of a whole new field of mathematics, are the surprisingly beautiful and complicated processes seen during the simulation. There are structures that remain stable, some that oscillate and some that start moving. There are blinkers and beacons, gliders and spaceships of incredible complexities that can grow infinitely (fig. 2.1). A run of the Game of Life can look like a time-lapse of the life, battle and birth of entire civilizations of blackened squares. A

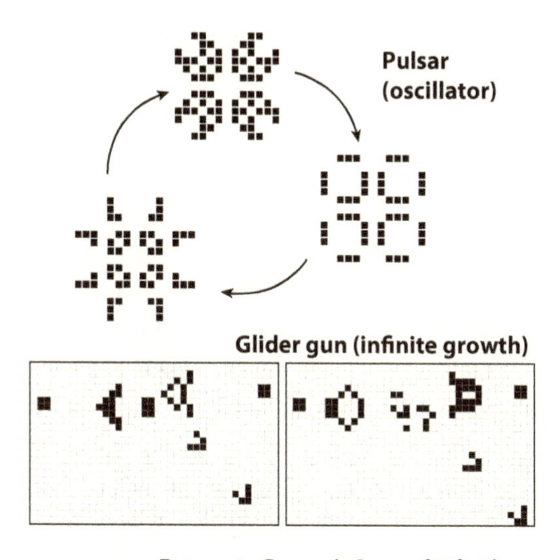

FIGURE 2.1. Patterns in Conway's Game of Life. The rules (adapted from https://en.wikipedia.org/wiki/Conway's_Game_of_Life) are: Any live (black) cell with two or three live neighbors survives. Any dead (white) cell with three live neighbors becomes a live cell. All other live cells die in the next generation. Dead cells stay dead. Top: The pulsar or oscillator infinitely loops between three configurations. Bottom: The glider gun infinitely produces moving "gliders" that coast away to the bottom right.

cult following has explored the behavior of the Game of Life ever since. In 2010 the first self-replicating structure was built in the Game of Life.[14] Dubbed "Gemini," it consists of two sets of identical structures that are the opposite ends of a structure called the "instruction tape." Over 34 million iterations of the four simple rules, the original "Gemini" destroys itself, while at the same time replicating itself into a new one. In 2013 the first structure was found that not only included its own copy, but also a copy of the "instruction tape"—and enthusiastic development continues.

The fascination with the Game of Life stems from the realization that simple, deterministic rules will create structures and behaviors that appear incredible lifelike. Strangely, the structures and behaviors the system produces cannot be predicted from the rules. Yet, to be sure, it is completely deterministic, i.e., running the simulation with the same input pattern and rules twice will produce the identical pattern both times.

This sounds confusing. How can something be deterministic and unpredictable at the same time? After all, I can predict the second run after the first. The point is that there is no other way of knowing what pattern an input and set of rules produce other than actually running the simulation. There is no formula, no equation, no shortcut. Any intelligence, be it natural or artificial, can only find out about the state of the system at, say, iteration 1,234 by actually going through the first 1,233 iterations. You can store and remember the output of a run, but there is no other way to compute it. Mathematically, the Game of Life is therefore called "undecidable."

This idea led the mathematician Stephen Wolfram (fig. 2.2) to propose *A New Kind of Science* in a book of the same title after many years of work on cellular automata.[2, 15] Wolfram systematically tested all possible rules for one-dimensional cellular automata, defined classes of rules and analyzed their behavior. In contrast to the two-dimensional Game of Life, the simplest nontrivial cellular automaton plays out in one dimension, on a line, starting with just one "live" cell. There are exactly 256 possible rule sets one can apply in this simple setting. The output is drawn in the next line, the rule applies on this new line, the next output is drawn another line down, and so forth. Remarkably, some of these rules produce completely unpredictable behavior, just like the more complicated, two-dimensional Game of Life. There are parts of the structures created by a run of rule 110 that appear to be stable patterns, but at the interface between patterns strange, unpredictable things happen (fig. 2.3). Matthew Cook and Wolfram proved in the '90s that the system defined by rule 110 is Turing-complete or what is also called a universal Turing machine.[16] This is a not an easy concept for the rest of us to grasp. In essence, a Turing-complete system is principally sufficient to perform any computation. Still not easy to grasp. Try this: despite the simplicity of rule 110 and the little information it contains, the patterns it deterministically produces when iteratively

FIGURE 2.2. Marvin Minsky and Stephen Wolfram, undated. Courtesy of Wolfram Companies and the Minsky family.

applied are so complicated that they contain any conceivable calculation of any level of complexity as part of the patterns.

But there is more: Rule 110, iteration 1,234 produces a partly repetitive sequence, but also a stretch of 0s and 1s that is mathematically indistinguishable from a random pattern compared to sequences of 0s and 1s at other iterations. This led Wolfram to speculate that, just maybe, the entire universe could be a rule 110-type automaton. From within this completely deterministic system the world would look indistinguishable from random noise. You may enjoy holding that thought, get yourself a drink and speculate on free will at this point, because I won't!

What can we really learn from rule 110? Let's start with the simplest lesson: a very simple system that can be described with ridiculously little information can produce a system of any conceivable complexity. And because it is deterministic, this means it precisely describes this complexity. For example, the information: "rule 110, iteration 1,234, cells 820–870," is a perfectly precise description of the following pattern of 0s and 1s: 10010010111101001001001100111100110110101101011010. Mind

FIGURE 2.3. The first 250 iterations of the rule 110 cellular automaton. As the pattern grows, it continues to change and produces infinite complexity with infinite iterations. Rule 110 is based on a very simple rule set and is the simplest known universal Turing machine.

you, the only way to know this result is by actually running the 1,234 iterations—again: no shortcut possible. The strictly correct mathematical term is *undecidability*: For rule 110, just as Conway's Game of Life, there is no analytical mathematical method to decide whether and where a certain pattern will occur based on a given input pattern.[5, 6, 16] You have to run the simulation, applying the rule, row by row. To precisely describe the sequence of fifty 0s and 1s above, we can either write it out in full, or write down the simple rule sufficient to run that simulation, to algorithmically grow the information content. The relative

FIGURE 2.4. Ray Solomonoff, sometime in the 1960s.
Courtesy of the Solomonoff family

simplicity of the algorithmic rule becomes dramatically obvious if we extend the sequence from 50 digits to a million 0s and 1s, a million iterations later. The rule to generate this information remains almost unchanged ("rule 110, iteration 1,001,234, cells 820–1,000,870"), while the information needed to describe each 0 and 1 individually explodes.

In the early 1960s, Ray Solomonoff (fig. 2.4), one of the core participants of the AI-defining Dartmouth workshop in 1956, mathematically described complexity as the shortest program or algorithm that could produce a given information content in the output.[9, 17] A sequence of half a million 0s, followed by half a million 1s is not as complex as a random series of a million 0s and 1s, because very little information is necessary to say "500,000 times 0, then 500,000 times 1." This type of complexity is now mostly known as Kolmogorov complexity, named after Andrey Kolmogorov, who, like his fellow mathematician Gregory Chaitin, developed the concept independently a few years after Solomonoff.[8, 18] Kolmogorov complexity also provides a good definition of randomness: a string of 0s and 1s is random if its Kolmogorov complexity, that is, any rule to describe the sequence, has to be at least as long as the sequence itself. Thinking about rule 110 is interesting again: when generating the pattern of infinite complexity row by row, the Kolmogorov complexity remains low because the information needed to generate it remains very little.

This leads us to another key difference between *endpoint information* and *algorithmic information*. In contrast to complete endpoint information, two more ingredients must be added to the mix to unfold an algorithmic rule to the full lengthy beauty of the endpoint: time and energy. An algorithmic description may be short, but to find out what it produces, what information unfolds, we must spend the time and the computational or brain power needed to decode it. The importance of this idea is apparent for rule 110: it encodes a perfectly deterministic pattern, but there is no shortcut to unfolding it step by step. Hence, there is a precise pattern in the millionth row, but a lot of time and energy has to be spent to find out what that pattern actually looks like. To account for at least one of the missing ingredients, Charles Bennett defined a version of complexity based on the time it takes to compute the output of an algorithm, rather than only the length of the algorithmic rule itself. Bennett called this concept the "logical depth."[19] For an increasing number of iterations of rule 110, its Kolmogorov complexity remains low, but its logical depth increases. And this definitely brings us back to biology: what is the genetic information in the apple seed without time and energy to grow it?

And yet, the relationship between information and energy seems to have fascinated physicists more than biologists to this day. The field of information theory started with Shannon using the term *entropy* to describe missing information.[20,21] Entropy had previously been described in the second law of thermodynamics. When you pull a divider between two compartments containing water, one cold and one hot, then you will soon have a container with water of a homogeneous, intermediate temperature. By contrast, water does not spontaneously unmix into cold and warm parts. Based on the work of Nicolas Carnot and Rudolf Clausius in the 19th century, we can calculate the energy flow based on entropy, the disorder of the system. The second law of thermodynamics says that entropy in a closed system always increases. There is more order in a separation of cold and warm water, and order does not spontaneously increase. I need to put in energy to increase order on my desk or to clean up the kitchen. I need to invest energy to separate cold and warm water. I need energy to grow a brain. Without addition of energy, order can only decrease.

James Clerk Maxwell, a contemporary of Clausius, had a funny idea. Temperature is the average speed of water or gas particles on either side of a divider. What, Maxwell asked, if there were a little demon that could effortlessly open and close the divider to let through individual particles depending on their speed.[22] The demon would only open for fast particles on the left side to go to the right, but keep the divider for slow particles. Conversely, the demon would only let slow particles from the right to the left, but keep the divider closed for fast particles. In this famous thought experiment, the temperature would rise on the right. If the divider is opened and closed effortlessly the process requires no addition of energy, in apparent defiance of the second law. Well, that can't be. But if order can increase in the thought experiment without adding energy, what has been added? Information. *Maxwell's demon* needs information to sort and thereby decrease entropy. *Boltzmann's constant* is a reminder just how fundamental this relationship between entropy and information really is. This single constant relates entropy to missing information just as it relates the pressure and volume of a gas to its temperature. The heat flow from my office to the outside can be calculated in bits, because temperature can be described

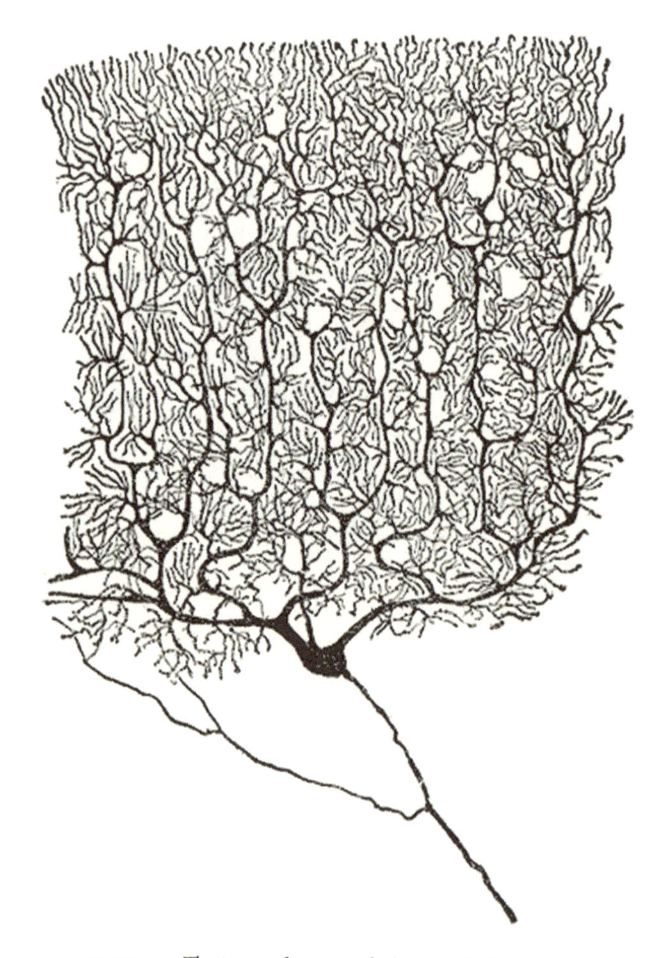

FIGURE 2.5. The iconic drawing of a human Purkinje cell by Ramon y Cajal, 1905.

with the dimension energy/bit.[21] Adding energy over time means adding information. Does this sound like something developmental biologists or AI developers should keep in mind? Let's see.

Back to neurons and their networks. Let's have a look at one of the most famous depictions of a neuron of all times, the 1905 drawing of a Purkinje cell by Cajal (fig. 2.5). Just looking at this picture, we are reminded of a tree. In fact, the branched structure is called the "dendritic tree." In the case of the Purkinje cell, the dendritic tree is a vast, largely two-dimensional branched structure that minimizes gaps and overlaps. Now imagine you want to precisely define this structure, including every branch, the branch angles, lengths, etc. Similar to the pattern produced by rule 110, the endpoint information required to describe the precise pattern is rather unwieldy. But is there an algorithm to generate this pattern, i.e., can the branched structure be algorithmically grown, step by step, based on simple rules plus time and energy?

In the late 1960s, the Hungarian biologist Aristid Lindenmayer devised a language based on the iterative application of simple rules that proved to be sufficient, with a few tweaks, to basically reproduce any branched structure observed in nature.[7] A basic Lindenmayer, or L-system, is deterministic. Once you introduce noise, you get stochastic L-systems that are commonly used to create digital models principally of any conceivable plant (fig. 2.6). But wait, now we have introduced noise. Let's look again at the dendritic tree of the Purkinje cell—is this a deterministic or stochastic structure? In fact, every dendritic tree, just like every apple tree, is different. It seems plausible that a simple program, allowing for some noise, could create such similar, but slightly different versions with every run. In the end, every apple tree looks a bit different, but it still looks more like an apple tree than any peach tree. There is a good explanation for both the similarities and the differences between apple trees: evolutionary selection. Many aspects of the appearance of apple trees are selected for functionality and usually well-adapted to the environment the tree finds itself in. However, selection may not favor a single branch a little more to the left over one slightly more to the right for either the apple tree or the Purkinje cell. In both cases, evolution selected for a structure that covers a certain space—for the apple tree to spread its leaves and collect most sunlight, for the dendritic tree to facilitate input specificity. Selection for the structure means: selection for the algorithm that created it. From an information theoretical viewpoint, the precise angle of an individual branch represents irrelevant information. An algorithm may permit noise during the iterative application of rules, as long as there is no reason not to. Selection pressure may exist to keep noise within certain boundaries, but below a certain threshold. Things become more interesting once noise

FIGURE 2.6. Rendering of a simple L-system. Generated with L-studio by Przemyslaw Prusinkiewicz. Lindenmayer systems are widely used to encode and render lifelike complicated plants and neuronal shapes based on a relatively simple code.

is not just something the system tolerates, but actually utilizes as part of the algorithm, but more on this later. For now, it seems a comfortable, plausible thought that a simple set of rules, run in iterations, can create more complicated structures, with or without noise. In this case describing the actual endpoint, including all the irrelevant little branch differences, seems pointless. Not only can the structure be encoded with less information, but a lot of the specific branch angles we see at the endpoint may actually be irrelevant and not encoded at all.

Let's delve a bit deeper into the nuts and bolts of algorithmic growth. In cellular automata, like rule 110, the same rule is applied for every iteration of the algorithm. In biological structures, the rules change as the algorithm runs. A good example for this are transcription factor

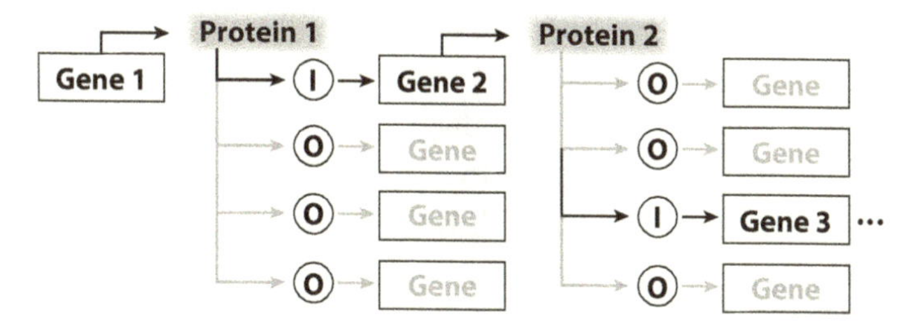

FIGURE 2.7. Schematic of a transcription factor cascade. A developing or functional neuron continuously changes its state in time. Under continuous supply of energy, the transcription of a gene for a transcription factor leads to the production of a protein that itself changes which genes out of many are transcribed a few hours later. In time, a cascade of such feedback interactions between the genome and its gene products creates changing states of the neuron. The system may never revisit an earlier state again. What proteins will be expressed three steps in the future may be unpredictable. This is a classic example for information unfolding in a time- and energy-dependent algorithmic growth process.

cascades. A transcription factor is a protein that, together with other proteins, binds to DNA and initiates the "reading" of a gene. Now imagine that the gene that will be read, or transcribed, encodes for another transcription factor, or TF for short. Let's say TF1 triggered the reading of gene X, which happens to encode the transcription factor TF2. TF2 has its own specific protein binding partners and DNA sequence it binds to, leading to the expression of a different gene. You guessed it, why should that gene not code for TF3? And so forth, to TF4, TF5 and TF6 (fig. 2.7). This is of course an oversimplification, as in real biological cells many other genes will be expressed that do not only code for TFs, but instead, for example, for metabolic enzymes that indirectly affect the action of the transcription machinery. It gets very complicated quickly. The point is that the products of the genetic code themselves provide *feedback* as to what parts of the genetic code will apply next. This process leads to a continuously changing state of the cell as long as energy and time are invested. This can happen during cell divisions, leading to more different cell types. In the same cell over time, this cascading process will lead to different states at different time points of the cell's development. For example, a neuron may not be in axonal growth mode for a long time as it runs through its transcription factor cascade, until the complement of genes that happen to be expressed at, say, iteration 1,234 render it competent for the first time to actually make synaptic contact with another neuron. If the growing axon would have met its partner too early, it might just not have been competent to do anything with it. Such is life.

If we think of the basic principle of transcription as a rule, then this rule is applied over and over again in an algorithmic growth process. If we think of the rules as something determined by all the proteins and tissue properties that influence what happens, the players of the game, then the rules change with the proteins that define them. For example, one of the genes expressed at iteration 1,234 may be an enzyme that modifies a TF or its interacting proteins, thereby fundamentally changing the way the whole transcription system works in a way that didn't occur until this iteration. There are 2,000–3,000 TF genes in the human genome,[23] but in combination of a dozen or so TFs a vastly higher number of TF complexes can occur, each with a different specificity for what part of the genome it can read.

When molecular biologists study a system like this, they put a flashlight at a specific moment in time and space. Imagine a neuron at an interesting developmental choice point. The biologist looks at the cell with a fancy method that provides a detailed view of the state of the system at this precise point in space and time. Currently it is fashionable to perform a method called single cell sequencing for many different cells or cell types at different time points, because it provides a more detailed and comprehensive picture of a cell at a specific point in time and space than ever before. With single cell sequencing we can reveal the "transcriptome", that is, a picture of the genes currently ready to be made into proteins. This is a powerful method and it provides a picture of the state of the cell at a given point during the developmental program, a snapshot of the growth algorithm at a distinct step. For this specific step, the transcriptome by itself, however, is only a very incomplete picture of the state of the cell. Proteins interact with each other and influence their own fates as well as those of other proteins. Some proteins hang around for days, others only live for minutes, and the transcriptome knows nothing about that. You may have guessed it—the "proteome" is another picture of the state of the system, providing complementary information about the set of proteins. Similarly, we might measure dynamic features of proteins that cannot be read from either the transcriptomic or proteomic information. There is no such thing as a complete picture of the state of the cell. However, there are more and less relevant aspects of the state of a cell, depending on the problem. For example, a neuron's decision to grow the axon more to the left or right will depend more on some factors than others. The biologist may measure the existence of a receptor protein on the growing tip of the axon. The presence of a certain receptor protein will render the neuron sensitive to a factor in its environment that it may like, causing it to turn that way. The biologist can then perform a perturbation experiment on the growing neuron. A favorite method is to literally knock out the gene for that receptor. Let's assume this particular receptor gene is at this point in time read out for the first time during the neuron's algorithmic growth, e.g., a TF complex at iteration 1,233 caused its transcription. The knock-out of the receptor gene will then not have caused any earlier developmental defects, but specifically affect the turning behavior of the axon at this choice point. If, of course, the same receptor protein had already been used at iterations 123 and 456 of the developmental algorithm, then the knock-out experiment may rather produce confusing results. In this case, the later defects are masked by the earlier changes and the knock-out experiment would not easily reveal the protein's function during the later step (fig. 2.8).

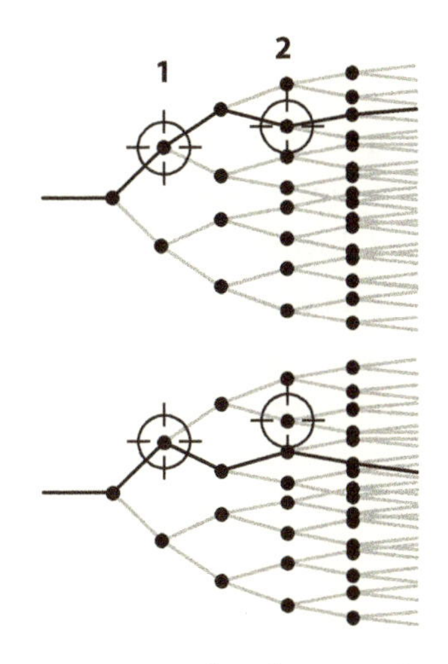

FIGURE 2.8. Algorithmic growth depicted as successive steps in a decision tree. Top: Many genetically encoded factors will function at several steps (target icons 1 and 2). Bottom: A mutation affecting this factor can change the decision at step 1 such that the opportunity for the second function never occurs.

Just like the rules of cellular automata, the genetic code does not provide endpoint information. And in both cellular automata and biological systems, algorithmic growth can lead to structures whose description would require much more information than the information laid down in the rules. For a developmental biologist this may sound like a rather trivial observation—nobody looks for a code for why a specific hair cell in my skin is exactly where it is on the back of my right index finger. Of course it happens to be there as part of the growth of my body—there is no "code" for its precise location; there is a developmental algorithm that results in it being where it is. We cannot read in the genome where that hair cell is going to be. Developmental biologists try to reconstruct how it came to be where it is by figuring out the underlying molecular and cellular mechanisms, going all the way back to what the genes encode. And the deeper we dig, the more we can understand each state of the system based on the prior state. The underlying molecular mechanisms are often fascinating and informative about how a particular step is executed in a particular system. And as far as we know, to understand how genetic information and molecular functions play together to create higher order structures, level by level during algorithmic growth, there is no shortcut. We must observe the actual program as it happens. There is no code that represents or explains the wiring diagram, there are only algorithmic steps, the rules by which they operate and the molecules by which they are executed.

As the developmental program runs, it will create changing states of the system, iteration after iteration, step by step. At one step, an axonal growth cone in search of a connection will express, for example, a certain set of cell surface receptors. These can be of the class envisioned by Sperry: molecules that provide chemoaffine tags. At the time and place such a molecule is placed, information has unfolded to that state. The molecule or the gene that encodes the molecule does not contain the information. The molecule in space and time is an output of the running algorithm at one step, and it becomes part of the input for the next step of development.

In biology, the information theoretical discussion is usually implicit without reference to the actual theory. Instead, developmental biologists have long referred to the information content of molecular mechanisms as *instructive* or *permissive*. In their words, a chemoattractant or guidance cue is instructive, because it tells the neuron where to go; a growth factor is permissive, because it is required as a survival signal for the neuron. Yet, the boundaries between instructive and permissive signals are not as clear as developmental biologists may wish. Let's first look at how the information theoretical categorization has been defined in experiments for biological network formation, and then try to relate it to artificial networks.

For many years, the best place to control and study information that could instruct a neuron where to grow was in a dish. And the most commonly used molecular chemoattractant for neuronal axon guidance in a dish is NGF, or nerve growth factor, first isolated in 1956 by Rita Levi-Montalcini and Stanley Cohen.[24] A growth factor can be required for survival, the ultimate permissive signal: you have to be alive to respond to an instructive signal. But by laying a path of NGF, it will define the path the axons takes. The information for the path the axon will take is encoded in the locations, the path, where the growth factor has been provided. The path is the instruction. This is the type of consideration that led Rita Levi-Montalcini to discuss NGF in the context of a "chemoaffinity in a broad sense" in an essay in honor of Roger Sperry in 1990.[25] From our viewpoint of information encoding, however, we haven't really gained much, because we now need to find out where the information came from to define the path containing growth factor in the first place. This is the type of information that algorithmic growth unfolds, given time and energy.

In addition to cell surface receptors and growth factors, there are countless more factors encoded by genes that contribute to neuronal development and wiring. As the developmental program runs and the neuron's state changes, step after step, the neuron will be receptive to some outside signal at one point and receptive to another signal later. It may grow faster now and slower later. It may be less responsive to tissue mechanics at first, but more sensitive thereafter. Combinations of permissive factors, like NGF, and permissive states, like tissue properties, can instruct decisions. Such *composite instructions* are the output and input of a preceding and subsequent step of development. They are part of algorithmic growth.

If every component of the composite instruction is relevant information, what is the threshold for relevance? If some factor only contributes to 5% of the composite, or only has an effect 5% of the time, is that the limit for relevance? How do we deal with an interesting choice point during algorithmic growth, say, the neuron's decision to make contact with partners on the left, but not on the right. In addition to the factor whose absence only causes problems 5% of the time, there are other factors that seem more relevant (say 60% of the time), and again others whose loss does not seem to make any difference. What if two different factors, whose individual loss makes no difference, suddenly cause problems only when both are gone? Clearly there are factors that are more important than others for a given decision. The metabolic state may play a role for the axon to turn, but it's probably less important than a signal from a growth factor, without which the axon won't grow at all. Tissue stiffness in the environment may be a factor, but a repulsive signal may override that and cause the axon to grow elsewhere. Those are the mutants with the high penetrance phenotypes biologists look for. For the experimentalist, the relevance of a factor expresses itself in the penetrance of the phenotype when absent. Whatever the role of gene X, if its knock-out doesn't cause a phenotype in the axon turning behavior, it appears to be irrelevant information with respect to that behavior. But is it?

Relevant information for composite instructions unfolds once energy starts to massage genetic information in time. The depth of relevant information, the factors needed to describe the system, depend on how deeply we want to study the system. A neuron in culture, with little context to worry about, will grow where the growth factor is. The same neuron during brain development will be subject to many more influences. If we now change the genetic code by knocking out gene X, a lot may change in the neuron, depending on the utilization of gene X as the neuron runs through its developmental algorithm. It may have quite nonintuitively altered properties affecting growth speed or sensitivity to a variety of signals. Depending on the depth of analysis, the "code in space and time" is rapidly in danger of becoming a rabbit hole of undefined depth. An infinitely deep code would be a code *ad absurdum*, a bottomless abyss in information theoretical terms, so we need to be careful when calling out the relevant and irrelevant bits. The idea of how composite instructions are encoded by relevant information remains a core struggle in genetics and developmental biology for which we will discuss some examples in the second session on "players and rules."

The code is a metaphor that works well for the genetic code or the rules of a cellular automaton. The code is a bad metaphor for the continuously changing states of neurons as they run through their algorithmic programs. Imagine looking for a code in space and time for rule 110, iteration 1,234, position 820–870. To study the mechanism that led to that state, we look at the point before, the relation of how it happened, in all available detail. Is that useful? Well, it can be if it leads to the identification of the underlying rule. In the case of the rule 110 automaton, the same rule applies everywhere, so the instance reveals something general about the whole. This is the hope of the biological experiment as well. But what happens if the rule changes with every iteration, as discussed for transcription factor cascades? To describe changing rules of algorithmic growth for every instance in time and space and in different systems is not only a rather large endeavor, it also suffers from the same danger of undefined depth. It is a description of the system, a series of bits of endpoint information, not a description of a code sufficient to create the system. The code is the "extremely small amount of information to be specified genetically," as Willshaw and von der Malsburg put it,[26] that is sufficient to encode the unfolding of information under the influence of time and energy. The self-assembling brain.

How can the genome, or any other code, represent an "extremely small amount of information" sufficient for self-assembly? For an answer we can look again over the shoulder of the problems in AI research: trying to understand, program and control every node, every parameter is the type of neat approach that just gets more complicated the deeper we dig, and easily out of hand. On the other hand, many successful approaches in machine learning today have given up on the details that happen within the network. AI researchers do not learn anything from the idiosyncratic details of a given gene or a specific guidance molecule in different anatomical regions of different animals—but neither do developmental biologists learn from the details of what synaptic weights are altered in a neural network that just learned salient features of an image or how to beat you at chess. Where we learn from each other is in the approach and in the algorithmic self-assembly nature of how information is encoded in the system. The idea of simple rules has nothing to do with how infinitely complicated the executing machinery in different systems may be as the developmental program runs. Simple rules are the common bits of information by which the neurons and molecules in each system operate. Similar solutions can be found in different implementations of the same problem in biology as well as in AI: an ANN can achieve the same confidence in handwriting recognition based on different underlying detailed strengths of connection weights. There are many paths in a decentralized network to reach a goal, just as the same patterning rule may be executed by different sets of molecules in different neurons and animals. In all these cases, the system was programmed by a fundamentally evolutionary mechanism, by selection of something that works. The underlying precise alterations are complicated in detail, but the rules they obey can be simple. This does not make brain wiring simple, but it gives us something to talk about between AI researchers and developmental biologists.

The quest for simple rules and algorithms is a classic reductionist approach. The problem with a term like reductionism is that it means something different to different people. What it means to me is based on various associations, a different history and a lack of clarity in its definition. A common dislike for reductionism stems from the idea that it signifies the reduction of something beautiful and complex to some simple components that contain none of the beauty. The simple components have none of the properties of the beautiful and complex thing; the parts are not the whole and tell us nothing about its wonders. This is an intuitively understandable concern. It is also the heart of algorithmic growth and the levels problem. The description of rule 110, iteration 1,234 contains none of the beauty and wonders of the pattern it generates. We have become familiar with lower levels not containing the information and properties of higher levels of organization. Yet, we must face the levels problem. If we want to understand how the beautiful thing comes about, we must understand the lower level code that can generate it. If such a code is sufficient to run a developmental algorithm that can produce the beautiful pattern—how beautiful is that? If we consider the higher order pattern or function beautiful, how much more beautiful must then be the simple yet in fact explanatory rules that are sufficient to generate it—this is not reductionism that takes beauty away, but explains where it comes from. Thank you.

1.2

Noise and Relevant Information

The Third Discussion: On Apple Trees and the Immune System

AKI (THE ROBOTICS ENGINEER): So, I guess the brain and the universe could be deterministic rule 110 machines then . . .

ALFRED (THE NEUROSCIENTIST): Don't we have an entropy problem? Remember the second law of thermodynamics: entropy, or disorder, can only increase in the universe . . .

MINDA (THE DEVELOPMENTAL GENETICIST): What are you talking about, Alfred?

ALFRED: Well, you and I are pretty ordered. But I guess we grew and it took some time and energy to become that way.

PRAMESH (THE AI RESEARCHER): We are indeed local low entropy anomalies. To keep us this way, we need to create some other disorder in the universe, like digesting food. In thermodynamics this is also called a dissipative system. Dissipation means local burning of energy to keep a thermodynamically open system far from equilibrium.

AKI: I like that! Pockets of order. Pockets of life. It is the ultimate question on information encoding: where did it come from and what happened to it in today's 13-billion-year-old universe? What was the 'genetic code' in the big bang that would lead to growing pockets of life, like us?

MINDA: Are you really comparing the genetic code and the brain with the big bang?

AKI: Your genetic codes and brains are kinda part of the universe. And maybe both are weird rule 110 machines, then both are deterministic and puff goes free will . . .

PRAMESH: I have to say this doesn't sound right.

ALFRED: That's an Aki specialty. But if you are inside a rule 110 system things do look random—and there is no way to find out. So I guess we'll never know. But it *would* make it difficult for you to do meaningful mutation experiments, Minda. If it looks random from the inside, who knows what a mutation does to the final wiring of a brain? Isn't that why most of your mutations really just mess things up unpredictably?

MINDA: That's not right. We learn from how exactly things get messed up. Of course there are many mutants that disrupt too much to be meaningful, but many are amazingly specific. We are looking for mutations that have really interesting, specific phenotypes.

ALFRED: But that's the whole point—most mutations should not give you a phenotype that tells you what really went wrong during development . . .

MINDA: Obviously there are good and bad experiments. If you knock out core components of the cytoskeleton or ATP metabolism, then the axon is not going to make the right connection—but that's trivial and not informative. On the other hand, if you knock out a guidance receptor that is required at a very specific place, then the phenotype can tell you exactly what instruction was missing.

AKI: Didn't we just hear about instruction as a composite of many factors and context, rather than a single factor?

PRAMESH: Right. But biologists prefer to look for simple, single instructions that define a connection to be here and not there. How do you deal with probabilities and noise, Minda?

MINDA: Every phenotype occurs with a certain penetrance, as we just heard. If you have 100% penetrance, then the mutation always causes a certain phenotype. If the penetrance is 50%, then only half the time. As I said before—biology is always noisy. It's a fact of life. All evolved systems have learned to deal with that.

AKI: But wait, if you are looking for instructions to make a synaptic connection at a certain place, then any noise is bad and you want to avoid that, right? So we are really just talking minimizing the effects of noise here. I can relate to that.

ALFRED: Does the apple tree try to minimize noise in its branching pattern? I guess only to a certain degree, at some point there will be no selection against a certain variability, so it doesn't matter . . . just thinking out loud here . . .

MINDA: Yes, that makes sense. It doesn't matter whether a leaf grows a little bit more to the left or right. It's irrelevant, so there is no selection pressure against it.

AKI: Fine, so maybe a certain level of imprecision is below the threshold. But noise remains the enemy of precision.

MINDA: Exactly.

ALFRED: Okay, then here is a thought: Many biological processes do not just try to avoid or minimize noise. They are built on noise and need it to function. The immune system, for example. It makes billions of unique cells, each expressing a random cell surface receptor. And you need that variety to fight random unknown invaders.

MINDA: These cell surface receptors bind very specifically to just one specific antigen, like a part of a protein of a pathogen. It's a lock-and-key mechanism to distinguish non-self from self. But isn't that a bit off-topic?

PRAMESH: The molecules may bind like locks and keys, but the cellular system requires a large random pool of variation to work. I can think of this as an algorithm. To make the immune system you need whatever developmental program immune cells go through. But then, at some point in the program, it can't be deterministic, because it needs to make crazy numbers of random receptors. I know this is actually very well understood . . . [1,2]

ALFRED: . . . so in this sense it is an example for how algorithmic growth can incorporate a noisy or random process. Without the random bit it wouldn't work.

MINDA: Sure. Evolution wouldn't work either if there were no random mutations to select from . . .

PRAMESH: Yes. So, I think the question is: to what extent is brain wiring specified by molecular functions that are based on avoiding noise to be precise versus functions that actually require noise and variability.

AKI: How would you require noise to make something more precise?

MINDA: Actually, I know of proteins with key roles in brain wiring that are variably and unpredictably put together similar to the cell surface receptor in the immune system. Dscam is a cell surface protein in *Drosophila* that has been celebrated for the sheer number of different proteins that can be made from a single gene—almost 40,000. The story was that any of these 40,000 different proteins can only bind to its own kind, so it's a key-and-lock mechanism. But every neuron expresses a random version.[3–5]

ALFRED: I know this story. It has actually been hailed as the long sought-after molecular identification tag for individual neurons. But I found it always odd that this identification tag should be random.

MINDA: Your passport number is randomly assigned to you . . .

ALFRED: Wait, the point of my passport number is that my government knows about it. There is a database. If nobody knew about my passport number, if I were the only one who knew what random number I had and I would never tell anybody, what kind of ID would that be? Certainly not an ID that would help in finding a partner or making a connection between matching IDs . . .

MINDA: Yes, it is correct that the Dscam ID is not an ID that other neurons know, but it is also not used as an address code. It is used for recognition of 'self' versus 'non-self,' just like the immune system. In fact, it has been proposed that Dscam might function in immune responses. It doesn't matter what random ID you have, as long as you know yours and therefore know who is not you.[6,7]

AKI: I get it. Of course you will only know who is 'not you' if everybody else really also has a different ID. The government makes sure that everybody has a different passport number by maintaining a big database. Your Dscam molecule makes sure that everybody has a different ID by having so many different versions that random choice virtually ensures you never meet somebody with the same code.

MINDA: Yes, that's the idea.

ALFRED: What does that have to do with Sperry's target definitions through molecular identification tags?

MINDA: Nobody said it did. It is a mechanism for self-avoidance, it changes growth patterns. But it doesn't define any targets.[4]

PRAMESH: . . . and it absolutely requires nondeterministic assembly of the protein as part of the growth algorithm. If the mechanism creating the proteins were precise, it wouldn't work. How does self-avoidance work, Minda?

MINDA: Well, the idea is that a dendritic tree grows by branching—and whenever branches from the same tree meet they stop growing, whereas other trees are ignored. This helps a single neurons to spread its dendritic tree. If you lose Dscam, the whole tree just clumps up.[3,8,9]

PRAMESH: And the branches are noisy, too. Like in the apple tree. But now you actually need the random branching, because you need something that grows and meets itself for self-avoidance to act on. If the branches were deterministic, then you wouldn't need self-avoidance to prevent them from overlapping with each other.

AKI: I'm not sure about this. You could also have a deterministic program that includes self-avoidance. It sounds rather like because of the random branching, the neuron needs self-avoidance to control the pattern. Isn't that like controlling noise?

PRAMESH: Hmm, maybe. The point is that both the dendritic tree and the apple tree have an algorithm in place that ensures spreading of the branches. For that algorithm to work, it must include noisy branching. I guess you can turn this around and say that, since it is noisy, the system utilized it to get the spreading to work. This gets us into evolution. Either way, the way the algorithm works, it is not trying to reduce noisy branching, it requires noisy branching.

ALFRED: So, Dscam needs to first be made in some random variant in each neuron to create a random 'self,' and then the neuron needs to randomly branch for self-avoidance to work. Fine, but what does that have to do with making specific synaptic connections?

MINDA: It's just part of how development works. The specification of synaptic contacts happens later. That doesn't mean that there are not earlier developmental steps that make the process easier. For example, having a spread-out tree makes it easier to be found by a partner neuron in the same region.

AKI: So, why are we talking about this Dscam then? It's just not a good example for a protein that specifies contacts!

ALFRED: Maybe a lot of specification is actually done by combinations of mechanisms that include things like this. This particular molecular mechanism is one of many factors that together create some composite instruction. Which brings us to the 'permissive versus instructive' debate—maybe instruction just happens when you put together enough permissive mechanisms in time and space. The question is how many molecules really need to function as IDs for synapses!

MINDA: We know that molecular specification by guidance cues and cell adhesion molecules is key in synaptic matchmaking. There has been so much work on this, it's something we know for certain. Their key roles in wiring the brain are just very, very well established![10, 11]

AKI: . . . and so said the Ruler of the Universe: 'You are very sure of your facts. I couldn't trust the thinking of a man who takes the

Universe—if there is one—for granted.[12] Or a woman, for that matter. Remember: all the early neural networks with random architecture, because nobody thought there would be enough information to specify all contacts in the brain? But now you tell me the molecules to specify all the synaptic matchmaking are well established. I think we are back to neats and scruffies!

ALFRED: It's not about whether there is random connectivity. For lots of connections we of course know that they are very specific, otherwise the circuits I am studying every day wouldn't work. The brain is not just randomly connected. The question is how do you get them to be specific. I like the idea that bringing the right partners together at the right time and place helps.

MINDA: One does not exclude the other. I am sure all these mechanisms occur. I never said everything is only synaptic matchmaking.

ALFRED: Sure. But wait . . . if a neuron can or cannot make random synapses, that would distinguish it, no? If contact is only established through matchmaking, then a neuron in a wrong place should not find its partner. You should not make a synapse if the key does not fit the lock. Actually, I know of lots of experiments that show the opposite—neurons by and large often make synapses with wrong partners when there is no other choice.

AKI: Even I know this! There are people who culture neurons in isolation, and if the neurons have nobody else to make contact with, they just make synapses onto themselves. They are called autapses. Totally not my field, but I remember this from a recent talk: 'Neurons want to make synapses. When they can't find the right partner, they do it with the wrong partner. And when they can't find a wrong partner, they do it with themselves.'

ALFRED: Yes, that sounds like something you'd remember . . .

MINDA: I don't think this argument is very strong. The way I see it is that the neurons do make the best choice they can. If the best code is not available, they settle for the second best. It is robust in this way, but certainly not random.

PRAMESH: But the question is how the information for all these contacts in the end is actually encoded. And I think the developmental algorithm idea is key. Part of the algorithm can be random patterning, as long as it is followed by some kind of selection to make it precise. In the immune system, one in a billion cells with the one cell surface receptor that recognizes the invader literally gets selected by the invader. The random branch gets selected by self-avoidance. Do synapses only form specifically during normal brain development, or isn't there also selection at work?

MINDA: There is. In some cases during development too many synapses are made initially, and then the correct ones are stabilized. Pruning the wrong synapses is an important part of development.

ALFRED: Yes, that's also the idea of activity-dependent plasticity. This is the famous 'cells that fire together, wire together' idea. It's true, for that to work the neuron must be able to initially make some unspecific synapses. Too many, actually.[13]

MINDA: Again, we know both these things are at work. There is a level of molecular specification, and then there is activity-dependent refinement. It's like a learning mechanism during development. Or the other way round: learning is life-long development.

ALFRED: But if the activity-dependent refinement happens prior to any environmental input, then it is actually part of the developmental algorithm—and that's genetically encoded.[14]

PRAMESH: Exactly. Making too many and unspecific synapses is part of the growth algorithm. Neurons becoming excitable is part of the growth algorithm. And the correlated activity that inevitably has to happen between excitable cells to identify the good connections, all that is part of the growth algorithm.

ALFRED: Incidentally, these activity waves in the retina have to be random. If they would always go from one side to the other, it wouldn't work. Do you want me to tell you how it works?[13]

AKI: Thanks. Maybe after the noise that's about to come.

Seminar 3: From Randomness to Precision

When you put a colony of *E. coli* bacteria in a dish with a spot of sugar, the bacteria will move towards the sugar source. The process is called chemotaxis, and it has been a prolific and fruitful research field since the late 19th century.[15] How do bacteria identify and compute the attractive signal? First, the bacterium has to sense the sugar, measure where the sugar concentration is higher and then activate a motor system to get there. All of this sounds a lot like a process for which we would use sensory organs, a brain and motor output. It also reminds me of the commonly used definition of artificial intelligence: "the science of making machines do things that would require intelligence if done by men."[16]

E. coli is an important bacterium found in the human gut and in molecular genetics laboratories world-wide. It is a simple cell, like all prokaryotes without a nucleus and limited intracellular compartmentalization. Any sensory system, information processing unit and motor machinery has to be organized at the molecular level. These molecular machines, and the principle by which they create chemotaxis, are well understood. Many scientists have been fascinated by the analogy to a cognitive process and dedicated years to figuring out this seemingly simple task performed by a seemingly simple cell. But let's not get into the debate of (proto-) cognition, and instead learn from how the bacterium actually does it.

Bacteria have a sensory system based on molecules that sit in the outer membrane, the chemoreceptors. In addition, there are a series of signaling molecules inside the cell that ultimately modify the function of a motor protein complex. The bacterial motor protein complex is famous in its own right, because it is the only known biological example of a molecular wheel that can spin round and round. The signaling in the cell can modify the motor to spin clockwise or counterclockwise. And finally, the motor is connected to a corkscrew-shaped hair on the outside, the flagellum. If you turn the corkscrew-shaped hair on the outside clockwise, then water is pushed against the bacterium. It is as if the flagellum tried to swim away from the bacterium. Because flagella all around the bacterium do the same thing, the bacterium just randomly tumbles. If the flagella turn counterclockwise, i.e., in the direction of the corkscrew, all flagella align and propel the bacterium straight forward, albeit in a random direction (fig. 3.1).

If all the movements are random, how does the bacterium move towards higher sugar concentrations? The chemoreceptors signal in response to sugar and the internal signaling events provide a comparison to the previous signal. Think of it as a memory of a single time step. This is the key bit of information the bacterium has: it knows whether the last movement was towards higher or lower concentrations. Of course, when it tumbles, it will randomly change direction again. The bacterium has no means of turning towards the higher sugar concentration. It also never learns where the sugar source actually is. The bacterium can only run in a random direction or tumble to change to another random direction. What's the solution to this riddle?

Well, the solution is simple enough, but strangely unsatisfactory, because it is not easy on our intuition. Here is what the bacterium does: if the concentration is increasing, it increases the probability for more running in a straight line by rotating the motors more counterclockwise. If the concentration is decreasing, it increases the probability of the motor to turn clockwise, thereby tumbling and changing direction. That's it, sorry. It's called a "biased random walk," and it just works. Think about it like this: if the concentration is decreasing, the bacterium is more likely to change direction, whereas increasing concentrations are computed as a good thing, and the bacterium keeps on running. That makes some sense, except when the bacterium arrives at the highest concentration, because it will keep on running a bit more. Since everywhere is worse than the spot with the highest concentrations, it'll start tumbling randomly soon again. On average, each bacterium will keep on running and tumbling around the place where the sugar is, without ever learning the place. Of course it keeps on sampling the environment that way, so if the location of the sugar source changes, the bacterium will change its biased random walk too. It's quite brilliant, has been described in mathematical detail and is absolutely dependent on the *random* bit of the random walk.[17–19]

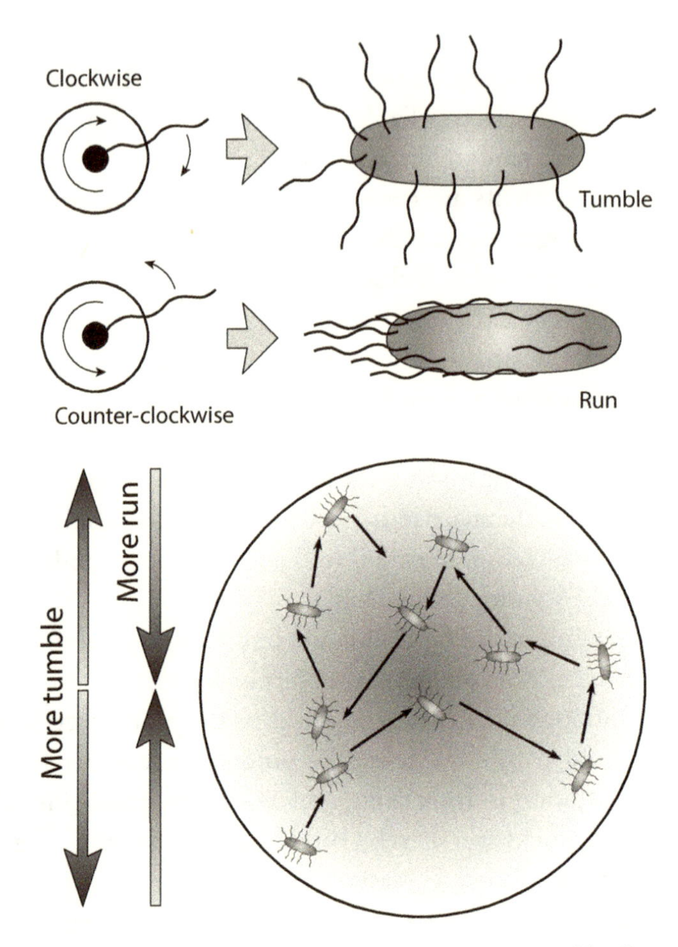

FIGURE 3.1. Bacterial chemotaxis. Top: Clockwise rotation of flagella causes the bacterium to tumble, orienting it randomly. Counterclockwise rotation propels the bacterium forward in a random direction. Bottom: The bacterium measures the relative change of sugar concentration for one time interval. If the concentration increases, the probability for run over tumble increases; if the sugar concentration decreases, the probability for tumble over run increases. This algorithmic process is called a *biased random walk* and ensures that the bacterium moves around the higher sugar concentration while continuing to explore its surroundings. The individual bacterium never learns the actual location of the highest sugar concentration; yet, the population exhibits seemingly goal-oriented behavior that would, arguably, require intelligence if done by humans.

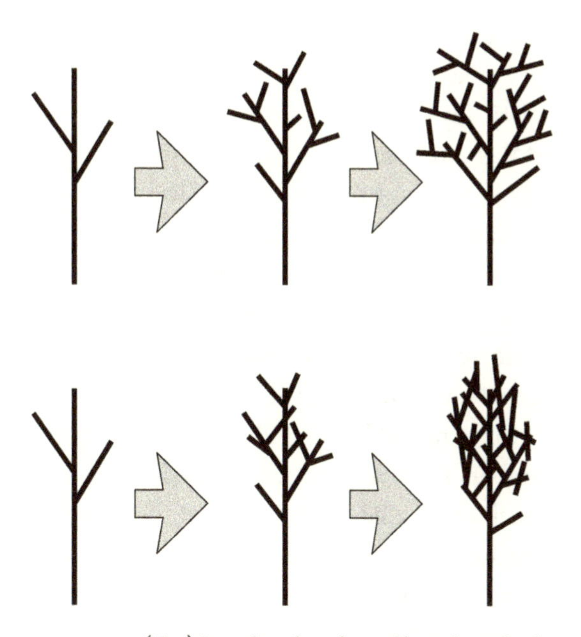

FIGURE 3.2. (Top) Branching based on self-avoidance leads to spreading of branches while (bottom) defective self-avoidance leads to branch clumping. Branch spreading is an algorithmic process during growth that is essential for functional outcomes—from apple trees to neurons.

Whenever there is a random process, the evolutionary principle is not far: you can think about the random directions as a pool of variation, from which "good" directions (towards higher concentration) are selected (by probabilistically increasing counterclockwise motor rotation). Here, noise is not the enemy of precision. Noise is necessary to make it all happen.

A pool of variation is also the basis of robust immune system function. And we can think about dendritic self-avoidance in a similar way. A neuron's dendritic tree may grow to spread in a certain direction. If the dendritic branches don't see each other, they will crisscross or even clump together. However, if they see and repel each other, then the only places to grow are away from each other, on the outside of the branched tree (fig. 3.2). Random branching events represent a pool of variation from which only some are selected, those that are not stopped by repulsion, thus allowing continued growth and further branching that spreads in empty areas.

How do "genetically encoded" neurons grow random branches? The neurogeneticist Bassem Hassan and his team are investigating this question for neurons in the fruit fly *Drosophila*. The fly brain, just like the human brain, is mostly mirror symmetric, and Hassan found neurons that have their cell bodies on one side, but send their axons all the way to the other.[20] Curiously, in the same brain, these neurons have independently varying numbers on both sides of the brain. The neurons on both sides are genetically identical, yet the process that determines their numbers allows for remarkable variability—anything from 15 to 45 cells. The precise numbers per brain half are unpredictable. All the neurons send axons to each other's side, but only a minority will form a particular type of branched extension. Again, the number of extensions with branches varies randomly, but does so independently of the varying number of cells on the left and right sides.[21] It turns out, the asymmetry between the left and the right sides has measurable behavioral consequences: the more asymmetric the branching between the two sides, the more "attentive," we might be tempted to say "obsessive compulsive" the behavior in a visual assay.[22] This is particularly remarkable, because it means that two individuals with identical genes will have meaningfully different behaviors due to a noisy developmental process. We'll talk about the consequences for selection and the relevance to monozygotic human twins in a few minutes. For now, let's focus on that enigmatic noisy developmental process. It turns out, the process is not simply noisy, but rather ensures a certain ratio of branching versus nonbranching cells. The cells play a competitive game with winners and losers. And interestingly, this kind of developmental game is very familiar to developmental biologists that study a particular molecular mechanism: Notch signaling.

Notch is a cell surface receptor that allows cells to engage in genetically encoded winner-takes-all games called "lateral inhibition." Two or more sparring partners compete by telling each other to give up using an activator protein that binds to the other's Notch receptor. Once a cell receives the signal to give up, it will itself express less of the activator

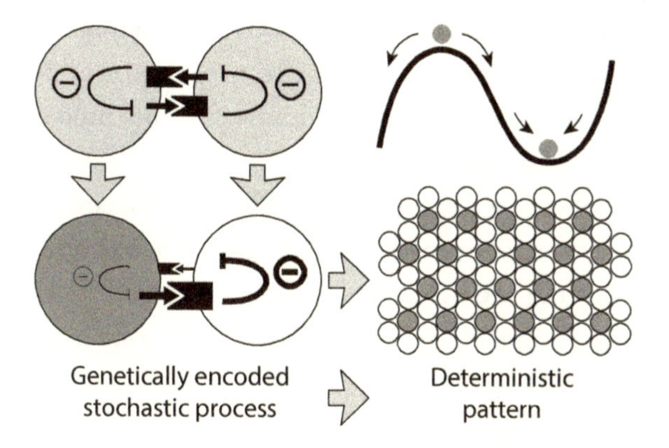

FIGURE 3.3. Symmetry breaking through Notch signaling. Top left: Two cells execute a competitive program based on the genetically encoded Notch receptor (rectangle receiving the arrow), its ligand (arrow), and other signaling components. Top right: The situation constitutes an unstable fixed point in which the slightest imbalance leads to symmetry breaking through a positive feedback loop. Bottom left: The broken symmetry leads to maximal signaling in one cell and shut-down of signaling in the other. Bottom right: Execution of this process in many cells in a two-dimensional array leads to a self-organized, precise pattern.

that would tell the other to give up, causing a positive feedback loop of . . . giving up. Winners do not tolerate a single competitor in their surroundings that has not given up (fig. 3.3). Amongst the cells that play the lateral inhibition game, it is impossible to predict which cell will win. But it is 100% certain that exactly one will win, and the rest will lose. Notch signaling executes a symmetry-breaking rule. As such, it is involved in countless contexts as part of cell differentiation and organ development across all animals.[23] Importantly, symmetry-breaking can allow for a well-defined amount of variability, as in the asymmetrically branched fly neurons. But it can also ensure patterns of absolute precision (fig. 3.3).[24] Algorithmic growth incorporates symmetry-breaking based on Notch signaling like a module—often repeatedly during subsequent developmental steps. Sometimes, a little bit of variability may be under the selection radar. Sometimes, a little more variation may actually be a selective advantage, as we will see in a minute. But sometimes, evolution has programmed the utilization of the random symmetry-breaking module to yield almost absolute precision. During brain wiring, we find all of these cases.

How does noise-based axonal or dendritic growth contribute to specific synaptic contacts? Spreading a dendritic tree has little to do with chemoaffine tagging of correct versus incorrect targets. However, it certainly increases a target area presented to an axon from another neuron in its search for the right partner. A fundamental idea underlies this notion: Not every axon will see every dendrite during its growth program. Depending on when and where it was born and when and where its axon grew, the choices for a neuron may turn out rather limited. There is a great potential here to facilitate the numerical matchmaking problem.[25]

The patterning of the axonal and dendritic structures have an effect on how easy it is for two partners to find each other because they restrict what axons "get to see" what dendrites at a certain time and place. In fact, growth patterns can facilitate partner finding to such an extent that they all but ensure two partners see each other only at the very moment when they are competent to make synapses. The interesting question then is: Will any two neurons make synapses, just because they are next to each other? I call this the *promiscuous synapse formation hypothesis.*[14, 25]

The answer is unlikely to be of the all-or-none type. Many observations suggest that neurons principally have the ability to make synapses with wrong partners, but they may not do so equally happily as with the correct one.

To understand the relevance of promiscuous synapse formation, it is useful to view the possibilities for synaptic specification in its extremes. First, let's envision complete synaptic promiscuity. In this theoretical situation, each neuron has equal drive and ability to form synapses with any partner. Partner specification therefore has to be ensured through correct sorting in space and time. The location of each axon and dendrite in space and time is a function of algorithmic growth. The rules of algorithmic growth may be deterministic (think rule 110 and transcription factor cascades) or stochastic (think dendritic self-avoidance). The growth patterns are flexible, meaning the growing trees adapt to environmental constraints.

The other extreme would be no promiscuity at all. Synapses only form between correct partners based on precise lock-and-key mechanisms. Note that this is not the same as a molecular lock-and-key: the cell adhesion molecule Dscam, for example, exists in thousands of specific variants that bind specifically just to the same variant following a lock-and-key mechanism, yet it does not function in synaptic partner matching.[8, 9, 26] The location of each axon and dendrite in time and space certainly helps to reduce the number of partners that actually encounter each other and thus need to be differentiated on the lock-and-key basis. Growth paths and synaptic partnerships encoded by locks and keys may have to cope with noise, but for them noise is the enemy of precision. In the limiting case, the coding is deterministic. The growth patterns are not flexible, the morphology rigid and precise. This is how Sperry envisioned it.

An obvious way to distinguish between these two models is to test the promiscuous synapse formation hypothesis. If a key does not fit the lock, it should preclude matchmaking. If, however, everybody's keys fit all locks, the promiscuity would be absolute. Or anything in between. So, to what extent has synaptic promiscuity been observed? A lot. There are abundant examples of neurons that do not encounter the correct partner and then form synapses with a wrong one. Sansar Sharma showed it for the retinotectal projections previously studied by Sperry in goldfish. When both target areas are surgically removed, axonal projections form ordered visual maps in nonvisual areas of the brain.[27] Similarly, if a fly grows eyes with photoreceptors where it should normally have antennae, then the photoreceptors will grow axons into the brain and make synapses there.[28] But the simplest example surely are autapses—synapses a neuron makes onto itself if no other partner is available.[29] Such synapses are less common in the brain—most neurons are supposed to connect to other neurons. But if a neuron is forced to grow all by itself in a dish, there is really just nobody else to connect with. A single neuron in primary culture from a mouse will make thousands of synapses with itself, which is a great boon to experimentalists who study neurotransmission in this system: they can stimulate and record from the same cell.[30]

The importance of neighborhoods for synaptic specificity was already highlighted by John White when he published his analysis of the first wiring diagram of an entire animal in 1985, the nervous system of the tiny worm *Caenorhabditis elegans*: "The low level of mixing within process bundles has the consequence that a given process runs in a restricted neighbourhood and has only relatively few potential synaptic partners. Neurones make synaptic contacts with many of these potential partners however. There is some evidence that neurones will still behave in this way regardless of what neighbourhood they happen to be in."[31] We will revisit this golden age of neurogenetics research and the advent of connectomics in seminars 6 and 7.

A thorough investigation of the synaptic promiscuity hypothesis has to include not just the principal ability to form synapses, but also the propensity to do so. Many studies found biases for and against certain partners. An adhesive cell surface molecule on a pre- and postsynaptic neuron can be sufficient to increase the propensity of two neurons to make contact in dense brain regions. In the vertebrate retina, Joshua Sanes and colleagues identified such a factor. The cell adhesion molecule Sidekick2 causes two specific subtypes of interneurons to preferentially make functional synapses more than ten-fold over other potential partners in a dense region of neuronal processes with dozens of cell types.[32] This suggests a high probability key-and-lock mechanism and provides an important example for molecular specification in restricting synaptic promiscuity.

Is there a case where promiscuous synapse formation itself creates a pool of variation from which the correct ones are selected? In this case we have to expect that promiscuity must be strong, as the random process itself becomes a necessary part of the growth algorithm. In the mid-'90s Carla Shatz (see fig. 1.6) and coworkers presented intriguing experiments in the vertebrate retina. They wanted to see whether there is any neuronal activity in the developing retinae of ferrets, where the retinae could be dissected out after birth, but prior to the photoreceptor cells being connected. This way, Shatz and colleagues could exclude any

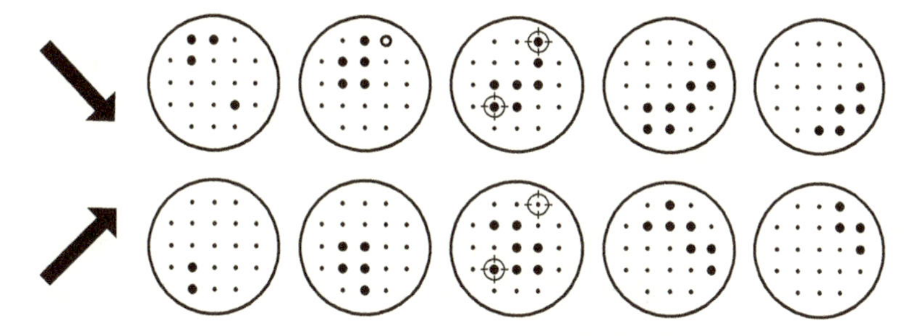

FIGURE 3.4. Schematic depiction of an experiment by Carla Shatz and coworkers in the mid-1990s describing spontaneous activity waves in the developing vertebrate retina. The arrows on the left indicate the direction of two random waves of activity that sweep across the same retina at different time points. Two target icons at the middle time point mark two separate locations on the retina. While the top wave from NW to SE sees both locations active at the same time, the bottom wave from SW to NE separates their activity in time. Through many iterations, random waves will correlate only those locations that are directly next to each other based on correlated activity patterns ("cells that fire together, wire together").

role of environmental input. To their surprise, the retina exhibited beautiful spontaneous activity waves that randomly occurred and crossed the retina on the timescale of seconds (fig. 3.4). They next wondered whether these activity waves played a role during the development of brain wiring. They injected tetrodotoxin, the poison you hope a certified chef in Japan has properly removed from a puffer fish before you eat it. Tetrodotoxin is deadly, because it blocks the conduction of neuronal activity, their famous "firing." Remarkably, they found that precise connectivity in the target area of the retina neurons that exhibited the spontaneous waves was lost after tetrodotoxin injection.[13] They concluded that neuronal firing is required for the precision of the wiring. Since the neuronal activity in the retina came in spontaneous waves, Shatz and colleagues came up with a brilliant idea: how about the fundamental concept of the Hebb synapse, the basis of McCulloch-Pitts neurons and all neural network architectures, "cells that fire together, wire together." When spontaneous activity waves cross the retina, only neurons that are directly next to each other will always fire together. Any given wave will also have cells fire together that are orthogonal to the wave, but the next wave orthogonal to the first will have the two cells' firing completely decorrelated. Hence, random direction changes of the waves are actually vitally important for this mechanism to work. And maybe most importantly: since all of this happens prior to any environmental input, from an information encoding perspective, it is all genetically encoded.[14, 33]

For Shatz's "cells that fire together, wire together" mechanism to work, the retinal ganglion cells that project into their first target region in the brain must initially make too many synapses across several layers. The activity-dependent process is not required to make synapses; it is required to weed out decorrelated loners in a pruning and fine-tuning process. I call this selection mechanism *post-specification* of synaptic partners: at first, synapses form promiscuously, and specification through selection only happens thereafter.[25]

The opposite case is *pre-specification* of synaptic partners: algorithmic growth and patterning may only bring certain partners together in space and time. The more accurately partners are sorted together, the more synapse formation *per se* can occur promiscuously. Again, a given system may employ a neat and scruffy combination: presorting can reduce partners, and probabilistic biasing between correct partners may further increase fidelity.[25] To explore these hypotheses in more depth, we will turn once again to visual systems.

The connectivity between the eye and the brain holds an important place in the history of figuring out brain wiring.[34, 35] Flies, like humans and Sperry's and Gaze's frogs, newts and fish, or Shatz's ferrets, see a picture of the world by mapping neighboring points in space as neighboring synaptic units in the brain. The peculiar eye-brain wiring diagram of flies was first described by Kuno Kirschfeld and Valentino Braitenberg in the 1960s,[36, 37] at the time when Sperry formulated the strong version of chemoaffinity, Sharma and Gaze obtained the first adaptive eye-brain mapping results in eye transplantation experiments, and Rosenblatt's perceptron, the first truly successful neural network implementation, had its heyday. Kirschfeld and Braitenberg worked in the small university town of Tübingen in Germany where the Max Planck Institute for Biological Cybernetics was founded between 1958 and 1968. Cybernetics as a field was defined by Norbert Wiener in 1948 as "control and communication in the animal and the machine."[38] Control of communication is fundamentally a question of information encoding and information transmission—and the field was fascinated by feedback loops and self-organization, which we will talk about more later.

The visual map of the fruit fly has a pretty low resolution of only 800 pixels, the equivalent of a 40 × 20 screen resolution. For comparison, in 2020, an inexpensive low end mobile phone display had a resolution of 1280 × 720 pixels. With its 40 × 20 spatial resolution, the fly does not see a good picture of the world. But it has a high temporal resolution, allowing it to process movement particularly well, a helpful feature when flying in three-dimensional space. The organization of each of the 800 pixels in the fly's brain is a copy of its neighbor; each synaptic unit represents one pixel in the brain and has the same types of neurons and synapses. But there is something funky and fascinating about the mapping of the eye to the brain in a fly. There are 800 single eyes that together form the compound eye typical for insects. Each of the 800 eyes has its own lens. However, underneath each lens there are six separate light-sensitive elements. If you put separate light-sensing elements at different positions underneath the same lens, then each sees a different point in space. Impressively, the angles between the unit eyes across the eye and the angles between the light-sensing elements underneath one eye are perfectly matched. In fact, the optic system is so close to perfect that six separate light sensing elements in six separate unit eyes all have the same optical axes; they see the same point in space, even though the six elements underneath one lens in one eye all see different points (fig. 3.5).[36, 37] Now imagine the cables to the brain: there are 800 bundles of six axons each, close to 5,000 in total, that must be sorted out in the brain. And sorted out they are: in what is considered an evolutionary and developmental masterpiece, it is exactly the six from separate unit eyes that are looking at the same point in space, that are sorted together in one synaptic ensemble in the brain, one pixel. Kirschfeld called this wiring principle "neural superposition."[36, 39]

From a cybernetics as well as developmental viewpoint, this is an interesting case study for information encoding. It seems obvious that there should not be 5,000 separate chemoaffine signals to connect each

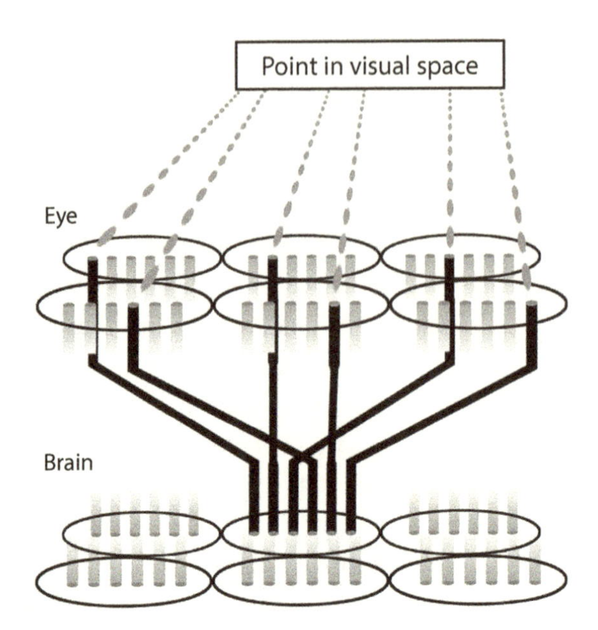

FIGURE 3.5. Neural superposition in the fly visual system. Six different types of photoreceptors in six separate eyes see the same point in visual space. These six carry the same information and are "wired together" in the same synaptic unit in the brain. This wiring pattern has been of great interest to cyberneticists since the 1960s, and the underlying growth algorithm remains only partially understood to this day.

photoreceptor neuron from the eye to its synaptic "pixel" in the brain. Yet, somehow, the close to 5,000 axon terminals all have to go "somewhere else" to meet up in groups of the correct six. Since the 800 units are all doing the same thing in a repetitive array, some sort of pattern formation should help with the information and synaptic specification problem. Experiments by Tom Clandinin and Larry Zipursky revealed important rules underlying the developmental program.[40, 41] For example, a mildly turned bundle is forced into the correct sorting pattern, revealing symmetry imposed by axonal interactions. In the early 2000s, major efforts in several labs were underway to identify the genes that regulate this exquisite sorting process. The idea was to identify genes required for the developmental process through the induction of random mutations in flies and genetic screening for flies with disrupted projection patterns. It turned out easy to find mutations that disrupted the correct sorting of the correct six axons for each pixel. The system is very precise, but obviously easily genetically disturbed. A detailed electron-microscope analysis of synapses for more than 40 such mutants with wrong connectivity yielded the surprising insight that synapses were by and large unaffected.[42] In fact, to this day, no mutant has been found that completely prevents the fly photoreceptor neuron from making synapses. So, what happens to the synaptic partnerships in developmentally incorrect visual maps? The answer is, photoreceptor axons make a precise number of synapses, even if they end up in the wrong target area. It really does seem that photoreceptor neurons really want to make synapses, and if the right partner can't be found, they will do it with a wrong partner. So, how promiscuous are these synapses formed—could a photoreceptor really make functional synaptic contact with any partner? We do not yet know. But this is a field in which it is worth digging deep to obtain answers to fundamental questions. For example, these photoreceptor neurons still form synapses in a mutant that makes them shoot through the area of the visual map altogether and end up in an entirely wrong brain area.[43] Maybe even more remarkable is the type of these synaptic contacts. Synapses at the axon terminals of these fly photoreceptor neurons form a special type of synapses where one presynaptic side is in contact with four protrusions of four different postsynaptic cells. The synapses in the wrong brain region are morphologically indistinguishable from the correct synapses in the visual map. It seems then that photoreceptors do not only want to make synapses, they really want to make this one-to-four type of synapse, no matter where they end up and who the four partners may be.[43]

Promiscuous or noisy synapse formation by itself does not contain information about brain connectivity. But synaptic promiscuity can be an integral part of a growth algorithm through pre- or post-specification in time. More than 70 years after Kirschfeld's and Braitenberg's work, we only understand parts of the developmental sorting process that contributes to pre-specification in neural superposition patterning. A first computer simulation provided a rule-based model for the sorting of axon terminals.[44] This step of the developmental algorithm concluded, the next step could, in principle, allow synapse formation with any partner found in the target area. However, even here more than a dozen partners are available. It is not yet clear to what extent during these later steps of the growth algorithm, Sperry-like target specification occurs. Probabilistic biasing for correct partners could put the final touch on pre-specification through algorithmic growth.[45] Alternatively, the timing and spatial dynamics of growth as well as time-dependent intrinsic states of synaptic competency could further facilitate or restrict what contacts are made promiscuously.[25]

Let's emphasize this again: promiscuous synapse formation can be a necessary part of the growth algorithm. Noise is not just the enemy of precision during brain wiring. Instead, random processes are abundantly utilized during the algorithmic growth of the brain, just as in so many other biological processes. Dietmar Schmucker and others originally discovered the importance of a noisy molecular mechanism for the remarkable cell adhesion molecule Dscam: In order for the receptors to be more or less randomly different in each neuron, the molecular process generating this diversity, alternative gene transcript splicing, must provide sufficient diversity.[5, 8, 9, 26] If the splicing machinery becomes more restricted through mutation, it leads to miswiring because too many neurons have the same Dscam version and fail to distinguish self from non-self.[46] Hence, the noisy process that produces stochastic variation of protein isoforms is essential for ultimate wiring specificity. By contrast, noise is incompatible with precise molecular target signals in Sperry's sense. To the extent that we see random processes as integral processes in brain development, we depart from the neat extreme to the scruffy side. We just don't know yet how far we need to go.

As we have seen, noisy developmental processes can lead to incredibly precise outcomes in brain wiring. So how about variability in brain wiring? Variability was already observed by Lashley in 1947: " the brain is extremely variable in every character that has been subjected to measurement. . . . Individuals start life with brains differing enormously in structure."[47] We have seen that behaviorally meaningful variability in brain wiring can even occur based on the same genetic program.[22] If a genotype produces several different behaviors, then a specific behavior cannot be selected for, right? How is this kind of variability helpful? Benjamin de Bivort tested the idea of genetic selection of variability itself.[48] The idea is that the *average* phenotype does not change, while widening the population distribution produces more variability to choose from.[49] In a wider distribution there are fewer individuals that actually exhibit the precise average phenotype, but many more divergent individuals on the left and right sides of that average. An increased phenotypic variability can be a selective advantage. Imagine an increased selection pressure, say an environmental change. The individuals are the messengers and survival machines for a specific genotype. If a specific genotype produces very similar individuals and the environmental change is substantial, all individuals will be on the wrong side of selection and the population dies. If, however, a specific genotype sends a noisy, highly variable army into the war, then some of those individuals at the far end of the distribution may survive the increased selection pressure, and the specific genotype still has a messenger in the game. It is at heart a "bet hedging" strategy (fig. 3.6).[50] We'll revisit the ideas of flexibility and robustness in the next section on autonomous agents.

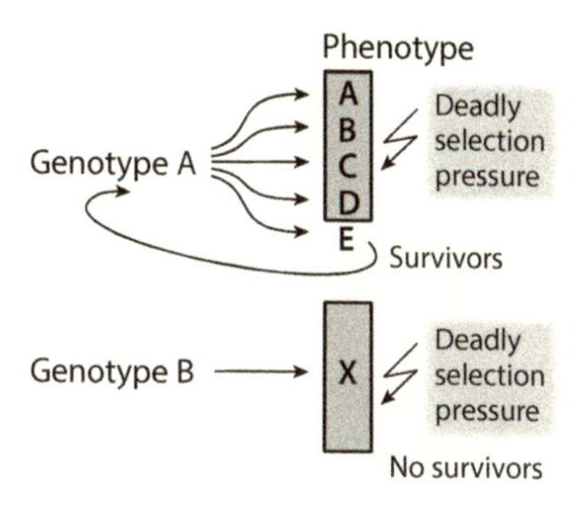

FIGURE 3.6. Imprecise genotype to phenotype mapping as a bet hedging strategy. Genotype A produces individuals with variable phenotypes that are differently affected by a given selection pressure. In the example, only individuals with phenotype E survive. These individuals will reproduce all phenotypes in the next generation. Genotype B produces a single phenotype exposed to—and selected against by—the same selection pressure. Since genotype B only produced this one phenotype with precision, there are no survivors.

De Bivort and colleagues identified a mutant that caused flies to choose their paths in a behavioral experiment more variably, without changing the average choice. They then asked when the gene in which they found the mutation needed to be active to affect the more variable behavior and mapped the critical period to the main time window during which synaptic connections are established in brain wiring. Maybe the biggest surprise however was the nature of the gene: It turned out to be the cell surface protein tenurin-a.[48] This exact gene was previously characterized as a key-and-lock synaptic matchmaker in the fly olfactory system.[51]

An important lesson from this work is that phenotypic variability can result from noisy development independent of genetic variability. There actually are genes whose function helps to ensure phenotypic variability even if all genes are identical. This adds an important dimension to the nature vs nurture debate: genetically encoded variability can be independent of both genetic variability (nature) and environmental variability (nurture).[22, 52, 53] Instead, variability is an in-built function of (brain) development and a limit to brain determinacy.[54] An emphasis of random factors and a warning against genetic determinism was the key advice the Harvard geneticist Richard Lewontin gave in an interview to the readers of the journal *Genetics* upon receipt of the prestigious Thomas Hunt Morgan Medal in 2017.[55] A similar consideration brought Kevin Mitchell in his book *Innate* to emphasize the role of stochastic developmental processes: "We really are born different and, in many ways, we get more so over time."[53] Think monozygotic twins: Here is a natural experiment of presumably zero genetic variability. How similar are such genetically identical individuals? They certainly look very similar. How about fingerprints? They are different. Not as different as between two randomly chosen people, but different. In fact, they are as different as your left and right index fingerprint.[56]

This brings us back to the concept of relevant information. The fingerprints are likely different because they are the product of developmental noise below the selection threshold. In this case, noise may just have been sufficiently irrelevant not to be selected against. Relevance is in the eye of the beholder. And the beholder is evolution. If the precise patterns of fingerprints are not under selection, than the developmental algorithm is free to leave the output noisy.

What is the role of the environment? Maybe, at the time your fingerprint patterns developed, you were lying a little bit to the left in your mother's womb. Your left hand happened to be in an environment where the temperature was 0.01 degree Celsius colder. That's a pretty

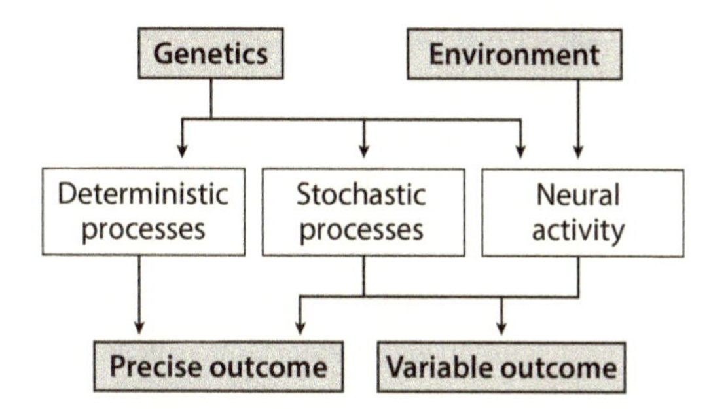

FIGURE 3.7. The nature vs nurture debate revisited. Genetic determinism has been associated with the false notion that genetics only encode precise and rigid outcomes, while only the environment and learning based on neural activity cause variability. This view is incomplete, because the genome encodes both deterministic and stochastic processes, and the genetic program can include neural activity independent of environmental information. Furthermore, stochastic processes can lead to precise outcomes.

irrelevant difference, much less than the normal body temperature fluctuations. But at the molecular level, any such difference could easily influence whether you had one groove more or less in your fingerprint. It's noisy—and it's irrelevant. Except that any such noisy process has at least the capacity of serving as a pool of variation. If such differences can be exploited, then the variability becomes relevant. This is the case for Dscam splicing. Like every biological process, splicing has an error rate, which is normally very low.[57] But if the errors create a pool of useful variability, and this variability becomes an important parameter during the execution of a developmental rule as the growth algorithm runs, it will be selected for. We are talking basic evolution here. DNA replication has an error rate. This error rate is normally very low, but it is never zero. If it were zero, there would be no pool of genetic variation to select from. Bacteria have a particularly high error rate, i.e., they replicate their own genetic code systematically with lots of random errors. That's how they develop antibiotic resistance. But the evolutionary principle does not only work at the level of DNA mutations that change the genetic code, it also works at many other molecular and cellular levels, within the running growth algorithm based on a specific genotype. So let's hold just this thought: the variability we observe may either be below selection threshold; it can be irrelevant noise. Or it may be relevant noise because it has been selected as part of intrinsically noisy developmental processes. We currently do not actually know to what extent two monozygotic twins are different because of environmental differences (nurture) or because of intrinsically stochastic development (the noisy part of nature) (fig. 3.7).

Neurodevelopmental disorders like autism and schizophrenia amongst monozygotic twins have a strong genetic component.[58] If one twin suffers from schizophrenia, the other has a 50–70% probability of having the disease. That's much more than the less than 1% of a random person on the street, but also not 100%. The difference could be made up by environmental differences or intrinsically stochastic development.

Each monozygotic twin, like each of de Bivort's or Hassan's flies, are unique individuals because of intrinsically stochastic development. But I want to end with a reminder that, as we have seen, a noisy process can also create order. We began this seminar with the example of bacteria that will chemotax with 100% precision towards the sugar source, as long as the biased random walk works. We will meet this idea again in the next seminar and the example of axonal growth cones: They clearly exhibit random dynamics that may be critical to "hit" the right partner. But as long as the random dynamics occur, the partner identification may well be 100% precise. We must look at how noise is utilized to understand when it creates order or variability. After all, noise is a fact of nature, and nature knows how to make it her own. Thank you.

1.3

Autonomous Agents and Local Rules

The Fourth Discussion: On Filopodia and Soccer Games

MINDA (THE DEVELOPMENTAL GENETICIST): I am not comfortable with the emphasis on random development. Of course I agree with the evolutionary principle though. It also makes sense that a growth cone must have randomly exploring filopodia to find the precise target. If I think about the one filopodium that hits the right target as "being selected," that's an okay picture.

AKI (THE ROBOTICS ENGINEER): So you say, but what does the filopodium know about the process?

ALFRED (THE NEUROSCIENTIST): Nothing, of course. The mutation in the bacterium knows nothing about antibiotic resistance either. It's evolution at work. This reminds me of Gerald Edelman's *neural Darwinism* idea. He proposed neuronal group selection as a mechanism for brain function, drawing on his earlier work on the immune system.[1]

PRAMESH (THE AI RESEARCHER): Yes, but that's function. How do we integrate the concept of mutation and selection with algorithmic growth? The mutations change the recipe, but how that changes the run of the growth algorithm may not be obvious or predictable. What matters is that the outcome is reproducible and selectable.

MINDA: I know the recipe analogy from Richard Dawkins. He said that the genetic code is like a baking recipe. It's not a code for the cake, it's just a set of rules for how to make it.[2]

ALFRED: But there is something missing here. The growth cone does what it wants—it must make its own choices whether to go left or right. There are no components of the cake that make any individually different choices where to go once the dough is mixed together and put in the oven . . .

PRAMESH: Hmm. The cake receives energy only from the outside, whereas the cell and the growth cone produce their own. This allows the growth cone to behave like what we call an autonomous agent. It does make its own local decisions. By putting many autonomous agents together that all have to follow the same rules and interact with feedback, you can create things like swarm behavior.

AKI: And I guess next the brain is wired by swarm behavior, probably noisy . . .

ALFRED: How about flexibility in an unknown environment? I think the reason why you need growth cones to make individual decisions is that they need to be flexible to find the right partner.

MINDA: Yes, we just explained that. Noise and selection allow the growth cone to find a partner with 100% precision, even if it is a bit more to the left or right.

AKI: Actually, you don't really need noise for that. I could program a perfectly deterministic search algorithm that equally probes the entire environment around the growth cone in a 360 degree circle. It'll find the partner all right.

PRAMESH: That's true.

ALFRED: But isn't that way too costly? I mean, the energy and resources to search everywhere equally would be enormous; that's not very efficient.

AKI: I can easily give it some priority rules, like search twice as much in this directions and not that, based on some growth information or environmental information or something.

MINDA: So, the key here is that the growth cone could not be flexible if it would not make its own choices. Even if the search mechanism were deterministic, it still had to do it on its own. If it were told where to go, it would not be flexible anymore. There is something more going on than the baking recipe. The genes have sent the growth cone on its journey, but out there it is on its own!

AKI: Wow, is that a new Minda speaking?

MINDA: Why? It's obvious that a genetically encoded organism has to fend for itself in the environment.

ALFRED: Yes, but now we are talking about this kind of behavior not for an animal or a plant out there, but for a growth cone during development of precise neural network connectivity . . . is that really the same?

PRAMESH: Yes, I think that's all true. That's definitely what we call autonomous agents. Each glider in Conway's Game of Life is an autonomous agent, it executes rules locally. Actually, the Game of Life is of course deterministic and yet the game is flexible, reacts to obstacles, etc.

MINDA: I am fine with genetically encoded development being flexible. And yes, if I can accept an organism to be sent on its journey by genes, and to fend for itself in the environment, then I don't see why I shouldn't accept a neuron to be sent on a journey by genes and fend for itself in what it encounters as its environment during development.

PRAMESH: But in contrast to the Game of Life or rule 110, the neuron, growth cone and filopodium are not run by a deterministic single rule. It is much more complicated. The local rules change as the growth algorithm runs. And as we discussed before, noise can be required to make it work.

AKI: Haven't we all long given up control? In my group, we *want* our robots to make their own smart decisions, that's the whole point. We want them to work with us, we just don't want them to take over and go wild. 'You wouldn't want your video recorder lounging around on the sofa all day while it was watching TV. You wouldn't want it picking its nose, drinking beer and sending out for pizzas.' Not mine, but pretty good anyway.[3]

ALFRED: You want them to work once you turn them on. You build them based on a designed blueprint. How about you start building robots by individual cells that make their own decisions. Grow your robot!

AKI: Ha! Actually, there was a hype time for nanotechnology when people dreamt of self-assembling cars. Bottom-up technology instead of top-down technology. Eric Drexler, another Minsky student, promoted the idea of molecular nanotechnology and self-assembling molecular machines in the '90s. I am not sure where that stands. But there continues to be a lot of discussion these days about the vision of little nanomachines that can be released into your blood stream to fix your body, kill cancer cells, this kind of thing. I like the idea![4]

PRAMESH: The problem is that the autonomous agents will be out of control . . .

AKI: Why out of control? They don't mutate, I can give them precise jobs. I can even program them to self-destruct after their job is done.

PRAMESH: But that's the whole point. If autonomous agents make local decision, then you can't predict what's going to happen once they start interacting with each other and the environment. Remember rule 110 just for starters, and that is without adding noise. But if autonomous agents are confronted with an unknown environment, interact with each other, and finally propagate, then you add evolution, I don't know . . .

MINDA: I think we are mixing things up again. Where are the autonomous agents in rule 110?

PRAMESH: I think the analogy is quite good actually. Rule 110 is really a local rule. Each cell of the automaton locally decides what to do by applying the rule. So think of Aki's nanomachines in the blood stream: With every interaction they have to apply whatever rules have been programmed into them. As time goes on, the state of the system changes—so the nanomachines operate in a changing environment all the time, with different challenges by the minute. Add feedback. You know exactly how a nanomachine is going to behave given a certain input, but you can't predict how the larger system with the nanomachines in it is going to behave.

AKI: I told you they could be programmed to self-destruct. And they really do not need to self-assemble or replicate—think of little machines that are sent to do a very specific job and then kill themselves. If those are the rules then how is it out of control?

PRAMESH: Well, if your nanomachines all self-destruct at a certain point then I guess that's a rather definite endpoint, whatever happened before. It doesn't mean you were in control while they did their thing though. But the evolution question is important. The application of the evolutionary principle does not require replication.

ALFRED: Of course, filopodia do not replicate either, yet they are selected. Immune cells are selected by a pathogen—I would call that an evolutionary principle. These were the examples where a noisy process is utilized to do something robustly, like detecting an imprecise target or unknown invader. But is that really a problem for the nanomachines? Maybe we are getting a bit paranoid after all?

PRAMESH: The point was that the autonomous agent will have to do its own thing, hopefully following great and safe rules. You program the rules; the agent does its thing. You give up local control. If a certain rule happens to be a bad idea in a particular situation, bad luck. Can you really predict all possible situations? In any case, it is much harder to predict how the system behaves on the next higher organization level. I understand the rules for the growth cone, but what does that mean for the network?

MINDA: The point of the growth cone, and I guess the immune system, is that they find what they seek robustly. When would that be a bad idea? They are flexible when and where it's needed—with precision!

AKI: Right, what's the disadvantage? Individual agents behave and react locally, that makes sense. What I am trying to understand is what can go wrong. Pramesh keeps on talking about something out of control.

ALFRED: Can we bring this back to neurons please? If the dendritic tree uses self-avoidance to branch, then the filopodia have to be the autonomous agents, I guess. And the precise branching pattern may be out of control, but the nicely branched dendritic tree can flexibly respond to its environment and grows robustly. Is that the idea?

MINDA: I have the feeling we are giving some rather trivial and obvious concepts fancy names again.

PRAMESH: . . . until you start to try and grow a car using something like dendritic self-avoidance . . .

AKI: Yeah, good luck with that. I can give you some specs for quality control.

MINDA: We are not here to build cars. We are trying to figure out how intelligent neural networks can be built. If biology is your template, then you need to grow them.

PRAMESH: If you have to grow it—and I am not sure about that—then the growth has all the issues we discussed. Individual components have to make local decisions, and even small environmental differences allow the system to adapt and change the outcome a lot. It's actually not really about control, it's about information again—what information do you need to provide to make a neural network. Something precise, but also flexible and robust.

AKI: What I want is the flexibility and robustness, but keep control.

ALFRED: I guess the growing brain does not have an Aki to keep control.

AKI: So, you think you can't grow a car?

PRAMESH: Well, maybe you can, but it won't be the car you like. Each individual car will be as different from every other car as apple trees are different from each other.

AKI: But if growth can also utilize noise to yield perfectly precise structures, can we not *only* use those bits of the algorithm that do that—and leave out the parts of the algorithm that would allow for variation in the outcome? As we said before, some noise may be left, but that can be below threshold and irrelevant. Every engineer will tell you that the machines they build are only precise up to a certain standard. What's the difference?

MINDA: Every cell has a primary program, it wants something. You could say a cell wants to live, for example. But then during development, other cells may tell it that it has to die. If that doesn't work, the cell is a cancer cell. The little worm *C. elegans* famously has 302 neurons, but more cells are actually made during development and then have to die through what we call programmed cell death.[5,6]

AKI: That's creepy. Why would I build a machine that grows while killing bits of itself?

MINDA: You may not get the right outcome otherwise. A cell's primary drive may be to live, but it is controlled by cellular interactions. Evolution has selected the survival of the whole organism, and that only works if certain cells die at the right time and place.

AKI: What's the primary drive of a neuron then?

MINDA: Well, I would say it wants to grow an axon. And it wants to make synapses.

AKI: I don't see how that idea helps us understand how to build the network.

ALFRED: Well, if the primary drive of the neuron is to grow an axon, then it will do just that if not prevented from doing so otherwise. If there is no specific direction for it to grow, it will just grow anyway. It's not like it will only grow if it is lured somewhere. And we talked about how a neuron really wants to make synapses. It doesn't need a very specific partner to make it do so, it will do it wherever you let it.

PRAMESH: That would mean that you mostly try to prevent the neurons from growing wrongly or making the wrong synapses. And whatever it is not prevented from doing better end up being the right thing.

MINDA: Here I need to respectfully object again. There are definitely a lot of repulsive signals, but also attractive ones. Maybe we can agree that the neuron has to respond to these signals, and that's definitely a local decision by the neuron. I agree this is not the same as if there was a director who simultaneously controlled all neurons to make sure everybody is making the right choices at the right time. The genome cannot be such a director.

PRAMESH: Exactly! That's the issue with self-assembly: the moment you give flexibility to the individual agents, you must give up some control. You can't have it both ways.

AKI: Could somebody tell me how all of this applies to AI? The issue here is learning, not growing. People don't grow artificial neural networks, not even in simulations.

PRAMESH: Both growing and learning are games with players—like a soccer game, really. If every player would do the same thing, it wouldn't be very interesting. Every player has to follow the same rules, but makes individual choices in context-dependent feedback with other players—and may do something you would not have predicted.

ALFRED: The question is whether we are playing the same game.

Seminar 4: From Local Rules to Robustness

Imagine you want to build a neural network that requires a certain ratio of neuronal components, say 46% of type A, 30% of type B and 24% of type C. The network architecture is flexible within a certain range. For example, good functional values for type A may lie between 43% and 49%. These are typical numbers for neuronal differentiation during brain development: ratios are stable around an average, but are variable within a certain range. How could this ratio be encoded robustly?

A good way to robustly produce this ratio would be to play a sufficient number of soccer games. The outcomes of soccer games in the German premier league in each of the last 10 years was 46% (+/− 2%) of all games ended with a home victory. The error of 2% is the standard deviation, meaning that 2/3 of all results fall within this range. No season ended with less than 43% or more than 49% home victories. Similar values apply to the two alternatives: 30% victories away from home and 24% draws. As a system to establish these ratios for neural network planning purposes, the effort of financing and running the entire German soccer premier league seems excessive, but it would certainly be robust. The number of goals per game is similarly precise: A soccer game in the first division of German soccer has had 3 goals per game since the 60s. There is some variability within the range of 2.5–3.5, but the average is within these boundaries for entire seasons.[7] On average, soccer is highly predictable and rather boring.

Individual games are a different matter. You may easily get unlucky with 0 goals or lucky with 5. Only from the perspective of the seasonal results the game is highly predictable, reproducible and robust. The rules of the game ensure that there has to be a winner or a draw; soccer is a zero sum game. If you are, for personal reasons, interested in the outcome of an individual game, then this is your game. If you are interested in the seasonal variation, soccer is not only boring, it is reliably and robustly boring.

What is interesting is in the eye of the beholder. From the perspective of the soccer fan, the outcome of every single game is interesting. The players and their performance are interesting. And interesting here means: unpredictable. From the perspective of the seasonal organization, advertising companies or stadium operators, the outcome of the individual games are a random bit of irrelevant information. They care about predictable seasons, numbers of games and spectators. The games just have to happen for their operations to function robustly.

What makes the individual game interesting? Soccer games have the right mix of being variable, flexible and robust. There are rules that ensure the game ends with a win or a draw and there are rules for what players can and cannot do. As the game unfolds in time, each player acts according to these rules, but makes his or her own context-dependent decisions. The players have their own drives, continuously interact with other players, and flexibly adjust their game according to the actions of other players. While the rules are relatively simple, the complexities of the ensuing dynamics are mindboggling. Every minute of every soccer game that ever occurred in history is a unique sequence of the movements of 22 players and one ball. Some minutes appear more exciting than others to the educated observer. Goals help. Some goal sequences may be more interesting than others. These we may indeed watch over and over again in time-lapse, where we analyze maybe one or two players central to the successful play. By understanding the detailed movements, the ploy of the attacker, the deception of the goalie, we can reconstruct how the goal happened.

That's what molecular biologists are doing. We want to understand the mechanism of a successful axon targeting event just as much as the mechanism of a successful goal. The growth cone calls the shot to the left or the right depending on the context of other players around it. The growth cone acts as a player that may compete with other growth cones for the same resource. The players move only in certain ways, constrained by rules. *As the game unfolds in time, each player acts according to these rules, but makes his or her own context-dependent decisions. The players have their own drives, continuously interact with other players, and flexibly adjust their game according to the actions of other players.* Yes, I did just repeat these two sentences. Both soccer players and growth cones act as *autonomous agents* in their respective games. Both are unpredictable in some aspects of their game. Both establish higher order phenomena that are robust. Whether the level of the unpredictable events or the level of the robust higher order system is interesting in a game of soccer or game of growth cone guidance is in the eye of the beholder.

We have now encountered two contributors to unpredictability and robustness: noise and autonomous agents. Let's analyze their contributions both separately and together.

Autonomous agents execute rules locally in interaction with their environment. The autonomous agent is not a lifeless puppet on a string with the puppeteer as the source of energy and the controller of actions. There is no puppeteer in self-assembly. Autonomous agents need time and energy *locally* to act. We can now look at the rule 110 cellular automaton in this light. Each cell locally applies the rule to decide the state of the cell in the next row. Rule 110 is not a puppeteer, it is locally executed with different results depending on the local environment. You could say each cell reacts to each other to decide the next time point. And in order to do so, it needs energy locally. The local cells behave as autonomous agents. And of course, rule 110 is unpredictable despite being deterministic. We can conclude that, even if a game of soccer or a biological process were deterministic, it could still be unpredictable if its components burn, or dissipate, energy and time locally in a self-coordinating, self-assembling process. And this ability of individual components to react to environmental differences is a key ingredient to render the system robust to perturbation.

From inside the system, neither the soccer game nor a biological system looks deterministic. However, note that two rows of a rule 110 automaton just a few iterations apart do not look deterministic either. In fact, they can be mathematically indistinguishable from a random process. From the inside a deterministic system can look noisy. We can calculate when events obey criteria for randomness—and measurably random processes abound, whatever their cause. Pools of variation, like immune cell variability, or the chemotactic biased random walk, or a growth cone's filopodia, all provide sources for selection. A rare variant in that pool may be key to robustness of the system. Think of the one immune cell that recognizes a pathogen, or the one filopodium that finds the partner. In other cases, noisy components in large numbers can simply average out to yield robust ratios at a higher level of organization. Think of the seasonal results of many soccer games or cell differentiation ratios. Both cases operate on the basis of individual components, or autonomous agents, that require local energy consumption and local rules.

What are the local rules? In the case of rule 110, the global rule at the outset of the game is identical with the rules applied locally at every iteration. Biological growth is more complicated than that. New proteins, including transcription factors, cell surface proteins and many others, enter the scene and change the neuron. These proteins add new properties and change the rules. Sometimes a rule is apparent by looking at a single molecular mechanism involving just one or a few proteins. In other cases understanding the rules requires more context and multiple contributing molecular mechanisms. Self-avoidance is again a good example. Here, a key molecular mechanism is the molecular interaction across cells through surface proteins like Dscam. However, interaction of Dscams on two membranes simply causes binding and this mechanism in isolation says nothing about attractive or repulsive consequence of the interaction. Repulsion is a response property of the filopodium, not the molecule. Similarly, self-avoidance is a response property of the filopodium, not the molecule. In these cases, the rules are composites of more than one molecular mechanism, plus properties of subcellular multi-protein interactions and structures. Self-avoidance is a property of a particular part of the neuron at the particular step as it runs its growth algorithm. The neuron must have a certain history. It must have initiated the growth of cytoskeletal and membrane protrusions. It must randomly branch and probe surroundings using filopodial dynamics. And it must have the right cell surface receptors localized to the right structure at that time. Only when all these variables, and many more, are in place, does the local rule of "self-avoidance" apply.[8] Remarkably, in a different context, a different compartment possibly even within the same cell, Dscam-Dscam interactions can be essential to sustain dynamic exploratory behavior of axonal sister branches.[9, 10] This exploratory behavior in turn is a stochastic process, which enables other molecular interactions to eventually select and stabilize a subset of axonal branches.[11] These diverse and context-dependent cellular responses are fleeting moments in the development of the neuron and the wiring of the brain. To call this "genetically encoded" is entirely correct, but what does it mean?

The knock-out of a specific gene may change local rules, but it may do so at several steps of the growth algorithm. Every next step depends on the previous steps, and changes propagate through space and time. To talk about a few thousand genes and how they encode brain wiring is to leave out information unfolding during development. The genes are a particularly interesting level, because it is here that both researchers and evolution perform their favorite perturbation experiments. But the consequences of a mutation that affects an early step during development are likely to create an avalanche of changes, modifications to many a rule, and ultimately an unpredictable outcome. Evolution can't predict that outcome any better than a molecular biologist. Evolution simply applies, tests and selects random—and typically small—changes. The final output is the subject of selection, however intricately complicated the developmental algorithm that led to it.

We are encountering the levels problem again. Autonomous agents may be playing their unpredictable game at one level, but at the next higher level the system may behave perfectly predictably. This is trivial if you are a city planner and have to estimate traffic. When I step out of my institute onto the street on a Tuesday at noon, I see a rather unsurprising number of cars forming a continuous stream with irregular interruptions that will allow me to cross within a minute, usually. There is an average wait for me to cross the street to go to the cafeteria on the other side. I can expect this wait, and while I sometimes get lucky and can cross quickly or sometimes have to wait longer, all of this can be expected. The driver of an individual car (our energy-consuming autonomous agent) is not thinking about her or his effect on the continuous flow, but rather about whether they will make it in time to pick up the kids from school to make it to the piano lesson. Hundreds of these individually motivated agents create a predictable, really quite boring, situation. Crossing the street every day for lunch is a predictable, worthwhile exercise because the noisy individual car movements exhibit predictable flow in large numbers. How about the more interesting idea that noisy individual autonomous agents provide a pool of variation for an evolutionary selection process in brain wiring?

In the late '80s, James Vaughn did experiments using the Golgi method on spinal motor neurons. Based on the static structures he saw, he proposed the idea of synaptotropic growth.[12, 13] The hypothesis built on the idea that dendrites and axons may utilize their filopodial dynamics and branching to detect where synaptic partners are located. We have discussed how branches spread based on rules like self-avoidance. But how do they make sure that they spread in the direction where the correct synaptic partners are? We can think of filopodial extensions from both partners as autonomous agents, scouts in search of partners. The presynaptic growth cone probes its surroundings with filopodial

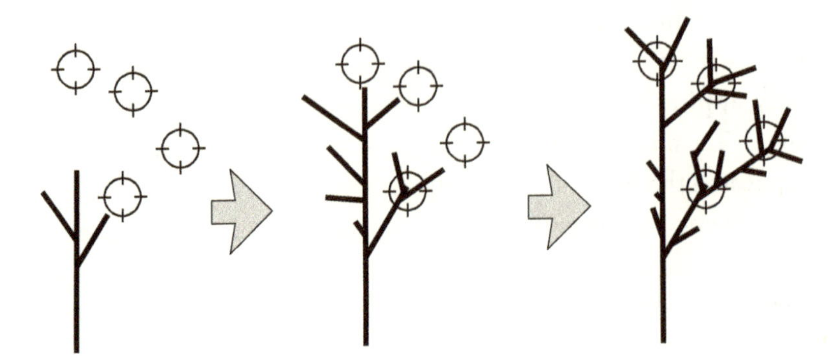

FIGURE 4.1. Synaptotropic growth appears as directed growth towards a target (synaptic partners). It is an algorithmic process based on stochastic branching, with stabilization on partner contact, followed by new stochastic branching from the contact sites. Based on synaptotropic growth, axonal or dendritic structures can find targets in an unknown environment.

exploration. The postsynaptic partner is spreading its tree. What happens when they first meet? Vaughn's idea was a two-step rule set: first, correct contacts are stabilized, while absence of contact leads to continued cycles of extensions and retractions; second, stabilized contacts become new branch points for further growth and another iteration of filopodial exploration (fig. 4.1). In this way, the dendritic tree will be seen growing towards where the presynaptic partner is, while regions devoid of presynaptic partners are ignored. The dendritic tree grows towards a "source of attraction."

Synaptotropic growth was later beautifully observed in both growing dendritic and axonal branches in living, normally developing fish by Stephen Smith and colleagues.[14] Many neuronal shapes are consistent with this type of growth, including the famous dendritic tree of Purkinje cells (see fig. 2.5). Synaptotropic growth is a prime example for the iterative, local application of a simple set of rules by autonomous agents. It leads to branching patterns that make every individual Purkinje cell as different from every other cell as snowflakes are from each other. The process is incredibly flexible and robust. But it is only one set of simple rules that comes into effect at a distinct step of the algorithmic growth program. Before and after the establishment of branches, the algorithm sees the neuron in entirely different states, executing different rules with different molecules. All of this is genetically encoded.

In AI, autonomous agents have featured prominently ever since Minsky and Edmonds built their first "crazy random design" neural network in 1951. Each neuron, each "rat" started to behave in unpredicted ways, and most interestingly so when interacting with each other. During the years when AI focused on formal symbolic logic, autonomous agents took a back seat. Instead, they thrived in the smaller, but dedicated community of artificial life researchers. Experimentation with cellular automata has never ceased to fascinate researchers since the '40s. Minsky found his own unique scruffy way to bring autonomous agents back in, while still focusing on symbol-processing AI, and independent of neural networks. In his 1986 book *Society of Mind* he presents a collection of ideas that all are designed to help understand how agents, not knowing anything about the bigger whole, can together create intelligence. In Minsky's words: "We want to explain intelligence as a combination of simpler things" (section 1.6); "I'll call 'Society of Minds' this scheme in which each mind is made of many smaller processes. These we'll call agents. Yet when we join these agents in societies -in certain very special ways- this leads to true intelligence" (prologue); "Unless we can explain the mind in terms of things that have no thoughts or feelings of their own, we'll only have gone around in a circle. But what could those simpler particles be- the 'agents' that compose our minds?" (section 1.1). Minsky refers to his agents sometimes as "principles," then "particles" and finally "tiny machines."[15] Hence, while the precise nature of the agents remains elusive, the agents always are a representation of the levels problem. Unpredictable actions of autonomous agents are not interesting, if it were not for the higher order organization of a system they constitute. Minsky's holy grail of higher order is the mind. Given our discussion it may therefore not be too surprising that agents can be very different things, depending on the process and level we are trying to explain. Minsky goes as far as describing something as an agent that need not be a "particle" or "tiny machine" in the sense of a physical thing: "The GRASPING agents want to keep hold of the cup. Your BALANCING agents want to keep the tea from spilling out. Your

THIRST agents want you to drink the tea. . . . Yet none of these consume your mind as you roam about the room talking to your friends. You scarcely think about Balance; Balance has no concern with Grasp. Grasp has no concern with Thirst" (section 1.2).[15]

In biology, ALife and AI research alike, an autonomous agent is an organizational entity that follows its own local rules, but has no knowledge of a higher organizational level. The filopodium doesn't know what its decisions mean for the growth cone, just as the parent on the way to pick up the kids knows nothing about the robust situation I encounter when crossing the street at lunch time. The growth cone doesn't know what its decisions mean for neuronal connectivity. And the neuron doesn't know what its decisions mean for behavior. An ant doesn't know what its behavior means for the colony. Yet in all these cases, the higher order organization exhibits robustness that can only be explained through the actions of autonomous agents that each had to individually fend for themselves. The individual leaf cutter ant has to have the ability to recognize a leaf in unknown terrain, decide where and how big a piece of the leaf to cut off and bring it back to the colony. From the ant's perspective, it finds itself in an environment that perfectly fits the rules according to which it is programmed to operate. It lives in a place full with positive (and sometimes alarming) olfactory cues, tracks to follow to where lots of other ants are going as well. There are leaves to cut where it goes that are exactly the type of structure that its mandibles are good at cutting, and the whole way it intuitively uses its mandibles just works great. There is food when and where it needs it. And if the termites attack, there is a rule book for that, too.

The growth cone finds itself in perfect surroundings, too. Just when it runs its filopodial dynamics program, there are other agents in the environment that provide perfectly fitting chemical and mechanical surroundings. In fact, neither would make much sense without the other. As a brain develops, all neurons run through their programs in an enormous coordinated and dynamic jigsaw puzzle. As each cell progresses through its individual growth algorithm, it finds itself in an environment of other cells running their algorithmic growth programs. As they interact, they change each others' programs, which in turn is part of what defines the overall growth algorithm. The growth cones effectively predict the interactions that must occur for the algorithm to run smoothly. Biologists make the distinction between cell-autonomous versus cell non-autonomous programs. Minsky, not being concerned with biologists much, noted: "The principal activities of brains are making changes in themselves. Because the whole idea of self-modifying processes is new to our experience, we cannot yet trust our commonsense judgments about such matters."[15] And yet, the beauty of the coordinated program of self-modifying processes during algorithmic growth is that each cell finds itself in the perfect environment for exactly what it is good at at the right time and place. At the time the neuron is really good at growing an axon, there is an environment that helps with the path. At the time the neuron is really good at making synapses, the right partners will be available, right then and there. The neuron lives in a dynamic normality that is characterized by continuous feedback with its surroundings and changing contexts that always seem to be just right.

The concept of *dynamic normality* applies to autonomous agents at any level. At the time and place the growth cone has to make decisions, it does so in an environment that, consisting of other agents running their growth algorithms, is the perfect setting for that decision. There is a sense here of a big flow, where every part has the fitting abilities at the right time and place, not knowing about the bigger whole, yet perfectly contributing to making it happen.

This is self-organization, a concept both beautiful and powerful. Autonomous agents, be they soccer players, neuronal filopodia or ants, can produce new, complicated behaviors through self-organization of agents that, each individually, follow simple rules, with nobody in charge. The autonomous agents don't even have to directly interact with each other. Like soccer players or neuronal filopodia, termites act as autonomous agents to build their impressive mounds. Termite behavior is characterized by each individual making its own, local decisions and following a common set of rules when moving about little mudballs. Each termite adds an attractive pheromone to the mudball, and it is itself attracted to that very pheromone. The individual termite explores the ground more or less randomly as it tries to find a spot where to drop the mud. If the ground is new and clean, the termite may drop the mudball randomly. If there is already a mudball lying around, the termite will be attracted to its pheromone and add its own mudball right there. A third termite in search of a good place to drop a mudball will be particularly attracted to a place with the double amount of pheromone from two mudballs on the same spot. As a few more million termites follow the same simple rule, a substantial termite mound grows. None of the termites are aware of the significance of the team effort. In fact, they need not be aware of the team or any team members: interactions are based on traces left in the environment, not between the autonomous agents themselves. In 1959, at about the same time as Heinz von Foerster thought about the principles underlying order from noise,[16] the French biologist Pierre-Paul Grassé coined the term "stigmergy" to describe the indirect self-organizing behavior of the termites through traces left in the environment.[17, 18] The concept has remained influential, for example in artificial life research and swarm intelligence,[19] but it is not widely known amongst developmental biologists, neuroscientists or AI scientists.

Discussions of self-organization are rare in "A causes B"-type studies prevalent in developmental biology. And yet, self-organization in biological systems has a long and rich history, including brain function. In 1947, the psychiatrist Ross Ashby (fig. 4.2) published a paper "Principles of the self-organizing dynamic system."[20] Ashby became a leading proponent of the new field of cybernetics, which he understood to deal with self-regulating systems. He built on Wiener's ideas on cybernetics[21] and applied them to the brain in his landmark 1952 book *Design for a Brain—the Origin of Adaptive Behavior*.[22] Many ideas that we are painfully rediscovering today as we hit the limits of linear causalities in development were already described by Ashby. For example, he clearly stated that each part of a system would have to adapt to an environment formed by all other parts or subsystems. We will discuss Ashby's contribution in more detail in the ninth seminar. For now, let's just note that he predicted self-organization of the brain, even though the actual self-organizing molecular and cellular components would only be discovered decades later.

FIGURE 4.2. W. Ross Ashby in his office in Stanford, 1955.

The idea that autonomous agents' actions underlie higher order behavior of a system is a representation and a property of what I call the levels problem. For every outcome of a neuronal differentiation ratio, there may indeed be an entire season of soccer games one level lower that give rise to the robust higher order property. Autonomous agents play such roles at an organizational level that allows them—or forces them—to react locally according to their abilities, without oversight from a controller at a higher level. The total information we would need to describe every single decision of an autonomous agent, every turn it takes, at every level of the algorithm, is fundamentally endless. There is always a deeper level or another layer of context to dig into. How this seemingly endless amount of information is coded into the algorithm is an analogous problem to rule 110, only more complicated, because the algorithm keeps changing the local rules, for myriads of different molecular to cellular agents, with every iteration.

At the moment we take a neuron out of its developmental context, it will be in a certain state of its developmental algorithm. How is the neuron going to behave? Once it is out of context, it lacks both inductive and inhibitory signals that would normally form part of the next steps of its developmental progress. If we culture at an early developmental stage, the neuron will exhibit the behaviors it is programmed to execute at that point. For example, the neuron is likely to begin growing its axon. At this time point, the neuron will not yet have the competence to make synapses. If we wait a few days, the cell-autonomous growth program has run its course to the point where the neuron starts to make synapses. This makes culture experiments both tricky and powerful. For example, we can observe the internal program (think transcription factor cascade) independent of external stimuli. Most neurons will indeed transition from axon growth to synaptic competency all by themselves. We also can control the external stimuli, for example by providing substances that the growth cone interprets as attractive or repulsive. But how similar is the behavior of the cultured neuron to its *in vivo* counterpart? Without the dynamically changing external stimuli, physical constraints, etc. the neuron has to run its cell-autonomous program, plus whatever an experimenter may provide. The neuron is not locked in its state, because time and energy alone are sufficient to let some version of its cell-intrinsic developmental algorithm progress. It is unlikely to be the same progression the neuron would have had *in vivo*, because here the growth algorithm itself changes based on interactions with other agents.

A cultured neuron displays mostly default behaviors for its developmental stage, behaviors that are normally constrained within the growth algorithm. Maybe we can look at it like this: The neuron just does what makes it most happy, its primary instincts, without anyone telling it not to do it. Axon outgrowth, for example. Addition of nerve growth factor in one corner will be sufficient to guide the axon towards the source of the substance it likes. This is interesting, because it shows that the default behavior—in the absence of any other signals or physical constraints—is to preferentially grow where the growth factor is. This is what makes the neuron happy. However, it does not tell us that guidance by the growth factor would play a role for axon growth decisions of the neuron at that time inside a brain. The growth factor might in fact serve a very different role for that neuron in the *in vivo* context of other developing cells. For example, limiting amounts of growth factor play roles in selective survival or death of neurons. Our neuron may happen to be particularly responsive to the growth factor as part of a mixture of competitive signals that allow some neurons to keep their axons and others not, or some entire neurons to live and others to die, while not carrying any information for axon growth directionality. In short, by taking the neuron out of the brain, we gained an experimental system to test parameters and roles in isolation, but we lost the growth program that will ultimately lead to wiring specificity.

Manipulations at the level of the genetic code can be revealing and a powerful way to understand the underlying rules. However, the problem with perturbations, as discussed before, is one of specificity in space and time. A loss of a component that specifically affects a process at level n, analyzed at level $n+1$, offers a good chance to reveal the exact function of that component. Just one step later, at level $n+2$, the changes of autonomous agent interactions and their effects on each other's growth programs may already be anybody's bet. And yet, both researchers and nature manipulate at the level of the genetic code to alter the outcome many levels up. How do you reprogram a dynamic, self-organizing jigsaw puzzle of the complexity of the brain through genetic mutations, if the effects of the vast majority of these mutations cannot be predicted after just two algorithmic steps? Evolution shows us how to program the development of neural networks successfully. To understand how it does it, we need to learn from biological examples and dive into the vast ocean of players and rules in the next session. Thank you.

2

Of Players and Rules

2.1

The Benzer Paradox

The Fifth Discussion: On the Genetic Encoding of Behavior

ALFRED (THE NEUROSCIENTIST): I really liked the idea of 'dynamic normality,' like, everything that happens to the growing neuron is supposed to happen just then and there, everything fits in the big flow of things. . . .

AKI (THE ROBOTICS ENGINEER): Yep, a big and dynamic jigsaw puzzle. Remember autopoiesis? A concept to think about how a living system behaves as a self-replicating and self-maintaining thing. Maturana and Varela were two Chileans who championed it. Not my cup of tea. Francisco Varela later spent much of his time meditating and linking his world of thoughts to Tibetan Buddhism . . . [1]

MINDA (THE DEVELOPMENTAL GENETICIST): It is always tempting to feel that there is something bigger out there, something fundamental that we just don't yet know or that is hidden.

ALFRED: Yeah, a feeling that something is going on in the world, something big, even sinister . . .

AKI: 'No, that's just perfectly normal paranoia. Everyone in the universe has that.' Slartibartfast, *Hitchhiker*. Cool dude.[2]

MINDA: Not again . . .

PRAMESH (THE AI RESEARCHER): I do actually think an approach can be a bit more holistic and still scientific. Autopoiesis is an interesting concept, it goes right back to the early cybernetics days: self-organization, the role of feedback, etc. The autopoiesis people had quite unique ways to express their ideas though; I think they tried not to be burdened with existing nomenclature and loaded metaphors. Some people felt that made their concepts harder to understand. But cybernetics also gave rise to similar ideas closer to both our fields. The introduction to self-organization in *Design for a Brain* by Ross Ashby in the early '50s, for example.[3]

MINDA: I do not see an obvious connection between what you describe as cybernetics and developmental neurobiology, certainly not historically. In biology, evolution selects for the phenotypic outcome. I appreciate that there may be self-organization during the development of the phenotype. But it seems to be just an abstract way to describe molecular interactions that drive development.

ALFRED: Wait, molecular interactions, yes. But there is a question what you can learn from such isolated molecular actions down the road, especially if the system is a self-organizing system. Remember, self-avoidance was nothing you could have predicted from looking at the DNA or even the protein's molecular mechanism, which is adhesion, I guess. That's the danger with molecules.

MINDA: That's not a danger, that's normal scientific progress. In our field biochemists and geneticists work hand-in-hand: the geneticist often finds the beautiful phenotype in the mutant and together with the biochemist they figure out the molecular mechanism.

AKI: That wasn't the point, was it? The question was how much a molecular function tells you about the complicated stuff that happens down the road through self-organization. I am okay with both molecular stuff and cybernetics, or what's left of it. But I'm not sure about the links to evolution and Buddhism.

ALFRED: As Minda said: Evolution selects at the level of the output, yet the mutation is in the genetic code. The problem remains how self-organization fits between what the genome encodes and what the output looks like. Of course the biochemist can figure out the precise interactions of a protein. But in different contexts, these properties can produce very different outcomes.

PRAMESH: Well, I think we can all agree that genes and molecules are the basis of what happens in any cell. But what Alfred points out is the levels problem: information at the level of a molecular mechanism may not be informative at a level only a few steps later during algorithmic growth.

MINDA: Why do you always talk abstractly when you are trying to explain something?

AKI: I have an idea, listen: we said when we can build something then we understand it, right? Remember Feynman's 'What I cannot create, I do not understand.' So let's ask ourselves what we would do to create something. Say, I want to create a neural network that makes my robot recognize a certain pattern and go there, dunno, an apple tree. Any apple tree, even though they are all a bit different, ideally. Let's see how each of us would build that. What do you think? Pramesh?

PRAMESH: Well, you know what I do. We simulate gene-regulatory networks in individual cells that can form a neural network, and then use evolutionary selection based on a performance criterion. Our goal is to evolve AI. So, I'd use the performance of the network to recognize apple trees as a selection criterion in our system. There are a lot of approaches out there for this kind of thing.

AKI: Okay, a simulation. Well, I would obviously take one of our artificial brains, powered by an ANN, and train it with ten million apple trees. I can do that before lunch. Alfred?

ALFRED: We don't really build things, you know. We look at neurons in networks and link neuronal function to networks and network function to behavior.

AKI: If you can't build it, you haven't understood it.

ALFRED: Oh dear, come on Aki! Obviously, the more we understand about the network, and the connectome, and the properties of the neurons in it, the better the approximation I could build . . .

AKI: Can you build it or not?

ALFRED: No. But I could train an animal . . .

AKI: Minda?

MINDA: Easy, actually. I would do a forward genetic screen in flies for 'apple tree walkers.'

AKI: You do what now?

MINDA: We do forward genetic screens. We mutagenize the animals so they have a higher mutation rate. Then you take thousands of individuals from the offspring, all with uniquely different random mutations, and perform an assay where the flies have to walk towards apple trees. The apple tree is a bit theoretical, but it would certainly work for simpler objects. Anyway, I'd take the ones that perform best and find out the mutations that caused the phenotype, the new behavior.

PRAMESH: Well, you're basically letting evolutionary selection do the work. This is accelerated evolution in the lab.

MINDA: Exactly. And when you have the gene you can find out exactly what it does.

ALFRED: Okay, here is where it all falls apart for me: There is not going to be an 'apple tree walker gene' in the fly genome. This just doesn't make sense.

MINDA: I didn't say there is a 'tree walker gene.' But there will be mutations, or combinations of mutations, that can change brain wiring and function to promote the behavior you select for.

ALFRED: So what if you find a mutation in a gene that codes for some metabolic enzyme in all cells? Is that possible?

AKI: What's a metabolic enzyme?

MINDA: An example for a metabolic enzyme would be a protein that is required to break down sugar or fat in cells. Different neurons and other cells often have different demands for such housekeeping jobs. Yes, it's true, there are a lot of surprises in forward genetic screens. But even a gene that is expressed in all cells can do something special in neurons. That's basically the story of Seymour Benzer. His idea was that you can find mutants that affect specific behaviors. Many people thought this was crazy, exactly like your reaction Alfred: there can't be a gene for every behavior! But Benzer went ahead anyway. He found amazing mutants and founded an entire field of research.

PRAMESH: I think we need to be careful not to mix up the idea of 'a gene for something' and mutations that lead to a selectable phenotype.

AKI: Isn't that splitting hairs?

MINDA: Exactly. For Benzer's clock and memory genes I think we are quite justified in talking about 'genes for something.' The mechanisms have been worked out in detail, and we now understand how these genes work.[4]

ALFRED: We are on thin ice, guys. I'm with Pramesh: We must not confuse a mutation that can lead to a specific behavioral change with the function of a gene. I think this has been the crux of our entire discussion. A mutation in a gene or some regulatory element may lead to a super-specific behavioral phenotype through an avalanche of cascading changes during growth, but the molecular function of the affected gene may not explain that behavior at all. Wrong level.

MINDA: It worked for the clock genes.

ALFRED: That's true. The clock stuff was a huge and cool surprise. The clock genes really do seem to do nothing else but run a molecular clock. There seems to be no growth involved. But many, maybe most, mutations that affect behavior will affect development, and then you have the problem with all the unfolding information stuff we talked about.

PRAMESH: Minda, can a mutation in a metabolic enzyme, as you suggested, be a gene for a specific behavior? Or do you find mostly the clock type of mutations?

MINDA: Well, as I said—almost anything has been found in forward genetic screens. People have worked less on metabolic enzymes, because they are comparably, well . . . boring. Maybe I should say less specific rather.

PRAMESH: You mean 'less specific' in the sense that the molecular function is not informative for a specific higher order behavior. But the mutation can affect a specific behavior through algorithmic growth. Think about it like this: Different cells may be differently sensitive to mild modifications of the same metabolic enzyme. A mutation can be weak, like only a small change in levels or function; such a change may be below the threshold for any measurable phenotype in 99.99% of all cells. But one specific neuron type may be just sensitive enough to change its properties. Any small change in the spatiotemporal jigsaw puzzle of feedback and algorithmic growth can lead to an unpredictable outcome that is nonetheless meaningful and selectable. Such a mutation will of course be heritable . . . so there you go, a truly specific mutation!

ALFRED: So, Minda: in this case it doesn't make much sense to study the molecular mechanism of your enzyme for brain wiring, don't you think?

MINDA: There you go again. Of course it makes sense. Every molecular mechanism is a piece of the puzzle.

ALFRED: But the puzzle does not have a fixed number of pieces. The pieces just become smaller and smaller the more you look, you cut them smaller and smaller simply by trying to understand the mechanism. Where do you stop?

PRAMESH: Right. When you describe molecular mechanisms you describe a specific snapshot in time and space. You can describe details without any knowledge of unfolding information. In terms of growth, you can maybe look at the step before and the step afterwards, but not more. And depending on the detail in which you want to describe it, you'll need a lot of mechanisms just for one step.

MINDA: What's the alternative?

AKI: You said you would just do the forward genetic screen and voila—you reprogrammed the whole thing to do what you wanted. You never said you would reconstruct all the molecular mechanisms. And you would never have been able to deduce, predict or mathematically determine what mutation you would find in your screen.

MINDA: The approaches are complementary. Once I have the mutant, I can figure it out. That's the beauty of the neurogenetic approach.

ALFRED: Okay, how about this: We know there are mutations in genes where the function of the gene product explains how you got the phenotype—like the clock genes. But for all of those where the gene product cannot easily be linked to the outcome—how do we know how many molecular mechanisms we will need to make the link? That's not an easy problem.

MINDA: I agree, this is not an easy question to answer. But how would you find an answer if not by figuring out the underlying mechanisms? Science is the art of the solvable. Some cases immediately make sense, and for others it may take dozens of labs to contribute puzzle pieces to put it all together.

PRAMESH: Science is the art of the solvable indeed—that's why you like to pick those mutations and phenotypes that are more obviously linked. The function of the metabolic enzyme is a puzzle piece all right, but a piece that needs a lot of context and probably many developmental steps to make the link. Now, putting many developmental steps of a growth algorithm back together on the basis of an undefined number of required puzzle pieces seems a pretty difficult approach to me.

AKI: . . . and it's not the answer any of us gave to my question: Minda would do her lab evolution screen, I would train a network, and so would basically Pramesh with his fancy evolutionary algorithms, and Alfred felt he couldn't build it at all, just find an animal . . . we all picked approaches that avoided putting puzzle pieces together. None of the approaches were bottom-up engineering. We all suggested an approach that selects out of some pool of random stuff; I did it using an untrained neural network.

MINDA: I am not convinced. We are getting further and further with the ability to design proteins or networks based on more and more detailed knowledge. The details are important.

PRAMESH: They certainly are. The question was just where the information about those details came from. By the way, you know how people are designing proteins today, right?

MINDA: Wasn't the 2018 Chemistry Nobel Prize for directed evolution of enzymes?

PRAMESH: Exactly! It turns out one of the best ways to design a protein with a desired property is to put evolution to work—selection based on a pool of variation. And selected are of course the final products—no need to be able to predict or even understand how a change in the genetic code would actually get you there. Unpredictability of rule 110 or the genetic code are not a problem for evolution; it is how it works.[5]

AKI: I also know of protein engineering projects using deep neural networks. Actually, in both cases the 'design' is not really based on understanding how exactly the protein is engineered. With both the AI and an evolutionary engineering approach you have just outsourced to a scruffy process. Not really neat design anymore is it?

MINDA: I am still not convinced. Just listen to the stories of Benzer and colleagues, all the amazing examples of genes and behavior. That's what the next seminar is about; then we can talk again.[4]

Seminar 5: From Molecular Mechanisms to Evolutionary Programming

There is a bloodsucking midge in Taiwan that the locals call "little King Kong." It is only a little bigger than a millimeter, and, depending on your age and need for reading glasses, you may not see it at all. You also do not feel it sitting on your skin. I know, because I had more than 300 bites on my arms just when I was preparing this seminar. Given my generous sampling of little King Kong's actions, I obtained an impressive map of where exactly it went to work. If I stretch my arms out and look at the back of my hands and lower arms, I see a few bites. If I turn my hands palms up and look at the underside of my lower arms, I see even fewer bites. Only when I twist my arm awkwardly or look into the mirror to see the back of my elbow, I see the extent of little King Kong's success. More than 90% of the bites occurred where I simply didn't see them. Is little King Kong really that smart? Clearly, it would benefit the midge if it were, just like any other predator or prey that has developed behavioral traits to become inconspicuous. Classic evolutionary arms races leap to mind. A stick insect did not need to be smart to evolve amazing mimicry—that's simply how evolution works. But how could evolution reprogram the midge's brain to recognize the far side of my elbow?

There are more prominent examples in nature for—literally—crazy evolutionary reprogramming of brains. *Cordyceps* is a genus of fungi that has evolved to infect specific host animals, mostly insects. In a well-known example, an infected ant will dramatically change its behavior and start walking upwards on local vegetation, just to fix itself with a bite on top to die. As the fungus sprouts out of its head, the spread of spores is better the higher the ant had walked before. Is the fungus really that smart? The question does not seem to make much sense: the fungus has no brain; it is a product of evolution.[6] The fungus' entry into the brain has been difficult to show in the ants, but it has been demonstrated in a similar example in flies.[7, 8] By processes that we do not understand, the fungus has "learned" to reprogram the ants behavior, likely by reprogramming its brain. This is remarkable. Amongst insect brains of similar complexity, the best studied brain by far is that of the fruit fly *Drosophila*. The field is currently figuring out the overall connectivity and the neural circuits underlying behavior with great success.[9–11] However, based on the known connectivity information alone, scientists are not able to reprogram the brain to make the fly stop what it is doing, start walking upward and then bite into a supporting structure to fix itself.

Let's talk about information again. We are obviously comfortable with the idea that a few thousand genes are enough to encode organisms such as *Cordyceps*, including the information it takes for the fungus to

FIGURE 5.1. Seymour Benzer at Cold Spring Harbor Laboratory, 1975. Courtesy of Cold Spring Harbor Laboratory Archives, NY

infect an insect brain and change the behavior of the prey insect in a meaningful way. Similarly, we are comfortable with a few thousand genes encoding the brain of an ant or a fly. A spider's brain is genetically encoded to make a unique net. A monarch butterfly's brain is genetically encoded to ensure a migration route of thousands of miles that takes several generations of butterflies, possibly even without proper navigational skills.[12] But what does "genetically encoded" mean? Surely, there are no genes for net-making, intercontinental migration or ant-brain-invasion. Yet, single mutations in the genome can change these behaviors in meaningful ways, in heritable ways and in selectable ways. One of the first scientists to recognize and experimentally test this idea was Seymour Benzer, the father of neurogenetics (fig. 5.1).

The field of neurogenetics took off in the 1960s, at the time when Rosenblatt's "perceptron" single-layer neural network made headline news and Sperry's idea of the specification of synaptic connectivity through chemical tags became widely accepted. The genetic basis of neurobiology and behavior was much discussed at the time, but experiments to generate and identify single mutations that altered behavior in a meaningful, selectable and heritable manner was not performed until

FIGURE 5.2. Sydney Brenner, François Jacob, and Seymour Benzer at Cold Spring Harbor Laboratory, 1985.
Courtesy of Cold Spring Harbor Laboratory Archives, NY

Seymour Benzer took up the study of flies. Benzer was a physicist who received his PhD in 1947, aged 26, for the characterization of germanium, a predecessor of silicon in semiconductors. The importance of this work led to several awards and patents and a professorship for physics at Purdue University the same year. Almost immediately he turned his interests to the completely unrelated and fundamental question of heredity and the genetic code.[4] DNA was only shown in 1952 to be the molecular substance of genes,[13] and the DNA structure was famously solved in 1953 by Rosalind Franklin, Maurice Wilkins, James Watson and Francis Crick.[14] Like several remarkable physicists at the time, Benzer was fascinated by Erwin Schrödinger's book *What Is Life?* Throughout the '50s, he worked with Max Delbrück, François Jacob and Jacques Monod, Francis Crick and Sydney Brenner (fig. 5.2), about whom we will hear more in the next seminar. Benzer made fundamental contributions to concepts of modern molecular genetics during this time.[4] In the

1960s he turned his curiosity to what may have seemed the ultimate higher-level outcome of what genes could do: behavior. This was not an obvious thing to do. It also required work with an organism whose behavior was amenable to both genetics and behavioral experimentation. So Benzer turned from phages and bacteria to an organism where the genetics he loved had been established, the fruit fly *Drosophila*. Thomas Hunt Morgan and Alfred Sturtevant had already in 1913 developed linear gene maps for *Drosophila* based on mutations.[15–17] Screening for mutants was well established, the genetics were known, but did flies have any meaningful behavior that could specifically be altered by mutations? Benzer's journey into the genetics of behavior started with a visit to Sperry's lab at Caltech in 1965.[4] Here at Caltech Morgan and Sturtevant had established their first fly room and revolutionized genetics. Morgan's student Ed Lewis was still there, spending year after year trying to understand developmental mutants. By this time molecular biology was the name of the game and genetics not *en vogue* anymore. Yet, rather silently, Lewis was on a path to seminal discoveries in developmental biology that would lead to a Nobel prize many years later. Lewis also provided flies for Benzer to see whether they (the flies) showed any behavior that interested him.[4]

The idea that a genetic change could lead to a meaningful, heritable behavioral phenotype was not at all obvious and a question deeply steeped in political and societal controversy. The debate is very much alive today and often difficult to separate from the science itself. Can there be a genetic basis for aggression, sexual orientation, intelligence or empathy? And if there are no single genes specifically for any of these, can there be mutations that specifically affect any of these behavioral traits? In the 1960s many researchers questioned whether genes influence behavior at all—and if they did, then surely manipulations at the level of the genes were too far removed from the level of behavior to yield any meaningful results. As Jonathan Weiner describes in his historical record *Time, Love, Memory*, Benzer extensively discussed his idea to screen for behavioral fly mutants with Sperry's lab, where members either liked or hated the idea.[4] Weiner left open what Sperry himself said on the subject. However, Sperry had spent the last 25 years reclaiming the development of brain connectivity from the domain of learning and psychology to establish it as firmly rooted in genetically encoded developmental biology. Already in 1951 he wrote: "with the foregoing pictures of the developmental processes, almost no behavior pattern need be considered too refined or too complicated for its detailed organization to be significantly influenced by genetic factors."[18] This was Benzer's project in a nutshell. As Benzer pointed out in 1971, they had no clear idea how genetic information could translate into behavior: "The genes contain the information for the circuit diagram, but little is known about the relationship between this primary information and the end result. How the tags of specificity are parceled out among the neurons so that they form the proper network, or even what kinds of molecules carry the specificity are, at present, complete mysteries. The problem of tracing the emergence of multidimensional behavior from the genes is a challenge that may not become obsolete so soon."[19] Here, Benzer adopted Sperry's wording ("tags of specificity") and the idea that "molecules carry the specificity," yet found it difficult to link this notion to "multidimensional behavior." But just because we do not understand *how* genes might do it, doesn't mean that they don't—so Benzer set out with an open mind to identify his first behavioral mutants.

Benzer started his quest for mutations and genes that affect behavior with a revolutionary, evolutionary experiment. He increased mutation rates (by feeding flies a chemical mutagen that is still used in labs today) and subjected the flies to artificial selection. His assay tested the flies' tendency to walk towards a light source, in the so-called phototaxis assay.[20] There are many reasons why a fly might not walk towards the light, from loss of vision to an inability to walk or laziness. And yet, or maybe because there are so many possible ways to affect this behavior, the genetic screen was a great success. Using this and other simple assays, Benzer and his colleagues supplied the scientific community with behavioral mutants that scientists are still trying to work out today.[19, 20]

Benzer's success showed beyond a doubt that identifiable mutations in specific genes cause behavioral changes that are meaningful, selectable and heritable. What is more, Benzer and his colleagues had indeed found some mutations in genes whose molecular functions could later be linked to specific behavioral changes. For example, they found mutations in a gene that could speed up or slow down the internal clock of flies; the same gene and protein's mechanism are fundamentally conserved in humans where they regulate our daily rhythm.[21] But there is also another class of mutations, those in less specialized genes, that were found over the years and decades that followed. Today, we know of a myriad of mutations, or single base pair differences, that can be linked to specific behavioral differences, even though the affected genes seem to be generalists serving basic cellular functions. Often these links come with a probability, or penetrance, of less than 100%, meaning that not every individual with the mutation really shows the same behavior. Most behavioral traits are of this type, including intelligence, empathy and countless other behavioral features that make us who we are. There are, beyond a doubt, genetic differences and mutations in genes that change the probability and predisposition for behavioral traits—yet, at the same time, for most of those behavioral traits scientist say their genetic basis is *complex* (code for "not understood"), and there does not seem to be a single responsible gene. In other words, a behavioral predisposition can certainly be caused by a single mutation, yet there may be no single "gene for that behavior." I call this the *Benzer paradox*. Let us approach the issue by looking at the actual mutations and genes studied by Benzer and the science that followed.

Ron Konopka was a graduate student with Benzer when he performed his evolutionary experiment: he mutagenized flies and looked for those that would eclose from their pupal case at odd times. In 1971, Konopka and Benzer published their findings describing three mutations that increased or decreased the daily rhythm of flies and proposed that they were all mutations in the same gene.[21] Konopka was awarded his own faculty position at Caltech following this remarkable discovery, but he did not do well in the science game of "publish or perish." While Konopka did not get tenured and eventually left science altogether, three other remarkable scientists picked up on his discovery of the gene aptly named *period*. In 2017, shortly after Konopka's death in 2015, Jeff Hall, Michael Rosbash and Michael Young won the Nobel Prize for figuring out the way the period protein and its interacting proteins regulate the circadian clock.[22] All three did their work in flies, but the mechanisms they discovered turned out to be very similar in humans, implicating the same genes and proteins. There is a genetic basis to our daily rhythms. If you are a night owl, like Benzer, it's "in your genes." How does the molecular mechanism of *period* help us understand the behavioral outcome? Beautifully, it turned out. The PER protein (encoded by the *period* gene) directly binds to the TIM protein (encoded by the *timeless* gene). The PER/TIM protein complexes accumulate and finally move into the nucleus at night, where they inhibit the expression of the *period* and *timeless* genes. Loss of *period* and *timeless* gene expression leads, with time, to loss of the PER and TIM proteins available to block their expression. It is a classic negative feedback loop that can run on its own time. It can also be entrained by light, as subsequent work found. The genes *period* and *timeless* are the heart of a molecular clock that determines our daily, or circadian, rhythm.[23–25]

Key to Konopka's and Benzer's initial discovery of mutations in *period* was that the mutant flies were viable and seemed fine except for their arrhythmic behavior. There are no obvious developmental defects. Mutations in *period* do not partake much in algorithmic growth, at least not prior to the onset of network function. More recent work has shown that the molecular clock may in fact be recruited late during algorithmic growth of the human neocortex, where network function is critical for a late phase of development.[26] But overall, the molecular clock in flies and humans is a functional, not a developmental mechanistic and exquisitely understandable molecular machine. Prior to its function, development must have created the cell, its nucleus and all kinds of molecular machinery for the clock mechanism to work. But once the cellular environment is there, it can start ticking. There is a true molecular clock that runs the behavioral clock, a single molecular mechanism for a single behavior.

Are all behavioral mutants of this type? Clearly, behavioral mutants must be viable and meaningfully affect behavior in order to be heritable and selectable. But this doesn't mean that they only kick in at the end of growth. In 1961, Kulbir Gill did his PhD on the developmental genetics of oogenesis in the fruit fly with Donald Poulson, a former graduate student with Thomas Hunt Morgan. Gill had performed mutagenesis studies, similar to Benzer's, in order to find interesting mutations. He stumbled upon a mutant that, bafflingly, didn't seem to cause any change in female flies, while changing male behavior dramatically: the male flies completely ignored females and started courting other males instead. As Gill noted in 1963 when he published his discovery, he had found a mutation that caused homosexual behavior specifically in males, and he tentatively suggested the name *fruity* for the mutant.[27] Jeff Hall (the same who was awarded the Nobel Prize for the circadian clock) later renamed the gene *fruitless* to keep the short gene acronym *fru*, with Gill's permission, as he reports.[28] The gene *fru* turned out to be a transcription factor. Males and females produce different versions of this transcription factor (splice variants). The male variant is necessary and sufficient to cause male courtship behavior. Regarding molecular mechanisms, transcription factors are conceptually easy: we know they are DNA-binding proteins that trigger the expression of other genes. We have discussed them in the context of transcription factor cascades as examples for how information unfolds through algorithmic growth. The context and the specifics are of course very important: the transcription factor Fruitless will only work in certain cells at certain times, dependent on the availability of other proteins (cofactors) as well as the accessibility of DNA sequences to bind to.

While the *period* gene is only activated at the end of developmental growth, *fruitless* is turned on early in the growth algorithm, at a time point when many cells already form the animal shape, but the brain has only just started its self-assembly. In 2005, the laboratories of Barry Dickson and Bruce Baker independently generated flies in which the expression of the male variant of *fruitless* replaced the female variant normally expressed in females. These females promptly started to exhibit all the key male behaviors of courtship.[29, 30] Hence, expression of this specific transcription factor variant is sufficient to encode and implement male sexual behavior.

Genetic sufficiency is always tied to a certain context. The human transcription factor Pax6 (eyeless in flies), is famously sufficient to induce eye development when expressed in undifferentiated cells of wings, antennae or legs of a fly.[31] But it can't do it when expressed in a differentiated cell, which is just another way of saying it has to be expressed at an early step of algorithmic growth. Both fru and eyeless/Pax6 are key regulators of subsequent growth. They are both transcription factors that initiate lengthy algorithmic programs, implicating hundreds or thousands of downstream genes, all expressed at the right time and place. This is why artificial expression of human eyeless/Pax6 in a developing leg of a fly will produce a fly eye, not a human eye. And this is why enforcing the male variant of *fruitless* will trigger an entire developmental "male" program, including male-specific brain wiring of neurons that normally do not exist in adult females. How about fru's molecular mechanism? A transcription factor's mechanism of action is obvious in a general sense: it will switch on an entire downstream cascade of algorithmic growth. But finding out all the genes that will be expressed just a single step down, in the next iteration of the algorithm, turns out to be surprisingly hard. A 2014 study using modern methods to identify direct target genes of *fru* indicated that both the male and female-specific transcription factors bind to dozens of places on the DNA responsible for the regulation of a plethora of different, and for the two sexes only partially overlapping, genes.[32] Many of these target genes encode proteins that are essential for neuronal development or function, from membrane receptors to more transcription factors. In the next iteration of the growth algorithm, two steps down of *fru*'s initial function, all these players will create a new state of hundreds of neurons, a new set of proteins that interact with each other in each neuron, and a different input for yet another step in the algorithm. The information unfolding implicit in this superficial description is already mind-boggling. The information unfolding of algorithmic growth becomes an information explosion.

Information explosion is a general problem in developmental biology and beyond—and a real problem when it comes to communicating relevant information to those who try to actually build intelligent neural networks. The renowned systems biologist Marc Kirschner noted on Christine Nüsslein-Volhard's developmental biology book: "*Coming to Life* is an antidote to the massive factual explosion in science that can obscure general principles, and thus it should be required reading for anyone wishing to understand where we stand in modern biology."[33] The "factual explosion" is what we get when we study the uncounted, and arguably uncountable, number of mechanisms of development and function at a distinct point in space and time of algorithmic growth. Most developmental biologists are not trained to think in terms of Solomonoff's and Kolmogorov's algorithmic information. At the same time, biologists are seeking *general principles*, but what are those? Is there an all-or-none criterion for generality?

A common idea of generality is that it should explain rather a lot, as opposed to being only applicable to a very special case. This is not as easy to define as it may seem. Arguably, the binding of a transcription factor to trigger the next gene's expression, and with that the birth of the transcription factor cascade, is a general principle. It happens again and again, a general concept or building block of algorithmic growth. But in order to explain any actual process during development or brain wiring, we need to consider a lot of specifics. For example, the types of target genes of the transcription factor that actually determine the properties of the cell. One of the dozens of those may be, say, a protein that promotes the ability to form membrane protrusions, ultimately growing axons and dendrites. The biologist will tell you that there is a general principle here too: the way this happens is again a module or building block, used over and over again in different cells throughout algorithmic growth. There are general principles for how proteins modify each other, how they control cell divisions, how they endow the cell with uncounted physiological and mechanical properties. There really are an uncounted, and arguably uncountable, number of general principles, and back to our factual explosion.

Another idea for a general principle could be: How far back can we trace the factual explosion to a principle so truly general that it explains everything that happens thereafter, making all the factual specifics of the explosion just details? Everything will automatically unfold to our eyes if only we knew the most general principle, the bits of information sufficient to create all that follows. This is of course the idea of algorithmic information. Here is what is trivial about rule 110: we should not be surprised that some very simple rules, applied again and again on the same simple stuff, create something of unmeasurable complexity. This is how astrophysicists explain the origin of the universe to us: very shortly after the big bang, there was a moment when the universe was very small in time and space and contained a lot of energy in a very small space. No atom existed back then, let alone complicated molecules—only some poorly understood basic particles—and energy, and a few, maybe just one fundamental, law(s) of nature, some rule(s). Arguably, from the perspective of our own planet, a factual explosion has taken place since then locally, with bilateral bipeds, brains and bipolar disorder. The problem with the unfolding, as proven at least for rule 110, is that the information content that invariably will unfold locally, given sufficient time and energy, is unpredictable. It helps little to say that complexity in the universe can principally (and seemingly) unfold from almost no information, because it doesn't predict anything that actually unfolds nor how the unfolding works. We still want to study the universe in all its detail.

This then is the hypothesis: energy dissipating pockets of the universe, the self-assembling brain and rule 110 all have in common that they unfold to unpredictable complexity. The simplest of the three is clearly rule 110. Here only eight simple rules are applied iteration after iteration to yield a 100% deterministic pattern. Stephen Wolfram (see fig. 2.2), who described properties of rule 110 in detail, came to the conclusion that this pattern is so complex that it is incompatible with evolution. In his 2002 book *A New Kind of Science* he wrote (p.393): "In a sense it is not surprising that natural selection can achieve little when confronted with complex behavior. For in effect it is being asked to predict what changes would need to be made in an underlying program in order to produce or enhance a certain form of overall behavior. Yet one of the main conclusions of this book is that even given a particular program, it can be very difficult to see what the behavior of the program will be. And to go backwards from behavior to programs is a still much more difficult task."[34] Here, Wolfram underestimated evolution's ability to select algorithmic growth processes and draws the wrong conclusion. Natural selection is not at all "asked to predict what changes would need to be made in an underlying program in order to produce a certain form of overall behavior." To the contrary: it does not matter how complicated the changes to the underlying program are in a *fruitless* mutant. As long as the behavioral change is meaningful and heritable it will be subject to natural selection. Evolution works by selecting based on output and has no need to predict how a growth algorithm got there. Natural selection is not only able to reprogram boundlessly complex developmental programs, it may in fact be the *only way* to meaningfully reprogram the otherwise unpredictable nature of algorithmic growth, precisely because of the impossibility to predict outcomes. It is understandable how Wolfram was in awe of the unpredictable complexity his simple set of rules could produce. The fact that rule 110 is to this day the simplest Turing-complete system and can perform any conceivable computation allows for uncomfortable thoughts: for all we know, the entire universe could be a rule 110-type automaton, including us and everything we know and do in it. From the inside where we live, things may seem random, and we might never know that we are part of a higher order system that simply runs some unknown algorithm. None of this may ever be measurable or knowable to mankind. But it is irrelevant for science: within the realm of the measurable and knowable, evolution is not only able, but may be the only way to meaningfully change algorithmic growth, including the self-assembling brain.

Given the unpredictability of the effects of most mutations during algorithmic growth, we should expect that selection for specific behavioral outcomes could be caused by mutations in many different genes. This is why, when scientists perform forward genetic screens, they are generally up for any kind of surprise.[35] The underlying mutation is initially unknown. If selection occurs for a highly specific phenotype, for example the color of the fly's eye, there will only be a limited number of mutations that can cause that specific phenotype. In fact, one of the first *Drosophila* mutants ever isolated was a fly that had lost its red eye color, found by Morgan in 1910/11.[36] The mutant gene, called *white* after the mutant phenotype, itself encodes an enzyme for the production of red pigment. Not much of a growth algorithm here. As in the case of *period*, the gene product functions specifically in the process that generates the selected phenotype. On the other hand, selection for behavioral phenotypes has identified uncounted mutations in a multitude of genes. There are the transcription factors, of course, but then there are also kinases that are known to be expressed in all cells, proteins that change how actin, a kind of skeleton inside every cell, assembles, genes required for metabolism, other housekeeping or cellular stress responses. Evolution can use any of these for reprogramming purposes.

Amongst other labs in search of mutants for neuronal function, Hugo Bellen's lab repeated Benzer's original phototaxis screen in the early 2000s. Because there are so many possible reasons and mutations that could cause a fly not to walk towards the light, they performed a secondary screen. A simple electrophysiological recording from the fly eye can tell whether the photoreceptor neurons are actually okay, or whether they cannot see the light, or whether they can see the light, but fail to transmit information to their postsynaptic neurons. Bellen's team selected specifically for the latter—flies that could walk and behave normally, their eyes could see the light, but somehow information did not get passed on. Even for this special case, there are several possible reasons. For example, there might be a molecular defect in releasing the chemical transmitter that alerts the postsynaptic neurons. Or, the flies might have no or wrongly wired synaptic connections. Bellen's team found mutations in almost any type of gene that one could (or could not) think of, affecting a surprising variety of different processes.[37, 38] In fact, they found mostly mutants in genes that nobody would ever have predicted, implicating virtually any basic cell biological process in neurons. But then there were also the easy ones, those that had been discovered and were at least partly understood already; for example, mutations in a previously described gene called *synaptotagmin* that specifically affect the ability of neurons to quickly release a burst of neurotransmitter.[39–41] This is such an important function for synapses that many researchers have dedicated entire scientific careers to figuring out the underlying molecular mechanism. The Nobel laureate Tom Südhof [illegible] of the prize for this and related work. Südhof and [illegible]ne, including Bellen, had figured out some key as- [illegible]min function. The molecule itself is a sensor for calcium ions, which provides the molecular link between the electrical signal passed along the neuronal axon and the release of a chemical transmitter at synapses. The beauty of this molecular mechanism is that calcium sensing is both exactly what the protein synaptotagmin does and the process that was known to convert the electrical signal into a chemical signal. When we say "synaptotagmin is the calcium sensor," then it has both of these meanings. Importantly, similar to the *period* or *white* genes, complete loss of *synaptotagmin* in *Drosophila* photoreceptor neurons, or any other neuron, does not interfere with algorithmic growth until the initial wiring diagram of the brain is done, and with it its entire growth algorithm up to the moment of neuronal function. In short, *synaptotagmin* is a molecular biologist's dream: we can study its loss in functional neurons, because it does not affect algorithmic growth and specifically disrupts function afterwards; and, maybe even more importantly, its molecular mechanism resembles exactly the higher order biological function that is lost in the mutant.

The team in Bellen's lab was happy to find a *synaptotagmin* mutant in their version of Benzer's screen, because it served the important function of a positive control: the screen worked, it lead to the isolation of meaningful mutants. However, most mutants were not of this type and remain more obscure to this day. Genes affected by mutations basically formed a list that reads like a random assembly of genes in the genome, from housekeeping enzymes to cytoskeletal modulators, membrane trafficking regulators to genes of completely unknown function. And then there were the mutations that could not even be assigned to a gene at all. So, what did they do? Science is the art of the solvable. They picked genes for which the molecular function seemed to make some sense at the time. Now this last bit is really what it sounds like: A highly subjective selection process, ultimately for what seems most likely to publish well. Almost none of the big genetic screens in yeast or worms or flies are ever published in their entirety—and the vast majority of mutations are never reported and never worked on. When we read the published work, we mostly see the result of our biased selection for *what seemed to make sense at the time.*

Some scientists are bolder and publish more surprising findings, especially when it comes to the genetic basis of behavior. In the earl-

2000s, Ulrike Heberlein, Adrian Rothenfluh and colleagues wanted to understand the genetic basis of alcohol sensitivity. Like humans, flies display reproducible behaviors when drunk. Depending on the genetic make-up, some flies are more resistant to the effects of alcohol than others. They literally collapse later. Screening for resistance to sedation, they found several mutants that had interfered with the function of the same gene. This experiment alone shows that there is a genetic basis for alcohol sensitivity in flies. They named the gene *white rabbit*, "for its diverse role in regulating responses to abused drugs as described in the song 'White Rabbit' by Jefferson Airplane."[42] The *white rabbit* gene encodes a protein that modifies structural elements in neurons, the actin cytoskeleton. This "skeleton inside the cell" has been implicated in virtually all cellular functions, from axonal and dendritic growth to dynamic signaling between the outside and inside of the developing or functioning cell. What does the skeleton inside a cell have to do with alcohol sensitivity? Despite great progress in the elucidation of its mechanism, this remains a difficult question. The gene *white rabbit* does not encode a calcium sensor for a biological process that is based on calcium sensing. The role of the actin cytoskeleton is known to be tightly regulated and highly specific in time and space. It looks like the overall growth algorithm is not obviously affected in this case, while other mutants isolated by the same team affect developmental signaling pathways as well as ethanol sensitivity.[43] However, even a purely functional defect is difficult to pinpoint if it depends on a complicated context, many other factors including molecular functions, physical constraints and the execution of its function as part of a coordinated series of events. We will revisit this idea of *algorithmic function* as an analog of algorithmic growth later. For now, it shall suffice to appreciate that even though a mutation does not affect the growth algorithm and its molecular function is known, it does not mean we have understood how it causes the meaningful, selectable and heritable phenotype of alcohol resistance. A genome engineering scientist would probably not have gotten the idea to mutate this particular gene to produce the specific behavior. Evolutionary programming, however, does nothing better than isolate those unbelievably rare changes that cause a highly specific behavior. Thank you.

2.2

The Molecules That Could

The Sixth Discussion: On Guidance Cues and Target Recognition

MINDA (THE DEVELOPMENTAL GENETICIST): Now you see why Benzer is a hero in our field! Evolution can select for mutants that produce all kinds of behavior. And that includes mutants that affect development, even if you think development is unpredictable.

PRAMESH (THE AI RESEARCHER): It wasn't that obvious for Stephen Wolfram. He took the idea of unpredictability in cellular automata and concluded that if nobody can predict the outcome without running the entire algorithm, then evolution cannot do it either.[1]

ALFRED (THE NEUROSCIENTIST): . . . but evolution doesn't have to predict the outcome, as we have seen. Cool story. The whole point of evolution is that it selects after the entire growth program is done.

PRAMESH: Yes, it sounds simple enough when you say it like that. Wolfram was right though that evolution could not violate the notion of unpredictability—even for a simple deterministic system like rule 110, much less for a system in which the rules change with every iteration of the algorithm and incorporate random processes on top of that. But evolution's trick is that it doesn't shortcut. It simply runs the full growth program to whatever end, and it does so in many different individuals to have something to select from. Selection happens at the population level.

AKI (THE ROBOTICS ENGINEER): Oh, we are back at the 'no shortcut' argument. Last time we had that on protein design we ended up realizing that the neat designer had outsourced the process to an evolutionary method anyway. Which kinda makes the whole thing scruffy. I guess evolution is scruffy as well, then?

PRAMESH: I'll just have to start to reject these poorly defined terms . . .

MINDA: What does an evolution-based method for protein design have to do with Benzer's behavior mutations?

AKI: Well, they are both top-down selection strategies, as opposed to bottom-up engineering. Benzer didn't try to make a behavior mutant by putting puzzle pieces together. Both are based on trial and error with unpredictable results, followed by selection. Which is also why we cannot build robots that way—I can't build thousands of random robots just to select one . . .

PRAMESH: Well, it's worse . . . you wouldn't build thousands, you'd have to let thousands grow to unfold sufficient information. . . . This is really where I see the analogy of rule 110 and brain development now. But the biological algorithm is much more complicated than rule 110, because rules change with every iteration of algorithmic growth. That's one difference. The other is that rule 110 is of course deterministic, while biological growth includes random processes during development.

ALFRED: What I still find fascinating is that the actual molecular function does not need to have anything to do with the selectable outcome. People look at successes like the genes *synaptotagmin* and *period*, but those may be the exception rather than the rule.

MINDA: Exceptions are what we often learn from most. Anyway, none of the molecules given as examples in the last seminar are what most of us are working on in brain wiring. We focus on guidance cues, proteins that are either secreted or sit on surfaces to guide where an axon should grow or where it should make synapses. We've discussed those in the context of noise and relevant information. Sperry's chemoaffinity idea was that the incoming axons and the targets have matching keys and locks. And here again you *do* have a direct link of molecular action to the outcome.[2]

AKI: Ah, then the question becomes: are these guidance cues the rule or the exception?

MINDA: It's not all black or white, and it's not exclusive. You can have mutations in transcription factors, in membrane or cytoskeletal regulators and cell surface molecules—all contributing to a phenotype. Cell surface molecules are particularly important, because they are the proteins that allow cells to interact with each other and serve as recognition molecules.

ALFRED: . . . and that's where chemoaffinity comes from, the idea that molecular keys and locks define some kind of synaptic address code. But Sperry did most of his stuff in regeneration rather than development, so he kinda threw the developmental algorithm out.

MINDA: That's only partially true: a regeneration process can be a recapitulation of a part of a developmental program.

ALFRED: . . . true, but the target is already there . . .

MINDA: . . . and Sperry considered that. He and his colleagues also looked into developmental timing. For example, he explained how axons in the visual system would form a compressed map when half the target was cut away.[3]

ALFRED: Wait, his explanation turned out not to be correct though, if I remember right? There was the problem that if half the precisely defined targets are lost, half the incoming axons should not know where to go. To explain why they instead form a compressed map, Sperry said if the ablation was early enough, a new compressed target code could develop. But really it turned out to be quite different, relative positioning, no target code. The systems-matching idea by what's-his-name?[4]

MINDA: Michael Gaze. Yes, that's actually true, maybe not the example I should have picked. But Sperry also predicted the gradients correctly that we now know play key roles in relative positioning. That was incredibly visionary given that literally nothing was known about the molecules involved and many people instead thought about mechanics and plasticity. Look, the most important thing was just this: Sperry was the first to propose specificity is achieved through the actions of molecules, not just some strange plastic force or process.

AKI: So he predicted the molecules, but not their mechanisms correctly?

MINDA: He described one important mechanism, and that was definitely correct. I would be happy to find just one important thing like this.

AKI: I have an idea. I liked the idea of your genetic screens—basically sped up evolution experiments. If you look for brain wiring defects you should get an unbiased sampling of whatever mutations cause a specific phenotype, right?

MINDA: Yes, it depends on how specific and good your assay is. But it is common to find all kinds of surprising genes. We just heard about that in the previous seminar.

AKI: Right. Your screens not only isolate mutations in 'Sperry molecules,' but in all kinds of other genes, the whole variety. What do you tell your students to pick when they look at the list of mutants?

MINDA: First of all, the phenotype needs to be really interesting. Next, we need to assess how hard it is going to be to study and what's most likely going to be successful. So if there is a mutation in a gene that encodes a cell surface receptor that nobody has worked on yet, that's simply more exciting than a housekeeping enzyme. You have to pick your battles.

PRAMESH: Interesting, and a cell surface receptor is of course more interesting because there is a clear hypothesis that it is a recognition molecule.

MINDA: Exactly.

ALFRED: Except that we keep on going back to the point that a molecular function may not tell you anything about the role the molecule plays as information unpredictably unfolds during brain wiring. So why restrict yourself to cell surface proteins? They usually turn out to do something more interesting than just 'recognition' anyway.

MINDA: I'll give you that: if you take a really strict version of chemoaffinity, then many recognition molecules have turned out to have some surprising and more diverse roles. Sperry just pointed out one important role these molecules have. We now have a much better understanding of chemoaffinity. Sperry was just a visionary to have started it.

AKI: With 'better understanding' you really mean a 'different definition' . . .

MINDA: A wider definition maybe, including many cell surface interactions and guidance functions during development.

AKI: So, is the guidance function then the rule or the exception for receptors on the cell surface?

MINDA: That's difficult to answer. The more we understand, the more complicated all the mechanisms become.

PRAMESH: Exactly! But the reason why you would advise a student to pick the receptor over a metabolic enzyme is the assumption that it actually is a simple guidance cue . . .

MINDA You have to start somewhere.

PRAMESH: Then why not start with the enigmatic enzyme? Sometimes it's better to start with no preconceived notion than to follow an assumed one.

AKI: 'Wonko the Sane' again: 'See first, think later, then test. But always see first. Otherwise you will only see what you are expecting. Most scientists forget that.'[5]

ALFRED: Thanks, Aki. I was waiting for that.

MINDA: Just because we pick something to start with does not mean we do not observe first. Listen, if I go to the lab and students have found a whole list of interesting mutations in genes, they already know what they want. They tell me they don't want to work on a metabolic enzyme, because that's not what people do in this field. They see what's in the big papers.

ALFRED: Basically, what you are saying is: since Sperry we have become so focused on cell surface receptors as guidance cues that now everybody wants to only look at those just because everybody else is, even if they are not guidance cues . . . I kinda do see a danger of seeing what you are expecting there . . .

MINDA: A recognition molecule is defined as a cell surface protein that has specific molecular interactions.

ALFRED: . . . but if the definition of a recognition molecule is that broad, isn't then every molecule that binds to some other molecule with some specificity a recognition molecule? Doesn't sound like something Sperry would be happy with . . .

AKI: . . . diluting a definition to keep a concept alive seems to me a pretty painful way of suffocating that very concept . . .

MINDA: Oh come on. Every field is like this. A hypothesis is great if you can build on it. Nobody needs to suffocate because of that.

AKI: Unless another, at least equally correct hypothesis is shut down. How about this other guy again . . . this relative positioning, systems-matching stuff. What was his name?

MINDA: Mike Gaze. What about him?

AKI: If he was the one who actually got it right, shouldn't people also try to find mutations and genes that work in relative positioning instead of address codes?

MINDA: Molecular mechanisms for relative positioning are now well established as well. But Gaze never proposed any specific molecules. He just proposed principles. I actually doubt that any of my students have ever read his papers or even heard his name.

ALFRED: Wait, that's terrible! He's forgotten just because he didn't propose some kind of relative positioning molecules even though he proposed the correct principle?

PRAMESH: I find this fascinating. The principle of relative positioning, or the 'sliding scale' idea they discussed, is an algorithmic process. You can't just propose a molecule for that. There have to be many molecular functions together, and then, step by step, produce the outcome. I think I got it, this is the key: Sperry proposed a type of molecule that would not require any complicated algorithm or feedback in time and space. He just said there are these chemical tags—a simple and clear hypothesis, and one that people could get their teeth into. On the other hand, if you go and say the process you are looking at in brain wiring is algorithmic by nature, then you start off with something nonintuitive and you cannot propose a single class of molecules. The process can still be genetically encoded, mutations will exist that create meaningful and selectable changes, all that. But you can't propose a molecular mechanism with a hypothesis that directly relates to the outcome! It's too complicated . . .

AKI: Cool. It seems easiest and safest to think in terms of specific contact information, like a blueprint. For our ANNs, I think what saves us is that we don't just have a blueprint, we let the network learn.

ALFRED: I think rules like 'relative positioning' are important because they are something we all can communicate across fields. Maybe then we can also find out whether aspects of your ANNs have something to do with self-assembly or not.

MINDA: But you need all the molecular detail to give evolutionary programming every chance of modifying it.

ALFRED: Ha, finally I can agree with Minda!

Seminar 6: From Chemoaffinity to the Virtues of Permissiveness

The quest for genes underlying behavior heralded the quest for the genes responsible for the wiring of neuronal circuitry that, so the connectionist view, regulates behavior. As we have seen in the previous seminar, a surprising variety of gene functions were discovered, from transcription factors to cell biological machinery. This large variety makes sense in light of evolutionary programming of the network; it is fundamentally unpredictable what mutations, in whatever regulatory elements or type of gene coding region, may yield a particular behavioral output, because algorithmic growth has to unfold that information. Only then, through the adult behavior, does the mutation become meaningful and selectable. Thus the precise molecular functions of the implicated genes' products need not carry the burden to explain the higher-order behavioral phenotype.

How different the quest for brain wiring genes! In the 1980s and 1990s, the field of axon pathfinding was driven by one motivation: to find the molecules proposed by Sperry. A transcription factor or actin-binding protein may be required for an axon to find its path, sure enough. But such molecular mechanisms were labeled *permissive*—merely something that needs to be there. If I stop breathing, I can't look at the street signs, navigate and walk all the way to the opera. Breathing is permissive, whereas the street signs are *instructive*. What are the really interesting, the instructive molecules? Sperry's chemical tags, molecules that either sit on the neuron's surface or are secreted ligands of receptors on the neuron's surface. The hunt was on.

One of the leading scientists excited by the prospect to find Sperry's molecules was Friedrich Bonhoeffer (see fig. 1.6). He had obtained a PhD in nuclear physics in 1958, but spent most of the subsequent two decades with seminal work on the molecular mechanisms of DNA replication. Bonhoeffer brought a fresh approach to the quest to find Sperry's molecules. He designed elegant in vitro choice assays in a dish, where he and his colleagues would, for example, offer retinal ganglion axons from different regions in the retina a choice between a piece of anterior or posterior target region tissue. Sure enough, axons from different parts of the retina exhibited surprisingly clear preferences, suggesting a substance, provided by the target, that is differentially recognized by the axons. Throughout the 1980s, Bonhoeffer and colleagues published groundbreaking work that brought Sperry's molecules closer than ever: his assays held the promise to isolate and purify the molecules that function as cues.[6–9]

Throughout the 1990s, molecular cues and their receptors were discovered at a fast pace. The labs of Bonhoeffer and John Flanagan at

Harvard identified two of the molecules that provided the target cues in the visual system studied by Sperry and Gaze. These molecules were later named ephrinA's.[10, 11] The ephrinA's are ligands for the Eph receptors (EphRs), which were found on the growth cones of the retinal ganglion cell axons.[10, 12] Surprisingly, these cues turned out to be repulsive, rather than attractive signals.[13] The study of ephrins and their Eph receptors blossomed into its own scientific field for the next 20 years.[14] The field produced a series of groundbreaking discoveries, from the early realization that ephrins and EphRs indeed form gradients as predicted by Sperry to unexpected modes of signaling, where not only ephrins would signal through the receptor unidirectionally, but also the other way round. More discoveries led to the appreciation of an ever-increasing, unexpectedly complicated ballet of molecular interactions. Bonhoeffer and others also realized early on that the molecular interaction not only occurred between the axon and the target, but also between the axons themselves.[6] Axon-axon interactions were thus proposed early on as key to the topographic mapping of neighboring axons.

The repulsive mode of ephrins' functions was surprising, because the quest had been one for the chemoaffine molecules, attractive cues for the axon to positively identify its target. A repulsive signal put the problem on a slightly different footing: now we need the axon to grow, with or without being attracted, to end up wherever it is least repelled. Work by dozens of scientists in the ephrin field revealed much of the beauty of how topographic patterning ultimately happens, and much remains to be discovered.[14] The quest had been for Sperry's: "gradients successively superimposed on the retinal and tectal fields and surroundings [that] would stamp each cell with its appropriate latitude and longitude expressed in a kind of chemical code with matching values between the retinal and tectal maps."[2] However, an absolute address code in latitudes and longitudinal coordinates, as unequivocally proposed by Sperry, has mostly not been supported in the decades that followed. Instead, the retinal axons achieve precise mapping through relative positioning to each other as well as the tectal cues, as development progresses in time. Neither the realization that the ephrins were repulsive nor the later work describing axonal competition and relative positioning instead of an absolute target address system prevented scientists from celebrating ephrinA as the discovery of Sperry's chemoaffinity molecules.[15, 16] After all, they were molecules, they were expressed in gradients, and they did guide the retinotectal projections. And thus, Sperry's visionary formulation of the chemoaffinity theory found its validation in an unexpected form.

The prediction of the existence and key role of such chemistry is Sperry's lasting legacy from that time. He brought to life the idea that molecular mechanisms execute developmental processes in the brain long before the advent of molecular biology in what he called the "basic guideline concepts."[17] And the time of molecular biology did come, probably even more forcefully and as a more sweeping victory than even Sperry had anticipated. In the age of molecular biology, the chemoaffinity theory has expanded its position as a cornerstone of modern developmental neurobiology. Thousands of research studies in developmental neurobiology have found specific molecular interactions and functions that are part of cell-autonomous neuronal development (differentiation), nonautonomous (inductive) interactions and gradients. To this day every other talk or paper on the development of neuronal connectivity cites Sperry as its historical foundation.

In contrast to Sperry's recognition, those who had proposed the idea of relative positioning as an alternative to Sperry's absolute address code in the 1960s and '70s, are largely forgotten.[4, 18] Michael Gaze (see fig. 1.5), in particular, was concerned about the applicability of chemical tags as absolute address codes in systems even more complicated than the mapping of the eye to the brain. What about the cortex? And thus he wrote in his 1970 book (p.217): "We may well ask whether, if we provisionally accept the hypothesis of neuronal specificity, it can be usefully applied to the patterning of synaptic relations in the more central parts of the nervous system—which become relatively larger and more important as we go up the animal scale. Sperry maintains that this is so. The local-sign specificity that is impressed on afferent fibres by the process of differentiation is presumed by him to determine the type of second-order neurons which the growing fibre-tips will find acceptable for synapsis; and this inference implies the existence of a similar refined qualitative specificity among the central neurons."[19]

To understand where we stand today, we need to become a bit more precise on the definition and meaning of chemoaffinity. Sperry, as we have seen, proposed a rather strict version of the idea of chemical tags, as unequivocally stated again in 1965: "Early in development the nerve cell, numbering in the billions, acquire and retain thereafter individual identifications tags, chemical in nature, by which they can be recognized and distinguished one from another."[20] He continues to make a prediction about the need for the identification tags to make "hookups" (synapses): "Lasting functional hookups are established only with cells to which the growing fibres find themselves selectively matched by inherent chemical affinities." However, as we have seen, neurons readily make "functional hookups" with wrong partners if the correct choice is not available, even if the only available wrong partners are themselves. Beautiful examples of what appear to be true matchmaking, Sperry-like proteins were eventually found nonetheless, as we will discuss next. However, their absence never seems to abolish synapse formation per se—neurons just find some other partner. In this context, the repulsion of ephrinA's makes sense: as long as the axon has the intrinsic drive to make synapses, relative positioning is elegantly achieved through guidance by exclusion. It requires probabilistic probing of an unknown environment and keeps the system flexible and robust. And, as we have seen in the case of ephrinA's, this is still done by chemical tags of sorts as well as gradients.

I will call the version of chemoaffinity that Sperry himself promoted *strict chemoaffinity* to account for his definition of the individual identification tags. By contrast, most developmental neurobiologists rather refer to what I will call *general chemoaffinity*, in which molecules may function in more creative ways than providing identification tags or address codes. From the perspective of molecular biologists, this distinction is sometimes considered to be merely semantic. From the perspective of an information theorist, the distinction can barely be overstated. Why is this? Unique identification tags map information one-to-one. They do not generate or unfold new information. By contrast, properties that allow axons to sort themselves out relative to each other based on molecular interactions carry information only in the context of a specific time and place during development. Given a distinct anatomical arrangement and developmental stage, axonal adhesion suddenly becomes part of the new information content of a rule governing relative positioning, as envisioned by Gaze. Sperry's idea of chemical tags as target definitions was the tip of an iceberg, where most of that iceberg represents chemicals that function in many different ways, allowing for feedback, growth and the unfolding of information.

Sperry was aware of the information problem. The strict chemoaffinity theory does not explain where the information for the placement of the affinity signals comes from. Somehow, the address code is read or transformed from the genetic code. The information content of a complicated wiring diagram is the same as for the precise code that describes it.[21] Sperry deals with this in his landmark 1963 summary of chemoaffinity simply by stating that the critique exists, but since he does observe this kind of specificity, the critique must be wrong. The observation that the specificity exists is the proof, deal with it: "The chemoaffinity interpretation also met objections on the grounds that there are not enough distinct chemical labels available in the embryo. The scheme requires literally millions, and possibly billions, of chemically differentiated neuron types, . . . This labeling problem, plus the further task of interconnecting in precise detail all the postulated millions of chemically specific neuron units into functionally adaptive brain circuits, also seemed prohibitive from the standpoint of information theory because of a supposed lack of enough "bits of information" within the zygote to handle all the developmental decisions involved in building a brain on this plan. Evidence obtained recently seems to provide a direct experimental answer to such objections. . . . In brief, we think we have finally managed to demonstrate quite directly by histological methods the postulated selectivity in nerve growth and synaptic formation."[2]

At the same time, in 1963, it was a common assumption in the AI community and amongst cyberneticists that, based on the apparent complexity of the neural networks, connections had to be random. Hence the random designs of the Minsky-Edmonds SNARC and the Rosenblatt Perceptron discussed in the first seminar. This assumption was based on the wrong intuition that there simply is not enough

FIGURE 6.1. Sansar Sharma, Lyn Beazley, Michael Gaze, and Michael Keating in Edinburgh, 1966.
Courtesy of Sansar Sharma, with permission of Lyn Beazley

information available to encode a nonrandom network of sufficient complexity. Sperry may not have explained how genes are supposed to lead to complex connectivity, but he had the vision, or doggedness, not to be daunted by the problem. Somehow, he felt, the genes simply must contain the information. He was correct in a way, except that strict chemoaffinity offers no solution to the information problem. Algorithmic growth explains the unfolding of information, and it requires general chemoaffinity, the observation that molecules both execute and generate new rules in space and time during the self-assembly of the brain.

Others at the same time focused on finding the rules of growth, whether they are encoded by molecules or not. Gaze and his colleagues Mike Keating, Sansar Sharma (fig. 6.1) and Tony Hope presented their systems-matching idea and a series of models, including computational efforts like the arrow model.[4, 22, 23] The models had to balance specificity and rigidity, not to say the power of chemoaffinity, with the flexibility and plasticity of development they observed in experiments. All these models demanded positional effects and feedback, often between axons relative to each other, in a process that happens step by step, based on the continuous supply of time and energy.[4, 19] Several fundamental ideas of these studies have proven to be correct, but they did not address how genes or molecules could encode or execute what was happening. Who could cross that bridge?

Most of the seminal studies in the 1960s focused on the visual system, which was of great historical importance in appreciating both the basic concepts as well as the intricacies of generating complicated wiring diagrams.[24] The debate on molecular codes of brain wiring continues unabated.[21, 25, 26] In the early 1960s, a contemporary of Seymour Benzer felt that visual systems, and a model like the fly, were too complicated to get a complete picture of the encoding from genes to behavior. Sydney Brenner (fig. 6.2) was a 26-year old postdoctoral scientist in Oxford when James Watson and Francis Crick invited him to see their newly constructed model of DNA in Cambridge before its publication in 1953.[27] Together with Crick and others, Brenner focused on the new field of molecular biology and played important roles in the discovery of the genetic code, including the stop codon.[28] In his own words, Brenner was searching "for explanations of the behavior of living things in terms of the molecules that compose them."[29] Just as Benzer had turned to *Drosophila* to link genes to behavior, so did Brenner soon get interested in the ways genes encode more complicated organisms and, in particular, the nervous system. In fact, the similarities of scientific scope and names led to noteworthy mix-ups, which the two men bore with some goodwill. Note Benzer's written note in a particularly remarkable mix-up in figure 6.3: "Sydney—it will probably always be like this! Best regards, Seymour."

Brenner started a search for an organism with a nervous system that would be as close to the ease of handling bacteria and phages as possible. He knew of Ellsworth Dougherty and colleagues in Berkeley, neighbors to where he had done his PhD, who had suggested the suitability of tiny microscopic worms for genetics research. The tiny nematode

FIGURE 6.2. Sydney Brenner and Seymour Benzer at Cold Spring Harbor Laboratory, 1975. Courtesy of Cold Spring Harbor Laboratory Archives, NY

Caenorhabditis elegans does not really have a brain, but only, and exactly, 302 neurons. Already during the first half of the 1960s Brenner envisioned that this should allow for a complete description of the anatomy and development of the nervous system much faster than in other, more complicated systems like the fly.[30] In fact, he described his masterplan in a few sentences to Dougherty in 1963: "To start with we propose to identify every cell in the worm and trace lineages. We shall also investigate the constancy of development and study its control by looking for mutants."[27, 31] In other words, right from the beginning, Brenner wanted it all, the genes, the mutants, the cells. In addition, Brenner knew he needed what he called the *wiring diagram* of the worm's entire nervous system—arguably the beginning of the field of *connectomics*. Brenner justified the need for the wiring diagram as quoted in Errol Friedberg's biography of Brenner (p.167): "If you explain to the skeptic: 'I have modeled this behavior and we've got this oscillator which interacts with this and it's coupled in this way and it does this'; and he says: 'That's very

NATIONAL ACADEMY OF SCIENCES

NORMAN H. GILES
DEPARTMENT OF ZOOLOGY
UNIVERSITY OF GEORGIA
ATHENS, GEORGIA 30602
(404) 542-5469

SECTION OF GENETICS

November 15, 1976

TO: Members of the Genetics Section

FROM: Norman H. Giles, Chairman of the Section NHG

Foreign Associate. To my chagrin, I must report that I missed an error in the report on balloting for foreign associate. The top nominee in the vote was Brenner not Benzer (who nominated Brenner). Consequently, unless Seymour wishes to renounce his U.S. citizenship (which I assume is doubtful!), I shall promote Brenner's candidacy as "vigorously as possible".

Sydney—
It will probably
always be like this!
Best regards,
Seymour

THE GRADUATE SCHOOL

AND

THE DEPARTMENT OF

DR. SYDNEY BENZER

MRC LABORATORY OF MOLECULAR BIOLOGY
CAMBRIDGE, ENGLAND

THE MOLECULAR GENETICS OF THE NEMATODE

DATE: Thursday, April 13, 1978
TIME: 3:30 p.m.
PLACE: Room 625, T-wing

FIGURE 6.3. Historic mix-ups of the names of Seymour Benzer and Sydney Brenner in the 1970s. Top right: Nomination to the National Academy of Sciences in 1976. Bottom left: A seminar announcement of Sydney Brenner at the University of Washington in 1978 made a complete job of it by announcing the hybrid "Sydney Benzer."
Courtesy of Cold Spring Harbor Laboratory Archives, NY

nice, but how do you know there isn't another wire which goes from this point, you know, goes right around the back and comes in at the side again?' You need to be able to say: 'There are no more wires. We know all the wires.'"[30] And thus Brenner and his team spent the second half of the 1960s and all of the '70s establishing the genetics and methods to perform genetic screens and map the wiring diagram of the tiny worm.

Key to the effort of obtaining the worm's wiring diagram were technological advances in electron microscopy and computer science. John White was trained in electrical engineering and recruited by Brenner to lead the project to map the worm's nervous system. To Brenner that meant: to identify every single synaptic connection. It took longer than White and Brenner may have hoped for, but the first fully electron microscopy-based reconstruction of a nervous system with all synaptic contacts was eventually completed for the little worm in 1985.[32, 33] Brenner, however, was reportedly rather disappointed, because the wiring diagram of 302 cells and about 8,000 connections between them did not reveal as much as he may have hoped "about how genes control the nervous system and behavior."[30] And yet, White's remarkable first account in 1985 already provided some key insights into the origin of synaptic specificity and developmental flexibility. Purely based on the analysis of a single worm's mature wiring diagram, he already postulated as one of the "three determinants of connectivity . . . the developmental mechanisms that generate particular neurone types at defined places, enabling their processes to select an appropriate neighbourhood."[32] The idea of positional effects in synaptic specification has lost and gained traction over the years, while the idea of molecular specification and "Sperry molecules" has remained a focus of the field. Yet, most scientists in the field would certainly agree that developmental mechanisms restrict what potential partners "see each other" in time and space, while the degree to which synaptic partners are ultimately molecularly matched remains an open question.[34]

Ever since the trailblazing work by Brenner and White, scientists have been refining and learning new features of the worm's relatively simple nervous system. New genes, ways of information processing, and behavioral nuances have been discovered in *C. elegans*, and continue to be discovered unabatedly, to this day. The hope has long been that with only 302 neurons and exact knowledge of how they are connected to each other, we might be able to run a computer simulation of what the nervous system does. What seems simple at first glance turned out to be surprisingly difficult. A remarkable community effort has come together over several years to develop the first complete "digital life form" based on the worm's 302 neurons and total of 959 cells. The *OpenWorm* project is an ambitious, open-source project to achieve this lofty goal.[35–37] The project provides an important platform for data integration as well as an example for what it may take to establish a fully understood neural network. A key question remains: Down to what level does the simulation need to simulate? Should a simulation consider the diffusion dynamics of individual molecules? A single molecule simulation of *C. elegans* is currently, and for the foreseeable future, a pipe dream. In the meantime, OpenWorm is designed to simulate all cells with certain properties. This seems to be the most reasonable compromise to shortcut the need for, as many hope, unnecessary details. We will revisit this fundamental question soon again.

Brenner published his seminal paper entitled simply "The genetics of *Caenorhabditis elegans*" in 1974, beginning with the words "How genes might specify the complex structures found in higher organisms is a major unsolved problem in biology."[38] In this work he not only established the genetics of *C. elegans*, but also performed a rather open-ended forward genetic screen for mutants. When Benzer did his first screen in *Drosophila*, he chose visual behavior to reduce the complexities of all the things that could go wrong in a complicated animal using a more specific assay. But what should Brenner expect from the little worm? Worms do not really do all that much. The most obvious behavior is the regular, snake-like movement in sine waves. When Brenner accelerated the worm's evolution by increasing the mutation rate with a cancerogenic chemical, much like Benzer, what mutant phenotypes could one even expect? There were some shorter "dumpy" worms, there were some longer ones, there were some that looked like they had blisters.[27] But by far the most common phenotype induced by his random mutagenesis were worms that didn't move quite right anymore. He called

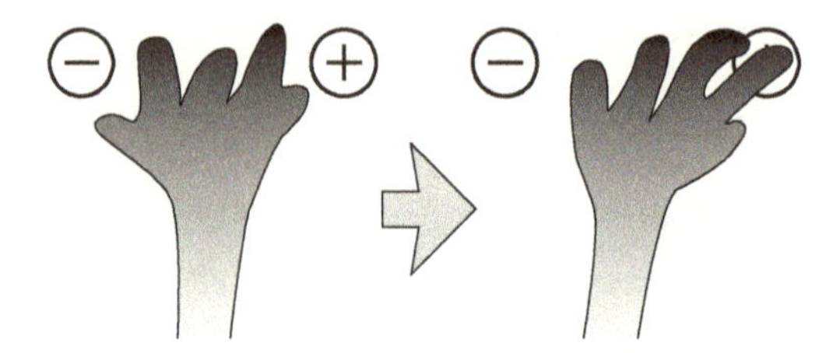

FIGURE 6.4. Attractive and repulsive growth signals guide axonal growth and provide an intuitive link between a molecular mechanism and network connectivity. Growth cones respond to environmental cues, avoiding repellent signals (−) and growing towards attractive signals (+). Many cell surface molecules and their ligands are initially studied with this idea in mind, but many more mechanisms have been shown for this class of molecules.

these "uncoordinated," and numbered them accordingly *unc-1* through *unc-77*. Brenner's initial *unc* mutant collection already included many of the genes that molecular neuroscience would focus on for decades. An analysis of *unc-6* in the late '80s, for example, revealed misrouted axons, indicative of a connectivity defect. It has become the classic textbook chemoaffine Sperry molecule, to this day: netrin.[39, 40]

Already in the late '80s, Marc Tessier-Lavigne, Tom Jessell and colleagues had suggested that cells from the ventral part of rat spinal cords secrete a substance that attracts axons growing from the dorsal part of the spinal cord.[41] In experiments similar to Bonhoeffer's choice assay, Tessier-Lavigne and colleagues showed that explants of the dorsal spinal cord *roof plate* exhibited undirected outgrowth of axons, while adding tissue from the ventral *floor plate* some distance away was sufficient to attract the axons (fig. 6.4). Apparently, the floor plate released some attractive chemical, just as Sperry predicted, but how to identify the chemical? The rat floor plate did not provide enough material to isolate the substance. Using their efficient axon outgrowth assay, Tessier-Lavigne's team looked for other sources and found attractive activity in extracts of ground-up, early developmental chick brains. A heroic, biochemical tour-de-force led to the isolation of sufficiently pure protein to

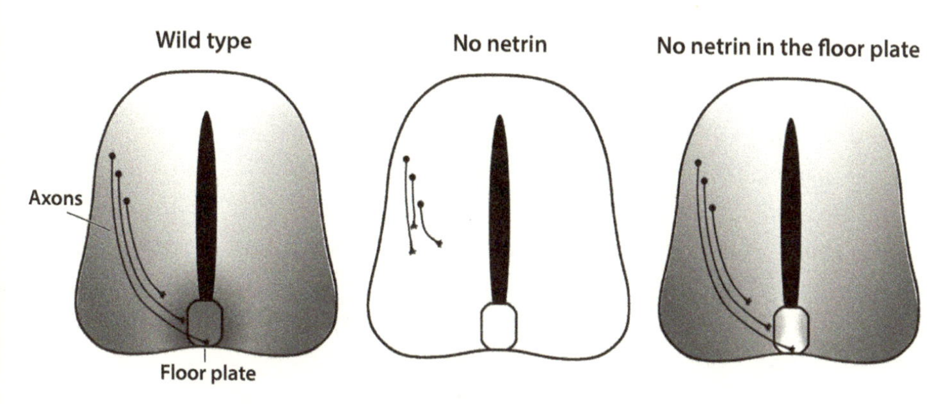

FIGURE 6.5. Axon pathfinding at the spinal cord midline and the role of netrin. Left: The floor plate is a central source for netrin. Middle: In a classic experiment, loss of netrin was interpreted as a chemoattractant for axons to grow towards the floor plate. Right: Selective loss of netrin from the floor plate does not cause strong axon pathfinding defects because netrin along the growth path is sufficient to "guide" the axons.

determine its identity. They found two proteins, called netrin-1 and netrin-2, both highly similar to the predicted protein product of *C. elegans unc-6,* which they published in 1994.[39] For more than 20 years, netrin became the textbook example of a *long-range attractant.* In addition, midline crossing of the axonal growth of commissural axons, i.e., axons that grow from the roof plate towards the floor plate, which they cross, became the favorite model for the burgeoning field of axon pathfinding.[42]

Midline crossing turned out to be an ideal system to study how growing axons make seemingly simple choices. It was much less complicated than dense brain regions, including the visual system (fig. 6.5). An axon first needs to find the middle and then cross the middle once. And netrin had what it took to find the midline: in addition to its proven activity to attract axons in culture, a knock-out of netrin in mice exhibited profound commissural axon pathfinding defects.[43]

Over the years, maybe not surprisingly, studies in different systems seemed to refine the general mechanism of the textbook long-range attractant. For example, in flies a nonsecretable version could still do a significant part of its job, leading to the suggestion of an additional short-range attractive function.[44] However, the experiment of abolishing netrin selectively in the floor plate in developing mice (i.e., removing the targeting signal from the target) was only published in 2017. To the surprise of the field, the laboratories of Alain Chedotal and Samantha Butler independently showed that loss of netrin only at the place to which it was supposed to attract axons had little effect.[45,46] By contrast, netrin along the path of the growing axon was clearly required. We are reminded of nerve growth factor: If a path is labeled with a growth-supporting chemical, then this is where the axon will grow. Even though it may be considered as a permissive factor, the information for the path is laid out in its spatiotemporal distribution. However, as Tessier-Lavigne and others have since pointed out, under the initial experimental conditions, including the guidance ability in vitro, the initial results hold true. In fact, for specific conditions, Tessier-Lavigne defended the long-range action of netrin as recently as 2019.[47] What is going on?

Clearly, the early interpretation that netrin is the one molecule that represents the information of a target "address" was an oversimplification. While the importance of netrin in this process has been undeniable since its discovery, the way it functions in space and time seemed to become more complicated over the years. It appears to serve several functions in sustaining growth at different times and places as the axonal cables are routed. Amongst netrin's functions there is now netrin along a labeled path that only exists as the result of a lengthy preceding development program. Then there is midline netrin, but here netrin turns out to function in a logical "and" with another secreted molecule called sonic hedgehog.[48] There is also a repulsive growth factor–like protein called NELL2 to consider, depending on when and where you look.[49] In fact, any midline role of netrin is highly dependent on the context of the particular vertebrate species under investigation and the exact section of the spinal cord where one does the experiment. All of this is important context for netrin, just as netrin's presence provides context for the function of, say, sonic hedgehog or NELL2. If we think of netrin as part of an address code or identification tag, then this address code seems to have become rather complicated after all—and that's just for the simple task of finding the midline.

To make matters worse, our list of factors that contribute context is certainly incomplete. We already have a sense that tissue stiffness plays a role. The axon itself may be in different metabolic states at different times during the growth process that indirectly affect its drive to grow. Metabolic states are influenced both by intrinsic properties of the developing cell as well as extrinsic factors. Both the neuron as well as its surrounding cells run their own developmental programs that change probabilities of how an axon may react at different time points of its life. If we are looking for a comprehensive code for the growth of the axons, which ones of these parameters can we ignore, and which ones are merely permissive and contribute nothing to the instruction? If all of these factors are part of the decision-making processes of the axon, how complicated is the code that we are trying to understand in molecular developmental neuroscience—maybe dozens or hundreds of critical parameters for every minute of the developing axon. In all likelihood, this is again only the tip of the iceberg, even for that seemingly simple axon pathfinding process of finding the midline, far away from the density of the brain. This is a problem we have to deal with if the objective is to decipher the "code in space and time" for the developing brain.

How about the notion that netrin provides an archetypical example for a *general* mechanism, namely short- or long-range attraction? The initially reported phenotypes were undeniably clear. A few hundred publications later this clarity, or the simplicity thereof, has somewhat faded. The insight that started with the idea of a general principle got more complicated the deeper we looked. We have encountered this phenomenon before in the discussion on relevant information. How many of the details that appear to make a mechanism less general are relevant? One could argue that the less detailed version of a mechanism is the more general mechanism. Unfortunately this also means that a mechanism that is more general because it is less detailed will more often not be sufficient to explain an instance of a problem. This is a bit of a conundrum: we can choose between fewer general mechanisms that explain more instances less, or many less general mechanisms that explain fewer instances better. Where we draw the line for "generality" is anybody's guess.

Every biologist will agree that a molecule only functions given a specific context. Yet, studies that focus on a single gene or protein are often titled "Gene X regulates process Y"—where "regulates" is interchangeable with: "directs, governs, orchestrates, controls, etc." Clearly, "Gene X contributes to the context- and feedback-dependent, multifactorial composite regulation of process Y" sounds a bit cumbersome. We have no simple words to express the concept. The danger, of course, is to believe what we write when assigning the control of a process to a single factor. The debate about netrin has never been about the relevance of its seminal discovery or its importance for axon pathfinding. Issues arise only when we pin the instruction on a single factor.

And this brings us back to the idea of the instruction, and its less exciting counterpart, the "permissive factor." In the field of synaptic specification, researchers are not typically excited by permissive mechanisms, like, say, cell biological mechanisms of receptor trafficking. We are looking for an instructive signal to tell the axon all it needs to know to reach a certain place. But who instructed the instructive signal? We like the idea of an instructive signal, because it provides all necessary information for the process. But a single factor can only transfer this information; it does not generate new information. New information only unfolds through interactions and feedback between more than one player in the game of algorithmic growth.

Long-range attraction and synaptic matchmaking are instructive concepts. Combining more than one factor to compose the instructions is often considered essential for specificity.[50] A combination of several positive and negative factors that bias the decision to varying degrees can work together by increasing and decreasing probabilities of individual synaptic pairs in different contexts. Feedback between the factors can make this "code" rather complicated and difficult to predict. If there are many factors working together, then the removal of any single factor may change probabilities of the outcome more or less—or not at all. In cases where the removal of a single factor does not seem to change anything (based on the outcome we are looking at), the removal of a second factor may increase the probability for a change. If the removal of two factors changes the outcome, while removal of neither alone does, the geneticist speaks of redundancy. Every single factor contributes to the

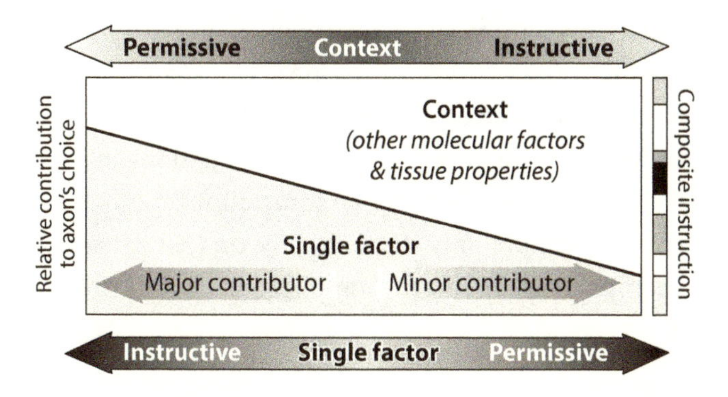

FIGURE 6.6. Instructive versus permissive functions of a single factor form a continuous gradient based on the relative contribution of other factors (context). Any "instruction" that directs, e.g., an axon's growth direction is a composite of more than one factor. One factor may be more dominant (left side of the gradient). Alternatively, a composite instruction can also be the result of many small contributing factors, each by itself "permissive" (cartoon on the right side).

outcome, yet no single factor determines the outcome. This is the idea of composite instructions.

Composite instructions can be dominated by a single factor (fig. 6.6). These are the cases biologists are focusing on in single genes studies. We can call the dominant factor "instructive," with everything else being permissive context. Netrin was originally described as a 100% instructive cue. Over the years, we have learned to appreciate its different roles, including permissive ones. We are sliding down a gradient from instructive to permissive functions for single factors. As more factors contribute their little bit to a composite instruction, there is more room for interactions and feedback amongst many players in a game that leads to the unfolding of new information. At the other end of the spectrum, a composite instruction may have no dominating single factor (fig. 6.6). Now, here is an interesting idea: A process may appear instructive at one level, yet represent the composite property of the set of molecules and their functions at a lower level, none of which themselves need to be a cue, and all of which may by themselves appear permissive. Each small contributor may function probabilistically, and its removal may cause small or no defects with varying penetrance. A combination of only such ones, of many permissive factors, means that from the perspective of every single factor, the remaining context is the major instructive contributor. We may not know how many instructions in brain wiring are of the "one dominant factor plus some context"-type, versus the "many permissive factors only"-type. But we do know that instruction is always a composite property that unfolds during algorithmic growth. Thank you.

2.3

The Levels Problem

The Seventh Discussion: On Context

AKI (THE ROBOTICS ENGINEER): I can't wait to hear what Minda has to say to that . . .

MINDA (THE DEVELOPMENTAL GENETICIST): Provocative, obviously.

ALFRED (THE NEUROSCIENTIST): But why?

MINDA: I have no problem with the idea of context-dependent gene functions. But I do not think every gene is equally important for brain wiring. I think you will find cell surface receptors much more often as major contributors to what we just heard as 'composite instructions.' Most mutations in random genes will cause no effect on brain wiring whatsoever, while mutations in guidance cues and receptors will have a much bigger effect.

PRAMESH (THE AI RESEARCHER): But it does seem curious to me that the behavior-altering mutations found in your genetic screens discovered so much more than 'Sperry molecules,' yet the study of brain wiring that's supposed to determine behavior is guided by searching just for these molecules?

ALFRED: Sperry's legacy. There is a lot of history to this. Gerald Edelman's *Topobiology*, for example, was based on the idea that you need to look at molecular interactions between cells. He also highlighted that behavior is a continuum based on development.[1]

MINDA: Exactly! In our experiments the cell surface molecules often have such beautifully specific expression patterns. They are exactly at the right time and place where they are required to mediate meaningful cell-cell interactions. Other molecular mechanisms like cytoskeletal assembly or membrane trafficking are much more similar in all cells.

PRAMESH: But the moment you see the specific cell surface receptor at a precise time and place, algorithmic information has already encoded the spatiotemporal specificity information by getting it there. The idea that the cell surface receptor is now instructive is the famous one-to-one mapping of information that first had to come from somewhere. As we heard, the information is in the growth process that got the receptor where it needs to be. When you see the beautifully specific pattern you are looking late in algorithmic information terms, at the output of the information encoding process.

AKI: I get that. Funny, why would a complete field focus on one-to-one mapping, the one idea that doesn't generate more information because it is already an output, the most boring hypothesis of them all?

MINDA: Not correct. The idea of molecular guidance is not just an idea of one-to-one mapping. When we look at a specific instruction at a specific place we are just taking a snapshot of developmental growth. I appreciate there are many levels. Nobody said it was easy.

ALFRED: Well, I remember when netrin was discovered, they definitely made it sound easy—like, look, here is an archetypical example for a long-range attracting molecule—and everything else is just details of different versions of this general principle . . .

MINDA: But the principle still holds, it is just more complicated than originally thought. That's normal for any scientific discovery. In the right context netrin does what netrin does.

PRAMESH: Aha! So, what is the right context?

MINDA: Well, all the factors that are necessary for its function. A minimal set of them will be sufficient for one function or the other.

ALFRED: This is classic genetics talk. In circuit neuroscience they now use the same terminology, which I find quite iffy.

MINDA: Why?

ALFRED: Okay, a gene or a neuron is called 'necessary' when its loss causes a certain phenotype. But that can depend on a lot of things. A neuron may only be necessary for a certain behavior if the animal is also, I dunno, hungry or something. It depends on context.

PRAMESH: Then you are just defining your phenotype with more parameters and more dependencies.

ALFRED: But it may also only be necessary 50% of the time—randomly. Sometimes you suddenly lose the behavior, sometimes not. What do you do with that?

MINDA: That's what we call penetrance in genetics. Many mutations cause phenotypes with reduced penetrance. And there are many reasons for that.

AKI: Well, if everything is context-dependent, then I don't understand the dichotomy of necessity versus sufficiency anymore.

PRAMESH: I would say the degree to which a gene or neuron is necessary or sufficient, like the instructiveness of a cue, . . . they all live on a gradient of context-dependency.

AKI: Classic Pramesh . . . if only we knew what he meant . . .

PRAMESH: But that's exactly what we just heard in the seminar. Of course instructions exist, because obviously axons make choices. The question was only whether we find sufficient information content for the instruction in a single molecular mechanism. Similar to a gene's necessity for a phenotype—it may only be 'necessary' when at least one other gene is lost. It may only be sufficient if at least three other genes are there.

MINDA: Yes, that's correct. In flies, most genes are not required for survival—under laboratory conditions, I should add. But once you start removing more than one, the effects are much stronger.

PRAMESH: Yes, that's the context-dependency I meant.

MINDA: But there are wild type and non–wild type contexts. Hundreds of both permissive and instructive molecular mechanisms have been described for individual genes for a wild type context. They all contribute to a complex code during developmental growth. Attractive and repulsive molecules are used and reused at maybe hundreds of different places and times during development. The problem of little information in the genes compared to lots of information in brain wiring simply turned out to be no problem at all.

AKI: . . . the whole point seemed to me that you just can't pin the information to a single molecule, that'd be too easy.

MINDA: Am I alone now? As you know, single molecular instructive functions have of course been described. Even if other factors play roles, let's say, the repulsive function of ephrins during eye-brain wiring is truly an instruction by itself.

ALFRED: Absolutely everything is context-dependent, isn't that even a bit trivial? But if the idea is that only the biggest contributor is the instruction, whereas the smaller contributors are just permissive, I mean, where do you draw the line?

PRAMESH: That's the gradient I talked about: We can think about specificity in brain wiring in terms of molecularly defined instructive mechanisms that tell axons and dendrites when and where to go and make contact. On the other end of the spectrum we have the idea that no such explicit instructions need to exist, because composite instructions continuously emerge from whatever permissive conditions exist at any given point in time.

ALFRED: The question is, to what extent could you wire the brain if all neurons could in principle make synapses with all others, purely by restricting who sees whom during algorithmic growth. Here unfolding of information really helps.

MINDA: To me, your 'unfolding of information' precisely means the unfolding of the molecular code in time and space. Of course only axons and dendrites that get to see each other during development can make synapses. But we also know that specific molecular interactions in dense brain regions can then positively or negatively bias the choice.

AKI: I think we are all happy with your molecules, Minda. The question was whether you could pinpoint the instruction to individual molecular mechanisms at every point in space in time, collect those, and when you put it all together you would understand how brain wiring happens during growth. That's how I had always thought about your field. But now I am not so sure anymore, so you guys have achieved something!

MINDA: I still find this a bit esoteric. Is this really not just semantics?

PRAMESH: Let's revisit the discussion we had last time. If you identify a mutation that changes a behavior, say, your predisposition for talking . . . not everybody is a born talker . . .

AKI: Hear, hear . . .

PRAMESH: . . . well, a mutation in some gene of some molecular function may affect that and thus become evolutionarily selectable. But do you now call this a 'talking gene'?

MINDA: I am actually with you on that. We have that problem with the genetics of intelligence, aggression and many other complex traits. I don't think there are any single genes for such traits. It's usually a collaborative effort of many genes.

PRAMESH: Exactly. Context-dependency is just another way of putting it. A certain mutant may only have that effect on your predisposition to talk if some other mutations are there as well. So, a mutation may only disrupt an instructive role in a certain genetic background. And a mutation may only be sufficient to cause a defect in a certain background.

MINDA: I can accept that part of the problem is that instructions are not usually all-or-nothing, and they are always context-dependent.

AKI: And peace was restored! How about the levels problem? If molecules have permissive and instructive mechanisms, how about a filopodium? In synaptotropic growth, if I remember this right, or in self-avoidance, filopodial interactions are the level of the game, right? The autonomous agents.

MINDA: What does that have to do with anything?

AKI: If composite molecular instructions determine what a cell or filopodium does, can composite filopodial instructions determine what the cell does? Like synaptotropic growth or self-avoidance?

ALFRED: That's cute. All molecular stuff in the filopodium produces some kind of composite instruction, so now the agent is the filopodium. That makes synapse formation a problem at a higher level, decision by filopodia, not molecules . . .

PRAMESH: Hmm . . .

ALFRED: The filopodium executes composite molecular instructions, but no single molecular mechanism may tell you what the filopodium does. Makes sense, for something like synaptotropic growth you need to describe the filopodia, not the molecules. You can describe synaptotropic growth at the level of the filopodia without knowledge of the molecular mechanism. That's actually how we work in neural circuit analysis. The molecular stuff is somebody else's problem.

AKI: Ah, an S. E. P.!

PRAMESH: A what?

AKI: A Somebody Else's Problem is how you can make starships invisible, just because of 'people's natural predisposition not to see anything they don't want to, weren't expecting or can't explain.'[2]

ALFRED: Let me guess . . .

AKI: Right, so here is the thing I actually find funny: In your field, Alfred, people don't care much about genes and molecules. You manipulate neurons.

ALFRED: Of course, if we want to understand how neural circuits work, we need to manipulate how neurons talk to each other. What's funny about that?

AKI: Minda, if you try to figure out how the same circuit that Alfred works on is put together, you manipulate genes and study molecules, right?

MINDA: Sure.

AKI: So why do you study circuit function—obviously—at the cell level, but circuit development—equally obviously—at the molecular level?

ALFRED: Oops.

Seminar 7: From Genes to Cells to Circuits

"The functional elements of the nervous system and the neuronal circuits that process information are not genes but cells." These are the introductory words of Gerry Rubin in a 2008 landmark paper on methods to study neural circuits.[3] At that time, Rubin had recently started a research institute with a focus on understanding how brain wiring produces behavior, the HHMI Janelia Research Campus near Washington, DC. Rubin obtained his PhD in the early 1970s in Brenner's department in Cambridge. He witnessed the birth of the neurogenetics of *C. elegans* firsthand and saw the initial effort to generate the first connectome from the get-go, including the development of advanced electron microscopy and computational analysis techniques in Brenner's lab at the time.[4] Rubin switched to the morphologically more complicated fruit fly early in his career. For many years, the idea of a complete connectome for the more than one hundred thousand neurons of the fly brain seemed an intractable problem. But with the start of Janelia in 2006, he had not only the vision but also the resources to follow in Brenner's footsteps. The goal was the fly connectome, and advanced electron microscopy and computational analysis techniques would feature again prominently. Prior to the work on the fly connectome, Rubin had codeveloped the first method to create transgenic *Drosophila* in the '80s and led the way into the genome era by spearheading the sequencing of the *Drosophila* genome just prior to the human genome. And then, in 2006, the Janelia Research Campus opened to much fanfare, and shortly after Rubin wrote the introductory quote above, followed by what reads like a manifesto for the road ahead: "we will need to be able to assay and manipulate the function of individual neurons with the same facility as we can now assay and manipulate the function of individual genes."[3] The message was clear: less genes, more neurons. What happened?

According to the connectionist view, the information for brain function is embedded in the connectivity between neurons. The burgeoning field of neural circuits progressed quickly in the 2000s on the shoulders of decades of systems neuroscience and accelerated by new methods to deactivate and activate neurons.[5] The idea was that the (de-)activation of neuronal function should yield insights into the workings of neural circuits (i.e., subsets of connected neurons in the brain) much the same way as the deactivation and activation of genes had uncovered the workings of developmental genetic programs for decades. If we can knock out genes to understand a transcription factor cascade, we should also be able to knock out neurons to understand a neuronal firing cascade. In just a few years, many new methods were developed to control and monitor neuronal activity. Most prominently, optogenetics became a key technique and buzzword. Several scientists developed methods to introduce genes encoding light-activated proteins to acutely activate or deactivate neurons.[6–9] This was a game changer. Why? Neuroscientists long knew that permanently silencing neurons led to changes of the surrounding network. Neural circuits are loosely defined as ensembles of connected neurons that serve a tractable function, like motion vision for example.[10] Such higher-order functions are under evolutionary selection to function precisely and robustly. Malfunction of a single component of the circuit should not, and usually does not, lead to malfunction of the whole. In an airplane, entire circuits are duplicated to make them redundant. But redundancy in biological and artificial neuronal circuits is of a different kind. If a neuron drops out, others can adapt their properties to minimize the effect at the level of higher-order output. Key to this property of compensation is feedback. Even single synapses show compensation of their sending and receiving sides: if the sending presynaptic partner is weakened, the postsynaptic side can compensate by increasing its sensitivity. The brain is full of feedback, just as the early cyberneticists envisioned.[11, 12]

Analogous to genes, the manipulation of a single neuron's activity comes with the hope that its loss or gain may lead to a change that tells us something about the function of that neuron. Compensatory changes mask such defects. However, most compensatory changes take time. Hence the desire to activate or deactivate neurons acutely, without warning and minimizing compensatory change of the surrounding network. The discovery of light-activated ion channels opened the door to genetically manipulate neurons such that light could be used to activate or deactivate them.[13] In recent years, these types of experiments have been driven to amazing levels of sophistication—flies, fish or mice are now being put in virtual, computer-generated environments where they are allowed to behave more or less normally, while light of one color can be used to activate or deactivate specific neurons in the brain, and light of another color can be used to visualize what other neurons are active in the brain. These are exciting times in neural circuit research.

The neural circuit field has fully embraced the notion that *functional elements of the nervous system and the neuronal circuits that process information are not genes but cells*, just as Rubin suggested. Of course important research on genes in neuronal function also continues. But if the question is at the level of the link between neuronal connectivity and the ensemble output that leads to behavior, then genes are at least one level too far down to be useful. The driving instructor does not teach the student atomic forces underlying friction in order to brake at the next red light. That doesn't make atomic forces irrelevant; they are just not helpful to understand higher-level action and not the right level for the student to know what to do next.

Circuit function is a question to be studied at the level of cells that constitute that circuit. How about the development of the same circuit? If the function of the ensemble is a question to be studied at the level of cells, shouldn't the same be true for its assembly? Curiously, this is not the perception in the field of circuit development: clearly, circuit assembly is a question to be studied at the level of genes, the "molecular mechanisms of connectivity." We have just spent the last two seminars debating genes, from the transcription factors to cell surface molecules. New mechanisms that molecularly encode specificity are the name of the game.

How is it that the study of neural circuit function is obviously a question to be studied at the level of cells, while the study of the same neural circuit's assembly is obviously a question to be studied at the level of molecules? A naïve observer might think that molecules are equally important in both. A neuron cannot develop without running its transcription factor cascade and expressing specific cell surface receptors at the right time and place for neuronal interactions to make sense. Similarly, a neuron cannot function without running its metabolic machineries and express ion channels and neurotransmitters and the molecular machinery that ensures their regulated release at the right time and place. Neurons, not genes, are the right level for perturbation experiments to understand circuit function. Similarly, neurons, their growth cones and filopodia are the players whose interactions lead to circuit assembly. Why then do the vast majority of studies on circuit assembly nonetheless focus on gene perturbation, not cell perturbation? There are at least two answers.

First, it is not obvious what the developmental analog of activation and deactivation of neurons is supposed to be. In order to affect developmental activities of neurons, the perturbation is likely a genetic perturbation. The functional (de-)activation of neurons through optogenetics is achieved through the expression of a gene that encodes a light-activated ion channel. However, here the gene is just a tool; the relevant manipulation is considered at the cellular, not molecular level. What is the activity of a developing neuron? Early in development, many neurons migrate—an activity that can be deactivated through genetic means but is harder to activate. Later in development neurons grow axons and make synapses—again activities that are not easily switched on or off with specificity, although promising approaches exist.[14, 15] For example, cell biological manipulations of membrane trafficking or cytoskeletal function through gene manipulations can decrease or increase axonal growth activity. However, since the manipulations occur at the level of gene manipulations, it is difficult to make the argument that the meaningful level of the perturbation experiments is a cellular developmental activity, rather than the function of gene products. The beauty of the light-activated ion channels is that they are genetic tools that do not normally exist in the neurons where they are used to manipulate a cellular activity. There is a separate field studying the molecular mechanisms of these channels, and another field using them as tools for cellular perturbation. There are few developments in this direction for the manipulation of developmental activities, and several are based on optogenetics.[14] Scientists studying circuit assembly have lacked technologies for acute cellular level perturbations for a long time.

The second, and probably more important reason why genes have remained the perturbation level of choice is that genes are where we expect to find the information that makes circuits. It is the same reason that led us to discuss genes in the first two seminars of this session: since the '80s, the molecular revolution has been delivering on Sperry's vision, in one way or the other. Genes obviously encode information, and the genome must somehow encode circuits. The idea that energy is information and that the developmental process itself unfolds new information under the influence of energy and time are, well, less obvious. Yet, it is exactly this process that creates new levels, between the genome and the circuit, through which structures like the growth cone or the filopodium attain properties that cannot be meaningfully studied at any other level.

Levels also separate the genome and the developing circuit in time. At the time a growth cone acts as an autonomous agent, it encounters a temporary environment whose existence cannot be read in the genome or deduced from the adult circuit. Let me give you an example. In their adult form, brains from flies to vertebrates exhibit highly organized anatomical features, including organization in layers.[16, 17] Specific neurons form synapses only in specific layers. Does that mean that neuronal growth cones once grew to identify those layers as target regions? If so, how did the layers come to be at the right place? In the fly, a brain structure called the medulla is the biggest neuropil, a region of synaptic connections, in the visual part of the brain.[18, 19] The fully developed medulla, similar to the human retina or cortex, consists of morphologically recognizable layers, where different types of neurons make contact. The classic neurogenetic approach in this well-studied system is to study a gene knock-out in a specific neuron type that exhibits a defect in what is commonly called "layer-specific targeting" (fig. 7.1). It is tempting to explain an experimental result where the loss of a cell surface receptor leads to the axon ending up in the wrong layer with a loss of the axon's ability to receive a target-derived signal in the right layer.

But this intuition does not take into account what the growth cone encountered when it grew. At the time when most neurons in the developing medulla undertake axon pathfinding, the layers do not yet exist. Instead, their own growth leads to layer formation itself.[18, 20] Rather than being a target region, the layers are the product of a self-assembly process. Remember our discussion of the pathfinding analogy in a complicated city network: the problem of circuit assembly is not one of finding a pre-existing address on a finished map, but a problem of finding a way and a target that itself develops, in interaction and feedback with other developing components. Layers in the neuropils of the brain come into existence because axons and dendrites grow to make them. The process has to happen in time, through relative positioning to whatever is there at a certain time and place during algorithmic growth (fig. 7.1). None of these targeting events could have happened at a different developmental time using the same mechanisms, because the situation each growing axon finds itself in depends on the context of when and where it grows, as it itself runs through its cell-intrinsic developmental program in a concerted ballet of feedback with its surroundings. The players in this game are the growth cones, their filopodial extensions and the interacting dendritic branches. We cannot study their behavior at the level of the genome. We can perturb the genome with care, making sure that the perturbation is acute or has no earlier, masking or compensatory effects. If this is the case, or at least a reasonable assumption, then we can study the changes to the growth cone *at that time*. It is risky to investigate the outcome at a later stage, because the changes caused by the mutant growth cone have repercussions in the subsequent game of algorithmic growth—from compensating to exacerbating effects. Any meaningful study of what goes right or wrong has to be at the level of the players at the time and place they act.

Let's briefly summarize where we stand on the levels problem. Mutations (genome level) that affect any of a number of different properties of the growing axon (cellular level) may cause a targeting defect in the adult structure (organismal level). Evolutionary programming can select a specific outcome with behavioral consequences at the organismal level through mutations in many different genes, including cell surface

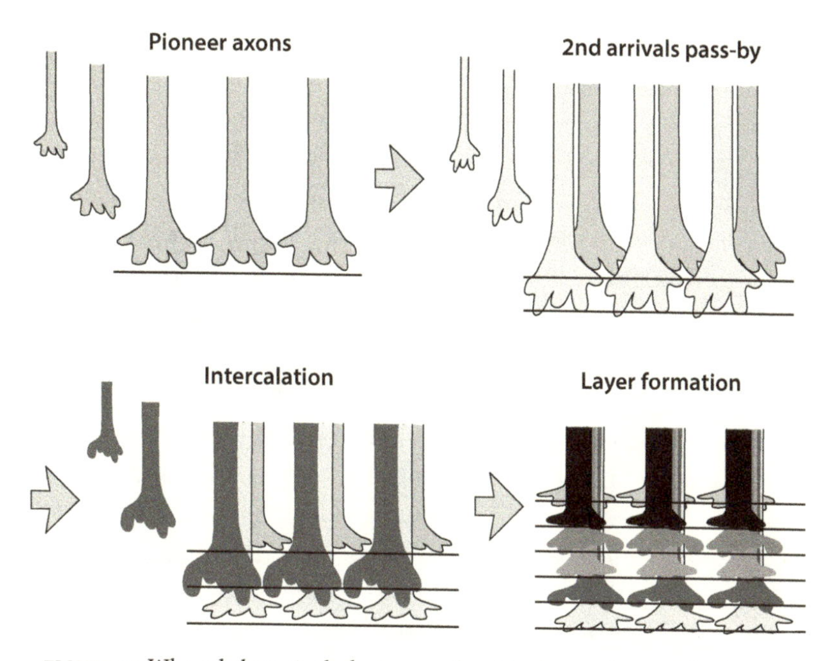

FIGURE 7.1. Where do layers in the brain come from? In the visual part of the fly brain, the medulla neuropil, layers form through successive arrivals of axons at a time when there are no layers yet. Pioneer axons form the first boundary, which is passed by a second series of axons from a different neuron type. Subsequent neurons intercalate between these two, all relative to each other, leading to layer formation. The neurons do not just find the right layers; they generate them in a self-organizing process.

receptors executing various functions that go beyond "marking a target with an X." Many functions affected by genome mutations affect cellular properties. Parameters like the speed or stiffness of a growing axon can be described independently of the molecular machinery that affect them. The axon is a player in the game, and it follows rules based on such properties. Both evolution and scientists can perturb the system meaningfully at the level of the genes, but these perturbations may play out as unpredictable properties of structures at a higher level. A specific gene product affected by a gene-specific mutation may critically influence properties of structures at many levels, even if the molecular function of the gene product is not indicative of or directly related to the properties at higher levels. At what level should we look for information that describes a system? The reductionist's dilemma is at heart a levels problem that has played out in science time and again: we try to find the smallest level to understand a system, but the properties of higher levels cannot be described by the properties of lower levels.

The schism between molecular biology and behavioral sciences runs deep. In 1944 the physicist Erwin Schrödinger published his famous book *What Is Life?* based on a lecture series that had kindled the hope that the time had come to study living matter at the level of physics and chemistry.[21] A few years later the discovery of the structure of DNA and its mechanism for replication and heredity in 1953 paved the way for a transformation in biology departments on both sides of the Atlantic. Physicists like Benzer and Delbrück joined the field of molecular biology in its infancy and brought with them the expectation and approach of an exact science that had fueled an age of successes and progress in physics and chemistry. When James Watson arrived at Harvard in the mid-1950s, a time began that the ecologist E. O. Wilson described in his autobiography as "The molecular wars."[22] The golden age of molecular biology that followed throughout the '50s and '60s led many scientists not only to disregard classical fields of biology like ecology, but even the field of classical genetics. "Genetics is dead!" as Delbrück said at Caltech, while Ed Lewis kept the fly room tradition of Morgan and Sturtevant alive.[23] But genetics was not dead; it became the basis to study higher order phenomena in organism development and function to this day. The questions scientists grappled with then as much as today are questions about levels and information: Do physical laws explain life? Do molecular interactions explain life? Of course all life is based on molecular interactions that obey the laws of nature. But the questions confront us with the levels problem: How can we bridge molecules and life, genes and behavior?

In this session on "players and rules" we encountered genes and their gene products as players in the game of algorithmic growth. The self-assembly of the brain is a game in which information unfolds based on what players are active at a specific time and place. The players execute rules that themselves depend on the context of specific levels. Think again Dscam: at the molecular level a cell adhesion protein that binds specifically to itself. If expressed in cells in a dish, this will lead to cell aggregation due to Dscam's adhesive properties.[24] By contrast, at the cellular level *in vivo*, in the context of stochastically branching dendrites, the properties of filopodia and many other level-specific properties, Dscam will execute the rule of self-avoidance.[25, 26] The resulting functional description is therefore a bit of an oxymoron in itself because it mixes the molecular and cellular levels: *homophilic repulsion*. Because of the possibility of many different level-specific functions, an assignment like "Dscam is a self-avoidance protein" can be tricky. Dscam is a key protein in the execution of the self-avoidance rule under specific conditions in the context of dendritic growth, but this is not its only function.[27, 28] Nor is Dscam the only molecule that can be employed to execute the "self-avoidance rule." For example, different cell surface molecules, the protocadherins, execute the self-avoidance rule in vertebrates.[26] Self-avoidance is a property of the cellular level, not a property of the molecular level.

Assigning the label of a higher level property to a gene or protein is a widespread phenomenon. A gene for an ion channel that is critical for the generation of action potentials (the signals that move along neuronal axons to function) may be called an "action potential gene." However, labels like "action potential gene" or "self-avoidance gene" are only meaningful in certain contexts: a specific animal, a specific cell type, and a certain amount of specific other contextual requirements must be met. Self-avoidance and action potentials are higher-order properties that rely on the function of many proteins and subcellular structures. As some of my systems biology colleagues would have it, the "gene may just not be the right *parameterization* of the system." What they mean by this, I think, is that the understanding of the higher order property may need to be described at a different level. There is a level where genes and molecules are not the right unit, just like atomic forces underlying the function of a brake at a red light are not helpful to the student trying to learn how to drive.

This brings us to the larger implication of the levels problem: players can be defined at any level. Neurons have been the players at the focus of neural circuit research. Yet, we have focused a lot on the level of the gene, because this is what the field of neurodevelopment has been focusing on for decades. Just as some scientists ask "What is a self-avoidance gene?," so can we ask questions of more obvious public interest: Are there genes for homosexuality, circadian behavior, alcohol-sensitivity, intelligence or aggression? But organismal behavior is an altogether different level from dendritic self-avoidance. How does the levels problem play out in these cases? The debate on behavioral traits typically sounds like this: We know a mutation that is associated with alcoholism or aggressive behavior 23% of the time. Hence, on average, flies, dogs or people with that mutation are more likely to behave aggressively. We therefore call this an "alcoholism gene" or "aggression gene." Why 23%, or maybe 45% or 67%? Context. Algorithmic growth. Levels. The gene affected by the mutation that correlates with the behavior may be some enzyme that plays a role during the development and function of the kidney as well as the brain, and in both cases may alter properties of subcellular structures, functional properties of the cells or properties that are specifically relevant only in the context of multicellular tissues. Let's look at the classic depiction of obvious levels in neurobiology:

Gene → protein → protein interaction network → neuronal properties → circuit properties → brain → behavior

This is a seemingly trivial hierarchy. Genes encode proteins, proteins encode protein interaction network, etc. But what does *encode* mean yet again? The gene contains the information for the primary amino acid sequence, but we cannot read the protein structure in the DNA. The proteins arguably contain information about their inherent ability to physically interact with other proteins, but not when and what interactions actually happen. Next level up, what are neuronal properties? A property like neuronal excitability is shaped by the underlying protein interaction network, e.g., ion channels that need to be anchored at the right place in the membrane. But neuronal excitability is also shaped by the physical properties of the axon, the ion distribution and many other factors, all themselves a result of the actions of proteins and their networks. Finally, neuronal excitability is shaped by the properties of the ion channels themselves. Genetic information had to unfold through all these processes to produce neuronal excitability. Now think circuit properties, brain and behavior. So, how much sense does it make to talk about an alcoholism gene? Our hope for an alcoholism gene is a hope for one-to-one mapping of information across levels.

We have discussed mutations that meaningfully change behavioral traits from circadian rhythms to sexual behavior—at least in flies. In humans such mutant analyses are more difficult. Experiments are mostly restricted to studies that correlate genomic sequences of individuals, your DNA, with any trait or phenotype of choice: from predispositions for neurodegenerative disorders to intelligence and aggression. If the data pool is large enough, one is bound to find spurious correlations[29] that are only funny as long as data-loving scientists do not fall into the trap. Since sequencing has become more and more affordable and produces enormous data pools, genome-wide association studies (GWASs) have become immensely popular.[30] Results from these studies are often daunting and suggestive, but they are not easily interpreted for several reasons.[31] For one, correlation is not causation; lightning does not cause thunder just because it precedes the other. Both are correlated because they have a common, separate cause. This is an enormously important discussion that would deserve a session by itself, but for now let's just keep in mind that we can principally correlate any two observations of unknown relationship.[29] GWAS analyses are entirely based on the correlation of information at the lowest (genome) level and a higher (outcome) level. To gain sufficient statistical power for such an analysis, we first need to look at large numbers of individual genomes and their individual outcomes. Only with a large number of these correlations under our belt can we predict for a certain genomic variant the probability of a certain outcome by comparison with the large numbers of previous observations. But remember, each individual case amongst these large numbers on which the comparison is based had to grow and unfold the information apparent in the outcome. We could not have predicted it otherwise; there is no shortcut.

In GWAS analyses, single letter changes in the DNA, also called single nucleotide polymorphisms (SNPs) correlate with a certain probability with a trait. The geneticist talks about the penetrance of a SNP or a mutation to cause a phenotype. For example, some mutations affecting a pigment for eye color will cause the change of eye color in 100% of all cases.[32] We say this mutation is 100% penetrant. But few mutations that affect animal behavior fall into this category, as we have discussed before.

When we look for associations between changes at the level of the genome and the level of behavior, context matters. A given SNP may contribute anywhere between 0% and close to 100% to the phenotypic outcome. The reduced penetrance of the effect of a mutation or SNP for the final outcome is influenced by other genetic factors that also contribute to that outcome. Additionally, the contribution of an individual SNP depends on whether it affects intrinsically imprecise developmental processes (fig. 7.2). Developmental biologists like phenotypes better the closer they get to 100% penetrance. A transcription factor like *fruitless* does what it does in most genetic backgrounds. So how about the genetic basis and variability of sexual behavior? Typically, the effect of the genetic background on behavioral phenotypes is substantial, and this is at heart a levels problem.

One of the best examples is the search for a human "gay gene." There is no such thing. And yet, as we have seen, sexuality has a genetic basis, the extent and relevance of which has proven extraordinary difficult to pinpoint in humans. In 2019, the largest GWAS-based effort to that date dramatically highlighted the difficulties involved in the attempt to associate single mutations with a behavioral trait like sexuality.[33] In an analysis of almost half a million (western) people, their sexual preferences and genetic variation, no single gay gene was found. The study analyzed SNPs in the close to half a million genomes. SNPs can be considered mutations, except that they are all normally occurring variants in a population. A few of these SNPs were significantly different in homosexuals compared to heterosexuals, but with very small effects. This means that these SNPs occur slightly more often in homosexuals, but they also occur in heterosexuals, while many homosexuals do not have them. Nothing black and white here. Based on this enormous dataset, the 2019 study also estimated the overall contribution of all such genetic variation as well as the predictability of sexuality based on this estimate of the genetic contribution. And here it becomes interesting. The

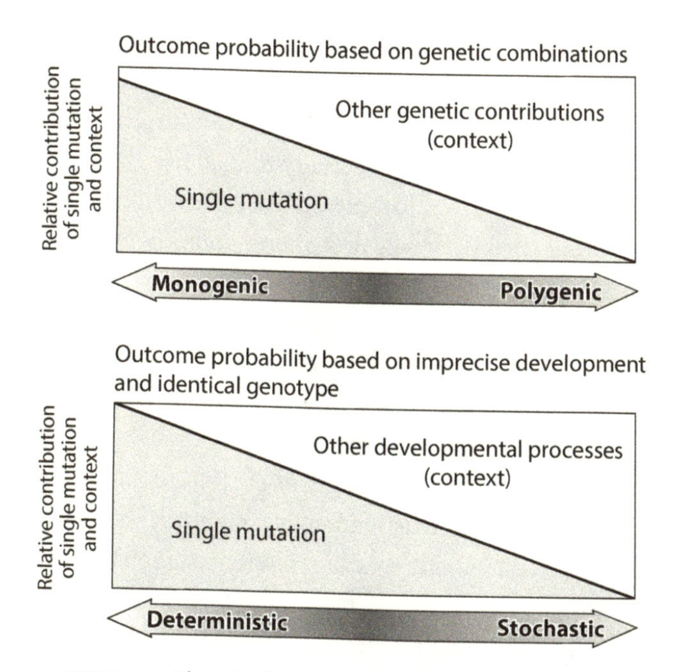

FIGURE 7.2. The role of context in monogenic versus polygenic processes and in deterministic versus stochastic processes. A single mutation leads to a precise outcome in 100% of all cases only if (top) the mutation is fully penetrant and other genetic factors play no significant role, and (bottom) the effect of the mutation is 100% deterministic. Alternatively, a single mutation may quantitatively contribute little to the final outcome if other factors contribute majorly or the developmental process introduces stochasticity. Both phenotypic strength and probability are a measure of the contribution of the single mutation to a composite instruction for the outcome.

overall contribution of genetic variation to sexuality was estimated to be 8–25%. However, when taking all these single mutations as a basis to predict sexuality, the scientists could predict less than 1% of individuals' same-sex behavior. In other words: not only is there no gay gene, but we also cannot predict sexuality based on the genetic code—at least not in these half a million people.[33, 34]

From the perspective of information unfolding during algorithmic growth between the genomic and the behavioral level this is what we should have expected. Hundreds of different gene products, slightly changed in their location or timing of expression, may all change the outcome, but are not likely to do so in a predictable fashion. As a thought experiment, imagine five different SNPs that cause homosexuality only if they all occur together. The probability that all five SNPs are present simultaneously is very low. However, one or the other of the five will occur in thousands of individuals without obvious effects. Hence, many heterosexuals will carry one or more of these SNPs, while many homosexuals will not—a low penetrance association. But this thought experiment illuminates only the tip of the iceberg if we assume that each of the five SNPs causes small effects that simply add up. Many, and probably the vast majority, of genetic effects are much more complicated: maybe two of the five SNPs only have their full effect if ten others do *not* occur. The number of possible interdependencies is countless—and that's not including environmental contributions. Geneticists call this "polygenic" and "nonadditive genetic effects." Polygenic means that the complex trait may be caused by a combination of mutations in many genes, and nonadditivity results from interdependence and feedback.

Here is the catch: Just because we cannot predict the outcome, or associate an individual SNP or a single gene function with the outcome, does not mean that the outcome is not encoded by the genome to a high percentage. In fact, the 8–25% estimated "aggregate contribution" of genetic variations for phenotypic variability may not represent an "upper limit" of the genetic contribution, even though it has been described as such.[34] The nonadditivity of maybe thousands of contributing SNPs that exist in as many combinations as there are individuals can predispose or even determine algorithmic growth to produce a specific trait. In other words, there may be as many different genomes that determine a specific trait as there are individuals. In this case, the individual's genetic code may not allow predicting the outcome (without growing it), nor would a GWAS analysis have much success in finding individual SNPs associated with the outcome.

How about a "gene for monarch butterfly migration"? In 2014, a remarkable GWAS-based investigation for these butterflies was indeed published.[35] Monarchs are originally migratory, but a few populations around the world have become secondarily nonmigratory. Monarchs

in which only a very specific neuron at a specific time is sensitive to the 5% reduction in the enzyme's function (fig. 7.4). In this way, a mild, modulatory mutation that principally affects all cells can cause a highly specific effect that can be meaningful at the level of behavioral output and is both heritable and selectable.

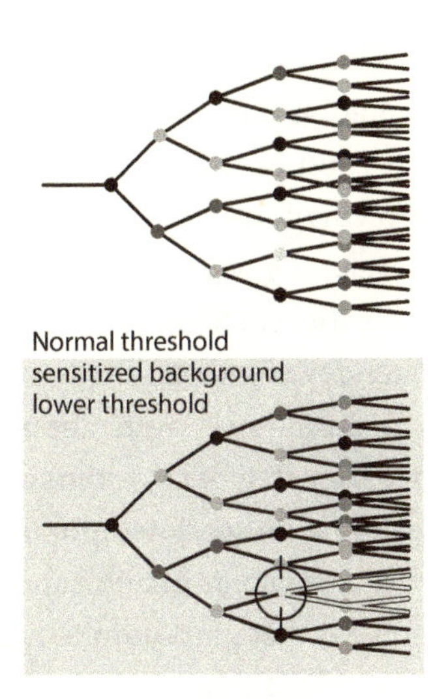

FIGURE 7.4. The effect of a sensitized genetic background. A mild defect (sensitized background, e.g., 5% functional reduction of a ubiquitous enzyme) can have highly specific effects during algorithmic growth. This is because every cell and decision point has a context-dependent different threshold (grey discs). In the example, the sensitized background in the bottom panel only has an effect on the one decision point marked by a target sign, making this defect appear highly specific.

Most mutations, even most single gene knock-outs, do not obviously affect organism development. The majority of the thousands of single gene knock-outs in *Drosophila* are viable, with no obvious defects. Official numbers are hard to come by, but my estimate is that maybe 50% of all fly genes can be completely knocked out without obvious effect on the fly under laboratory conditions. How is that possible? Imagine a mutation in a cell surface receptor gene was below threshold at a particular step in the developmental algorithm. Similarly, a mutation increasing the speed of the filopodium may not be sufficient to cause obvious changes, because faster filopodia may not stick quite as well, but still adhere enough to targets. However, both changes together may create havoc: Faster filopodia that additionally have reduced adhesion to targets may easily create a situation where no sufficient adhesion to targets is possible. The filopodium now operates to a new rule, maybe finding another type of contact that would normally have been competed out. A filopodial speed regulator, maybe an actin assembly factor, and the cell surface adhesion factor have no common molecular mechanism, yet appear genetically redundant. It's a question of genetic background, and that means context.

Context is everything, also for the discussion of whether a gene or mutation is necessary or sufficient to cause a behavioral phenotype. Expression of the transcription factor *eyeless* is sufficient to trigger the growth of an eye, but only if expressed in pluripotent tissues like the early developing wing or leg tissue of a developing fly. Expression of *eyeless* is not sufficient to induce eye development in any differentiated cells during later development. The type of rule change caused by *eyeless* has to happen during an early iteration of the growth algorithm. Flipping the switch at a later step of algorithmic growth may have different or no effects, depending on spatiotemporal context. This is because the eyeless protein contains no information about an eye. The eye represents an information content that must unfold through algorithmic growth from undifferentiated cells. Both *eyeless* and undifferentiated cells are necessary to trigger the growth of an eye. In terms of sufficiency, we could just as well say that undifferentiated cells are sufficient to create an eye, given the context of eyeless presence. From the information-theoretical viewpoint of algorithmic growth this statement is equally valid, however nonsensical it may sound to a developmental biologist. A very similar argument can be made for the expression of the transcription factor *fruitless* and the unfolding of neural circuitry for homosexual behavior in flies. In both cases, the penetrance is 100%, given a specific context. Netrin can be described as a guidance cue with 100% penetrance for *a specific context*. In this case, at least one other gene (*sonic hedgehog*) needs to be gone as well, and we have to look in a specific animal at a specific place—and voila, netrin is the long-range attractor we always wanted it to be.[38] For most mutations, the specific context creates a gray zone of reduced penetrance (fig. 7.2). To call out a gene for a specific behavior always has to come with the relevant context. The depth to which this context needs to be described is arbitrary, but can be meaningfully restricted based on its contribution to penetrance. A SNP or mutation that reduces the penetrance in one out of 10,000 cases may not appear worth describing. Besides, if we go down that deep, the amount of context that requires description blows up the description. Yet, it is exactly these "details" at the genomic level that form the basis of evolutionary programming at the level of neural circuit assembly through selection of meaningful, heritable behavioral change.

A practical way to meaningfully talk about a system without disregarding important detail is to focus on the players and rules at the level of the phenomenon we are trying to describe. A *rule* is a set of instructions sufficient to predict the behavior of a player, the autonomous agent. For example, the rules for synaptotropic dendritic growth could be (see also fig. 4.1): (1) Branch, extend and retract stochastically new filopodia (the agents), (2) stabilize a filopodium that makes contact with a target and (3) restart rule 1 at this target contact point. These rules can be written down and understood independently of the molecules that execute them. Moreover, different sets of molecules can execute the same rules (just like self-avoidance can be executed through different cell surface molecules). This does not diminish the importance of the molecules; they are the executive power.

The simple set of rules described above applies to the dendritic tree and its filopodial protrusions as autonomous agents. The rules for a specific structure at a specific time during development are the product of the molecular complement and physical surroundings. If we knock out one of the molecular components that contribute to, say, target adhesion of the filopodia, the rules change according to which the filopodium operates. A missing cell surface receptor may have been part of a partner recognition mechanism by several such receptors. Loss of just one type of receptor could change the specificity of the filopodium to preferentially make target contacts with a different cell in a different location. Algorithmic growth would go on, synaptotropic growth would continue, only with a different output of this particular algorithmic step with consequences for all downstream steps.

Similarly, a metabolic change, or the downregulation of a cell biological mechanism that is required for fast filopodia, may be sufficient to change the rules for an algorithmic step. Filopodia that are slower can have an increased probability to make a contact. If such contacts are possible between several partners (as opposed to only a specific pair due to a key-and-lock mechanism), then filopodia speed alone can be a mechanism to make certain partnerships very unlikely, while others can be favored through either increased filopodial probing or reduced filopodial speed.[39] Hence, it may truly be the entire molecular composition of the filopodium at a specific time and place that affects connectivity in the neural network. As the cell progresses through the developmental algorithm, the composition changes, filopodia may cease to exist altogether and new molecules define new rules for synaptic maturation and ultimately function.

Next level up, function itself is a higher-order property of neurons that plays an important role in the self-assembly of neural circuits. Once a neuron arrives at the part of its developmental program that renders it excitable, spontaneous functional activity may occur. And the information of correlated activity may then be used to select appropriate connectivity based on initially exuberant synaptic connectivity, as proposed for the vertebrate visual system in the pioneering studies by Carla Shatz and coworkers.[40] Of course genetically encoded components—ion channels, transmitter molecules and all the rest—define the functional properties of neurons, and correspondingly these properties can be altered through manipulations at the genome level. But the phenomenon can be, and needs to be, described one or two levels higher, where the "functional elements of the nervous system and the neuronal circuits that process information are not genes but cells."[3] This realization allows us to describe the higher level property independently of the molecular details. McCulloch and Pitts could describe neuronal connections without knowledge of the molecular details. Herein lies the hope of artificial neural network design since its inception: our ability to shortcut the molecular details by describing higher order properties at "their level." For many applications, these higher order simplifications are not only sufficient, they can actually be superior to the messy, unnecessary complicated biological reality. So, how deep do we have to go in our consideration or simulation of the details? What shortcuts can we take? This is the problem the OpenWorm project faces, and it is only concerned with a simulation of mature function. In order to implement the functional simulation, or most ANNs, the favored shortcut is this: we put it together by hand instead of letting it grow. The entire growth process is rather a big thing to leave out. Can it be done? Let's try to find out—in the last session. Thank you.

3

Brain Development and Artificial Intelligence

3.1

You Are Your History

The Eighth Discussion: On Development and the Long Reach of the Past

ALFRED (THE NEUROSCIENTIST): It's really interesting how this happens: suddenly cells are the focus of circuit function and molecules the focus of circuit assembly. As always, it should depend on the question, but fields have their own dynamics. Maybe you should focus more on cells than genes, Minda . . .

MINDA (THE DEVELOPMENTAL GENETICIST): As you said, it depends on the question. Of course we are studying circuit assembly at the level of neurons as well.

AKI (THE ROBOTICS ENGINEER): So you really do loss- and gain-of-function experiments with neurons?

MINDA: Well, that has mostly been done in the early days. Classic ablation and grafting experiments. Like cutting away half a tectum, as we have seen. These classic experiments were pretty limited. We can now ablate specific neuronal populations during development with more precision, but the results are typically still quite dramatic.

PRAMESH (THE AI RESEARCHER): As we would expect if you take out an entire team during a game of algorithmic growth. But can't you now take out individual neurons with more spatial and temporal control? The question of cellular dependencies seems interesting to me, independent of the molecular mechanisms that execute these dependencies.

MINDA: The molecular mechanisms are unavoidable. Maybe Alfred should rather be worried. Simply making neurons fire or preventing them from doing so may not capture all neuronal functions either, right? How about neuromodulator regulation and diffusion, for example?

ALFRED: I totally agree. But even though we may be missing some circuit functions, it is still a good idea to test how far we can get by manipulating neuronal activity.

MINDA: Well, people in my field are asking both cellular and molecular questions. If I submitted a paper on cellular dependencies alone, I think that paper would be lacking molecular mechanisms.

AKI: Not if I would review that paper. Maybe we should start sending papers to people outside the bubbles of their own fields . . .

MINDA: For that to work we would need a common language. You wouldn't like all the genetics terminology. It's gotten rather complicated.

AKI: I would not, indeed. Probably never will. So, after all of this, we may have become better at appreciating each other's bubbles, but have we come any closer to how we can talk to each other?

ALFRED: I say, we are all still here . . . it doesn't feel like it was all that bad . . .

PRAMESH: Well, it seems to me, for biological brains we have some consensus that the wiring diagram contains information that has to unfold during growth. Development needs to happen for the endpoint to contain the necessary information. And only evolution can reprogram such a system.

AKI: Ok guys. I don't want to grow robots. I want to build them. Minimally, I'll need a lot of shortcuts.

PRAMESH: Well, for biological systems, shortcuts don't seem to work very well.

ALFRED: With 'biological systems' you mean 'brains'?

PRAMESH: I am thinking of the OpenWorm project we just heard about. It fascinates me. Only 959 cells—that doesn't even sound too hard. The question is obvious: How much do you have to simulate to make it work?[1]

AKI: It depends on how much you want it to work. A simple artificial neural network of 959 cells is a piece of cake.

MINDA: *C. elegans* only has 302 neurons; the rest are cells that make up the rest of the worm.

AKI: Even easier then. But I guess they want to simulate the entire worm, not just its nervous system. What I meant was this: You can perfectly simulate chemotaxis of a population of thousands of individual bacteria without simulating the metabolism of the bacteria. That's an unnecessary detail at the level of the simulation and should be cut short. The question is again what you really want to simulate.

PRAMESH: They say they want to simulate the first real digital life form.

MINDA: Then I can tell you right here and now: There is no shortcut. Anything short of a simulation of every molecule throughout the entire developmental process will fail to recapitulate fundamental properties of the whole.

AKI: Minda! 'Properties of the whole,' is that still you?

MINDA: Why? I've told you all along you need the details of all molecular mechanisms.

ALFRED: Wait, I'm with Aki on this one. You don't need all the detailed molecular mechanisms to simulate a neuron! Think of the McCulloch-Pitts neuron. Or an action potential based on Hodgkin and Huxley. We actually understand these things better if we ignore the molecular details.

AKI: Exactly. We can simulate great neural nets without simulating synaptic molecules. We actually think about the machinery just as the hardware—it doesn't matter what hardware the neural net is running on.

MINDA: If you want to understand the behavioral output of a biological circuit, you'll need things like neuromodulators, state-dependent functions as we call them, even those based on metabolism. The hardware becomes part of how it works.

ALFRED: That's true. But again, neuronal activity is a good way to start. Pretty often we can explain circuit function just based on architecture and excitatory and inhibitory neurons, actually.

PRAMESH: I think we got stuck on two questions here: First, what level do we need to understand or simulate? Do we really need every molecule and when can we use shortcuts? And second, do we need development to understand or simulate a network? Do networks have to grow to be, well, intelligent? Here, again, the question is: When can we use shortcuts?

MINDA: History has a long reach, and it required all the detailed happenings to happen. Development has a long reach too, and it required all the molecular and cellular details to happen. There is no presence without memory of the past.

AKI: Minda, this is getting positively deep! As the 'Ruler of the Universe' said about the past: 'a fiction designed to account for the discrepancy between my immediate physical sensations and my state of mind.'[2]

PRAMESH: Indeed. Ray Solomonoff, the inventor of algorithmic information theory said much the same: 'Memory is what you invent to explain the things that you find in your head.'[3]

ALFRED: I love that, but when we study circuit function we are really not interested in how it came to be, just in how it works now. We are not interested in the precise machinery the system runs on either. That can be replaced, just as Aki said.

AKI: Definitely true for AI in robots. We can do away with all your molecular details as long as we have the essence of synaptic strengthening and weakening in ANNs that do the job. And it doesn't matter in what language it is implemented, that can be replaced.

MINDA: That reminds me of where we started—your artificial neural networks are stuck in the pre-Sperry days. I think the key argument has to be evolutionary programming: you need all the details during development to produce selectable outcomes.

Maybe the lengthy training period of your artificial neural networks can resemble or recapitulate at least a little bit the developmental period? I consider learning continued development.

PRAMESH: An interesting thought. The programming of an ANN is really the training, one data point at a time. It takes lots of energy and time and the order, the memory, matter.

AKI: If training is development, then I'm cool. But this one may need some more thinking through . . .

ALFRED: Hey, even if a certain circuit architecture only looks the way it does because of its evolutionary and developmental history, that's fine. I still only want to know how it works. When I look at function, I just don't care about how it evolved or developed or how it was programmed by evolution . . .

MINDA: It's not that easy. When you describe a circuit, you always make the point how all connections need to make sense for the thing to work.

ALFRED: Of course.

MINDA: But how about connections that don't actually make sense? Are there no connections or architectures or features that you have always wondered about, but that do not actually make functional sense? I am looking for the historical burden of development. Since you only report those circuits and connections that work, you are biased in what you look at . . .

ALFRED: This sounds like payback time for your bias in analyzing cell surface molecules!

MINDA: Be my guest! Again, I bet many circuit architectures are not the way you would build them if you could design them from scratch.

ALFRED: I guess we are still trying to figure out the actual role of cortical columns . . .

PRAMESH: Really?

ALFRED: Yes, but it's as always—just because you haven't understood something, or it seems dispensable in certain cases, doesn't mean it is not important. Remember, lots of neurons can die and the cortex still functions fine, but that doesn't mean that those neurons had no function. We just don't know yet.

PRAMESH: Hmm. The question is how the developmentally grown structure impacts function . . .

ALFRED: . . . and even if it does, as it well may, the function is what it is and can be studied no matter how it once came to be . . .

MINDA: But then you may not understand why it functions the way it does. We like to assume that a network is the best solution to a problem. But it is only the best solution given the evolutionary and developmental baggage. The burden of its history.

PRAMESH: Very nice—there is algorithmic unfolding of information in all of this. Development is growth, and function is growth as well.

ALFRED: Say what now?

PRAMESH: Well, the order of input matters. If we train an artificial neural network with input data in a different order, the network will be a different one. That's obviously the case in the brain. Every new bit of input you receive is compared and sorted and processed according to what is already there. If I tell the same story to Minda and Alfred, different things get stored in each of your brains, don't you think?

AKI: Yeah, they are both sure to confirm their biases . . .

ALFRED: But how is that algorithmic?

PRAMESH: The state of the system at some time point t is the basis for input-specific changes at time point $t+1$. Information gets stored in a sequence. Information is also retrieved in a sequence. Can you say your telephone number backwards?

ALFRED: No, but . . . hmmm. Decidedly hmmmm . . .

MINDA: Well, this is just another way of saying that developmental history matters. You are your history.

PRAMESH: And an artificial neural network is its history, too. Both function based on their histories.

ALFRED: But when I try to understand neural circuit function and behavior, I do not want to understand your personal, idiosyncratic history. Nor the personal history of a lab rat. I want to understand the general way it works.

MINDA: Unless the general way it works is the outcome of its developmental history.

Seminar 8: From Development to Function

The human eye is a remarkable structure. From an engineering perspective, it represents unusual solutions to several challenges. For example, the lens is marvelous. When human engineers build cameras, they make the lens moveable towards and away from the detector in order to focus. The alternative is to change the curvature of the lens, but that's much harder: the curvature has to be very precise and evenly changed in order to bundle light rays in a single spot. And yet, that's what our eyes do. The more straightforward solution exists in biology too: fish for example have evolved lenses that can move forward and backwards, just like in our engineered cameras.[4] So, why does our eye work the way it does? Biologists like to answer: if it evolved that way, it must be good for something. And biologists will usually find something if they search long enough. "The thing exists; therefore it must do something that evolution favored" is a biology mantra. But evolution has to work with constraints that can lead to surprising results. Selection can only operate on changes of existing structures. These changes are typically small, leading to the idea that there must be "paths of gradual change" for evolution to produce a certain phenotype or structure. The eye in particular has featured as a key example in a debate that is as old as Darwin's idea. Richard Dawkins made it a central discussion point in *The Blind Watchmaker*. Can selection of gradual changes lead to a structure like the eye, or is the eye such an intricate functional unit that it can be described as having "irreducible complexity"?[5] This is not an open question. We now have overwhelming evidence that there is no such thing as irreducible complexity in evolved biological structures. Rather, we are dealing with our own brain's irreducibly failed intuition.

We have the same problem with our intuition when we think of a mutation that affects a specific behavioral change. The key to our argument was the deeply nonintuitive concept of unfolding information through algorithmic growth. And this brings us to the algorithmic nature of evolutionary change: the output of the previous state has to be the input of the next gradual change. There are no mutations that can change our metabolism to stop depending on oxygen or bring back the fins. There are mutations that can cause you to have more or less fingers, but they will still be fingers. If tomorrow's selection pressures favor something you can swim with, while fingers are suddenly a terribly bad thing to have, whatever change may lead away from the fingers will have to operate on changing the way they look now. The outcome of that change may look like nothing on earth and may be very well adapted to whatever new demand has arisen, but it will also still carry a history of what it once was. In fact, our embryonic development retraces key remnants of our own evolutionary history. Famously, early embryos of very different vertebrates like fish and chicken look very much alike (fig. 8.1). We all start with similar vertebrae, including tails. Biologists like to quote Ernst Haeckel: "Ontogeny recapitulates phylogeny."[6] The spine of our ancestors is preserved in our history. Your spine is still subject to gradual change, but it won't turn into an exoskeleton any time soon.

Evolution has to work with the output of the previous state, no matter how inappropriate that state may be for what is currently under selection. This simple observation can lead to strange phenomena that will have a human engineer's eyes rolling. The eyes, in particular. The light rays that go through your squeezable lens are focused on the detector of your eye, the retina. But the retina is not only a layer of light-sensing cells. Instead, the retina is itself an astonishingly complicated neural network at a level of complexity similar to what we find in major brain regions of insects.[7] The retina is not a CCD sensor; it is an entire vision-processing computer. As a consequence, the signals sent from the eye to the brain are not a picture of reality, but instead substantially pre-processed data. We do not see the world as it is, but modified data that emphasizes certain features, for example contrast enhancement. The evolutionary assumption is that we see the way we see because it has been selected for being successful, not for being a faithful representation of the world.

FIGURE 8.1. A part of Ernst Haeckel's classic comparative embryo development picture from *Anthropogenie*, 1874. The common early ancestor of vertebrates left its trace in the early development of embryos: the early stages of fish, salamander, turtle, and chicken look almost identical. This common state of the developmental program must serve as input to subsequent developmental steps. Any fish- or chicken-specific evolutionary solution to adapt their bodies, physiology, and brains must develop from this state.

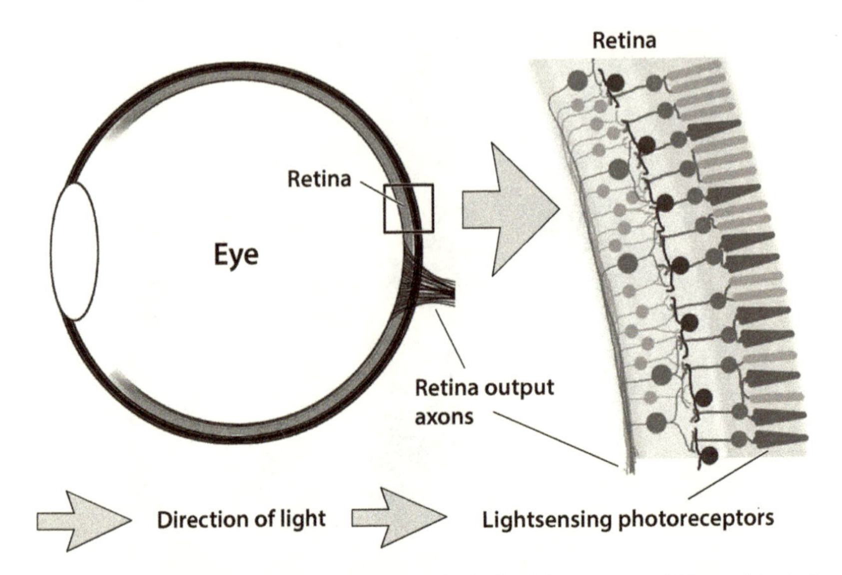

FIGURE 8.2. The vertebrate retina. The retina at the back of the eye is a multi-layered intricate neural network. The photoreceptors are at the far back end, thus requiring light to go through the output axons and several layers of neurons before it reaches the light-sensing elements. This is not for functional reasons, but because of its serendipitous evolutionary history.

The neural network of the retina is a beautifully intricate and practical system to study both the development and function of a tractable neural network architecture. The retina consists of several layers of cells, starting with the photoreceptors themselves, several types of interneurons and two separate major layers that contain the synaptic connections of all these different cell types.[7] The picture is reminiscent of a two or three layer ANN architecture, including the stochastic distribution of input neurons. There is only one glaring problem: the layers are the wrong way round. The input layer, i.e., the light-sensitive cells are all the way back. Light coming through the lens has to first go through thick bundles of axonal cables from the retina output neurons, then through all the layers of interneurons and the two major layers containing all the synaptic connections, then through the non-light-sensing bits of the photoreceptor cells themselves, until the light finally reaches the light-detectors (fig. 8.2). This seems crazy. The architecture is a bit like a design for electrical wiring for ceiling lighting, but instead of hiding the cables in the ceiling, they all dangle in the room in thick bundles in front of the lights. This design not only requires all the light to go through the cables, it also has the additional problem that the cables need to go back through the ceiling somewhere to connect to the rest of the wiring in the walls. This is why we have a blind spot: there is a substantial area of the retina that cannot have photoreceptors or any other cells, because all the retina output cables need to exit the illuminated retinal space to connect to the rest of the nervous system. The only reason why people with two eyes do not continuously see a grey, blind area in their visual field is because the brain fills in the missing information of the blind spot with parts of the visual scene seen by the other eye. Why would evolution come up with this difficult, clearly suboptimal design, including having to solve new problems like the blind spot, in the first place? What is it good for?

Nothing, really. "*What is it good for*" is arguably not even the right question to ask, because you can always come up with some explanation if you try hard enough. But whatever reason we may find will be a secondary rationalization. The primary reason for the existence of the crazy eye design is not because it is good or bad for anything. The eye has the light-sensing elements on the wrong side simply because of the *burden of its evolutionary history*. The human eye looks the way it does because of the algorithmic nature of evolution.

What is algorithmic about evolution? Evolutionary change happens in steps, however small, following a principal output-input relationship, step by step, in a necessarily sequential manner. And the process takes, of course, energy and time. Lots of those. We are our evolutionary history as much as we are our developmental history.

The evolutionary biologist Stephen Jay Gould was fascinated with evolutionary outcomes that cannot easily be explained with the Darwinian idea of adaptation through mutation and selection. Of course the eye is an adapted structure and indeed it is fully consistent with its evolutionary history. In fact, evolution is the only concept that actually explains why it looks the way it does based on observable data. But adaptation has to operate on previous outputs in an algorithmic process. The reason why the human retina is the wrong way round is a result of an embryonic tissue folding event that was 'locked in' by evolution in our ancestors. Like our metabolic dependency on oxygen, this change was so fundamental that evolution found it hard to reverse. There was simply no evolutionary path open to gradual mutational changes to flip the retina back. If only that ancestral change would not have been, then it could look like the camera eye of squids and octopuses. In a textbook example of convergent evolution, the ancestors of these animals started the algorithmic path to eyes that look almost like ours, except that the retina is the right way round. It can be done.

When Gould looked at features produced by evolution, he wondered: if evolution had to work with what was available, then some things may just occur as side effects, without really being themselves an adaptation. Both evolutionary and developmental constraints can lead to structures looking a certain way, not because it is the best way to build that structure, but because this was the path evolution took based on what was there before. This powerful idea led Gould to coin two important terms: *spandrels* and *exaptation*.[8, 9]

When the Italians built the Basilica di San Marco in Venice, they employed the ancient tradition of building self-sustaining arcs. Whenever an arc is built to support a roof, the arc's curvature leaves a triangular space between itself and the rectangular surroundings, called a spandrel in architecture (fig. 8.3). Gould and his colleague Richard Lewontin used the metaphor of the spandrel to describe side-effects in evolution.[8] Over the centuries, an art form developed to use the architectural spandrels of buildings like Basilica di San Marco, otherwise useless spaces, as canvasses for a new form of art. The artists didn't invent the spandrels and they didn't ask for them. They used them because they existed. This is what Gould called exaptation. If something comes to be as a side-effect of something else, then it will be a necessity under selection by proxy, a structure reliably available for further evolutionary change. In other words, the spandrel may have come to be without itself being useful or a selective advantage. Yet, its mere existence is a basis for new developments that themselves may well become a selective advantage. For example, the lens of the human eye is perfectly clear, because it is made of proteins called crystallins. Evolutionarily, crystallins are

FIGURE 8.3. Spandrels (asterisks) are an architectural side-effect. The "empty" spaces created by placing an arch inside a rectangular frame has historically served as an opportunity for artful elaboration. Stephen Jay Gould drew the parallel to how evolution can exploit (exapt) opportunities that originate as evolutionary side-effects.

derived from enzymes that happened to be at the right time and place to be *exapted* for a new function.[10]

Brain wiring is no less a product of evolution and development than the eye's lens and retina. The connectivity pattern of the brain is both an adaptation to functional requirements as well as a consequence of its idiosyncratic evolutionary and developmental history. Here again, the study of visual systems has featured prominently in history, as we have seen in the last sessions. In the third seminar we encountered a remarkable wiring pattern called neural superposition, where six photoreceptor neurons that see the same point in space are wired together and make synapses with the same postsynaptic neurons in the brains of flies.[11] From the perspective of the postsynaptic neurons it does not matter where exactly the presynaptic photoreceptors are in the eye, how complicated the entire arrangement is and where exactly the cables are, as long as they are available as partners. The synaptic unit where the six input-equivalent axons connect to the shared postsynaptic neurons is called a cartridge, and it has a surprisingly stereotypic architecture (see fig. 3.5). The six input axons are arranged in a precise order, with type 1 neighboring type 2, followed by types 3, 4, 5 and 6 finally neighboring type 1 to close the circle. Functionally, the six photoreceptors could have any arrangement, since they carry the same input information and make the same connections. So, why is the arrangement so much more precise than it needs to be? Just because we do not understand it and most likely have incomplete knowledge does not mean that there is not some functional reason somewhere for the precise arrangement. It is possible. However, we may do well not to be caught in the thinking of "it exists, so it must be good for something." It is also possible that the precision of the seemingly functionally irrelevant arrangement is a necessary side-effect of the way the wiring pattern develops. Indeed, a few years back a simple set of rules were described that would produce the precision of the arrangement as a side-effect of the developmental algorithm.[12] But even if this proves to be the case, remember that the arrangement may still be exapted for some functional feature, if only in future evolution.

It is currently unclear how important developmental constraints are for the field of connectomics, the study of wiring diagrams in the brain. Thorough analyses of the efficiency of neural designs have led to the suggestion that the developmental burden cannot have been all that bad and spandrels should not be too common.[13] The temptation will always be to assume functional significance for specific connections or topological arrangement in a wiring diagram. There is an engineer in all of us. But the biological wiring diagram, in contrast to the electrical counterpart, is not an engineered product, but a consequence of its growth history, its developmental algorithm. The biological wiring diagram is not encoded by endpoint information, but the outcome of unpredictably unfolding information. Interestingly, variability in the final, adult biological structure can reveal something about how the information is encoded. But a precise outcome can also hide a principally imprecise developmental path that imposed structural constraints on the final network topology. Brain connectivity is a consequence of both its evolutionary and developmental history.

One the best known structural elements in our own brains are cortical columns. As we have seen before, much of the brain is organized in a columnar and layered fashion, and the neocortex is a prime example. Columns were described by Vernon Mountcastle as structures that looked very much alike whether he looked at the visual, auditory or somatosensory cortex.[14, 15] Cortical columns were the subject of Lashley's worrying equipotentiality of cortical function. And cortical columns became textbook knowledge to all neuroscientists when David Hubel and Torsten Wiesel performed the Nobel-awarded experiments on the plasticity of the visual cortex. Hubel and Wiesel could show that a patchwork of neighboring regions in the cortex responded selectively to the left or the right eye in cats.[16, 17] Remarkably, preventing input from one eye led to an expansion of the cortical columns responsible for the other eye.[18] If you don't use it, you lose it—and if you use it a lot, you gain. These experiments have become a staple in undergraduate neuroscience education. But what happens when there is no input either from the left or the right eye? The surprising answer: the columnar architecture still forms—and is basically normal.[19] Activity is not required to make columns per se. Plastic changes only occur if there is more input from one side than the other.

The beauty and importance of cortical columns lies in their visualization of plastic changes in the brain, and the fundamental nature of this discovery has shaped neuroscience to this day. However, when it comes to the actual function of the columnar organization, the situation is less clear. The problem is this: it is not obvious whether the structure is functionally really needed. Squirrel monkeys have wonderful vision that seems to be equally good between individuals. However, when a squirrel monkey's brain is dissected, one may or may not find anatomically recognizable columns. Bafflingly, there is no way of predicting whether the structures are there other than opening the head and looking, leading some scientists to openly question the functional necessity of cortical columns in general.[20] Not surprisingly, the debate is ongoing and suffers from the conundrum that we cannot exclude something based on incomplete knowledge.

Whatever the functional significance of cortical columns, we must respect that the anatomical structure is a consequence of its developmental history. And once development is part of the equation, other arguments come into play. For example, the ease or speed with which synaptic partners are brought together at the right time and place to facilitate synapse formation between neurons can be an evolutionary advantage. In zebrafish, a mutation in a cell surface receptor called robo2 (a classic example of a protein considered a guidance cue) leads to loss of the layered organization of a specific brain region.[21] Surprisingly, the functionally important wiring in this brain region, direction selectivity that ensures correct visual behavior, is unaffected by this gross aberration of the anatomy. In this case, the researchers found that wiring specificity is still achieved, albeit at slower developmental speed. This example is consistent with the idea that neuronal arrangements observed in the adult structure, in this case the appearance of layers, may be a side-effect of a more efficient developmental process. Developmental speed is certainly a selectable advantage, and the layers may be said to serve the function of speeding things up. But the layers themselves may not be strictly functionally required—and their interpretation only makes sense in light of their developmental history.[22]

The brain, just like anything else in biology, is what it is because of its history, both in evolutionary and developmental terms. The problem with history is that it is messy. How do we know what exact parts of the anatomy, the physiology, the biochemistry and the wiring specificity is really functionally important? And if there is functional importance, is it really the best solution or just another weird biologically messy retina the wrong way round? Something along those lines may have been on the minds of the computer enthusiasts trying to develop AI in the '80s, when they took pride in ignoring neurobiology. "We can engineer things much better," they may have thought, so why learn about the nonsensical solutions biology had to put up with? An engineer can design something from scratch, whereas evolution and development cannot. Evolution is constrained by the need to work through gradual changes of existing structures, however inappropriate they may be for a new task. Developmental growth similarly builds new things only on the basis of what was previously built, step by step, and partly retracing its evolutionary history. It is all beautiful and interesting to study, but it certainly is not always straightforward and likely to contain unnecessary complications.

This kind of messiness is not commonly embraced, certainly not in engineering. But then, the very messiness of the developmental program, the very unpredictability of unfolding information of the feedback-rich interactions of molecules and cells that act as autonomous agents are what made Benzer's mutations possible. The idea that a mutation in some widely expressed gene can reprogram the brain requires the full unfolding of information during algorithmic growth. How much messiness do we need? How much messiness can we take?

Jeff Hawkins describes his quest to understand the brain as motivated by the hope that its intricacies may hold the key to building truly intelligent machines. In his book *On Intelligence*, he describes his attempts to join the AI community at MIT in the '80s with the idea that, surely, there must be lessons to learn from the brain. He describes how he met with widespread disregard for the biological messiness of real brains in the AI community at the time, which led him to seek his fortunes elsewhere.[23] Fortune came, allowing him to revisit his passion many years later by founding first his own research institute, and later a company—both with the twin goals of understanding and exploiting how the brain, and more specifically the neocortex, functions.

Hawkins put forth a proposal for the function of the neocortex as a *memory-prediction system* that exhibits remarkable similarities to our discussion of the adult structure in light of its history. Hawkins' idea is that the brain does not really *compute* solutions. Computation, that is the way all human-engineered computers work based on the von Neumann architecture, easily requires millions of steps for a complicated problem. As Hawkins points out, this leads to a whole series of incompatibilities with how, and especially how fast, the brain operates. Instead, Hawkins' idea is that new input is compared to existing memory, which continuously allows predictions about the next thing that is likely to happen. When you run down a familiar flight of stairs, the memory-prediction system keeps you safe based on the memory and the prediction that there will be another step. If, unexpectedly, the next step would be an inch further down, the surprise of a failed prediction kicks in. Hallmarks of cortex function according to Hawkins are therefore that memories are stored in a *sequence* (try to sing a song the wrong way round, or say the alphabet from Z to A), memories are *auto-associative* (a little bit of the stored memory is enough to recall the whole thing), and memories are critically based on *invariant representations* (as any fuzzy input is enough to recall the full, invariant memory).[23]

The idea of the sequential memory storage and access is fundamentally an algorithmic phenomenon: I often cannot access a certain memory without the preceding step in the sequence. This leads me to rephrase Hawkins' idea and enter a discussion about the relationship of development and function: maybe memories are stored as algorithmic information, as algorithmic rules, that can "recreate" the (invariant) representation. In other words, the brain may not store the memory as such, but the algorithmic information required to recreate it. In this model, memory storage (input) and memory retrieval (output) are difficult to separate; they are both parts of the same process. An input is required to trigger the retrieval, and this input simultaneously changes the memory, as it is itself an input that feeds into the same network. The same exact memory cannot be recalled twice. Some aspect of the memory, its invariant representation, clearly remains robust and similar every time I recall it, but any particular input influences what and how I remember in detail. There may also be a case against absolute invariance of memories. The longer different inputs change the network, the more a memory even of my grandmother will change. There are certain aspects that will remain more robust than others, like my grandmother's facial features, but I doubt they are truly invariantly stored.

If I envision every input to set the entire "talking-to-itself" machinery of the brain (neocortex) to be set in motion, I cannot help but have immediate associations. I immediately have a full picture of what a little fragment reminds me of (the invariant representation), but I also cannot help but simultaneously form a new memory that mixes with the new input. This process happens every minute, every second, with everything I see, hear or smell. Now, all these uncountable new bits of memories do not get laid down as well packaged memory units, or bits of information, somewhere in the strengthening or weakening of specific synapses. Synapses clearly do get strengthened and weakened, but these are the same synapses that are likely involved in retrieval—and they are the same synapses that will be part of uncountable other memory storage and retrieval processes. Every new input that comes through our senses may be treated more like a wave that propagates through the network and alters an unknown number of synapses that all are part of many other memories. This does not mean that all synapses get activated all the time; the "unknown number of synapses" may yet belong to a very low number of neurons in the vastness of the cortex. Certain neurons are activated more than others during the formation of a specific memory. Remarkably, we know that activation of a single neuron can be sufficient to trigger a memory, at least in mice.[24, 25] However, loss of that same neuron is not likely to affect the memory itself. In this kind of memory system, information storage must be at least somewhat decentralized and no component can be absolutely necessary. This does not mean there is no change. If a neuron is lost, some minute modification may occur to when and how exactly a memory is recalled or associated, as any new wave (input) propagates through the slightly altered network. But the invariant representation remains robust. Finally, the input does not even have to come through the senses, but can also be triggered through the workings of the network itself. The brain cannot help but talk to itself, and this is arguably most of what it does continuously, with sensory input only providing occasional modulations.[26]

So, what does this excursion into *algorithmic function* mean for our problem of how "to get information" into the brain? How do you *program* the network? Well, it is impossible to store new information without changing other information in the brain through new associations and reassociations. Information *has* to enter by changing the sequential, auto-associative network, which means it must change algorithmic information—it changes the rules. If memories are not stored as entities at all, but only as algorithmic rules sufficient to recreate them, then the basis for robustness, flexibility and variability differs entirely from the von Neumann computer architecture. We are reminded again of the rule 110 automaton, where the information (and in fact the only way to retrieve the information) is not stored, say, for the state of the system at iteration 123,548, but instead in the set of rules that are sufficient to recreate that state given 123,548 iterations. Of course rule 110 automata are deterministic and comparably boring, whereas the algorithmic storage/retrieval process in the brain is probabilistic and plastic. New information can be embedded in a network of other memories. This associative embedding process of new memories bears similarity to sending the wave for unpredictable propagation on a journey into the network. The input will self-select the associative memories it triggers, the neurons and synapses that will be particularly activated, and what synaptic weights may change. Storing is changing is retrieving. Such a mode of algorithmic network function would have serious consequences for the storage of a four-digit PIN code in the brain. The memory of the PIN changes, its associations and retrievability for example, although one hopes the actual four digits do not change too much with time. We try to memorize, reiterate and keep the four digits as robust as possible, although even this fails at times, especially if we have not used the PIN for a while. Associations with the PIN change as well, a fleeting glimpse of a picture of the card or an ATM where I used it last or a sound or feeling of keys I pressed when I typed it last. I cannot prevent these associations; they are interwoven parts of the memory of the PIN.

Experiences change the way we think. Think of how having a child changes a parent's adult brain over time. Every day sculpts the brain to integrate the information content "child"—to the extent that its existence influences every decision in an often imperceptible way. A "grandmother neuron" does not need to contain the information of a grandmother to be activated as part of a network effort recreating the memory. The information "child" or "grandmother" becomes an integral part of the network itself. There are no single neurons that can be killed to remove the "child" or "grandmother" information content. The memory is not stored in a data bank separate from the processing unit (the von Neumann architecture). Rather, the information becomes part of how the processing unit processes information itself. The information cannot be separated from the processing unit—it is not data that we recall and compare when making decisions; it is part of how our individual brains have wired and rewired themselves over time.

We create and program information content in the brain through usage. The more I keep on remembering (simultaneously storing, changing and retrieving) in many different contexts, the more robust and multiply-associated the memory becomes. This also means that feeding an input into the network will not simply create a robust memory of that bit of information. Just reading the PIN once may or may not be enough for me to be able to robustly retrieve the number later. There is no shortcutting of time to train the network. Similarly, there is no shortcut to many memories other than going through a sequence. I can still play a set of five notes on the guitar that are part of an etude from Bach that I learned to play some forty years ago. But the only way for me to retrieve the memory is to take the guitar and play. I actually have to hold and feel the guitar in my hands and my fingers have to go through the motions. I have no other access, my fingers "remember" the piece, and the five notes are only accessible as part of me playing the piece up to the point when they occur in the etude. If I try to just play the five notes by themselves, I fail. In fact, if I stop in the middle of the piece, the only way to again reach the place where I dropped out is to restart at the beginning and hope that this time the sequential retrieval of all necessary memory for the playing of the entire etude will unfold without interruption.

Programming the brain is training the brain, is learning, is using it. And training is changing the many parts of the network through alterations and embedding of new associations in algorithmic chains, or rather interwoven networks. The leap is to understand what and how certain neurons and synapses get altered and activated when new input information sweeps into the network. An interesting picture I mentioned briefly is that the wave self-selects how it propagates through the network. What kind of selection are we talking about? Scientists have shown that single neurons in a special cortex area can "recognize" specific faces.[27] The single neuron knows nothing about the face, yet it is a key part of the invariant representation of that face in the face recognition area of the cortex. The neuron was there before the monkey saw the face, but the way the wave of information self-selected its propagation through the network, the single neuron became a node and part of the memory. We can say the neuron got *selected,* very much in an evolutionary sense.

Gerald Edelman built his theory of neural group selection, or neural Darwinism on very similar ideas. Importantly, Edelman applied his evolutionary thinking based on the functioning of the immune system to both development and function sequentially. First, during embryogenesis he described *developmental selection,* and next, when the neurons start to function, *experiential selection.* But the heart of neural Darwinism is the third part of Edelman's theory, the idea of *reentry.*[28] Edelman did not mean feedback in the sense of an error correction signal back to the input level, as it is routinely applied using backpropagation in neural networks. Instead, Edelman envisioned a selectional system based on a recursive process of changing synaptic strengths in a massively parallel manner. This leads to synchronization and selection of some subcircuits over others. Edelman linked his theory to both evolution and the function of the immune system, work for which he had previously received a Nobel Prize in 1972. Like one out of the billions of randomly produced antibodies that gets selected by a random antigen (think an unknown bacterium), Edelman saw circuits in the brain under selection. In 1978 Edelman wrote: "It is clear from both evolutionary and immunological theory that in facing an unknown future, the fundamental requirement for successful adaptation is preexisting diversity."[28] We can think of the face-recognition neuron, in the state of the network, given its associations, etc., as the one that gets selected by the face's picture in the brain of a monkey that had never seen that face before. In a way, the face was already in the network, just as the antibody-presenting cells were already in the body, before it got selected through activation of the system.

Why should the ideas of Hawkins' memory-prediction system or Edelman's neural group selection be restricted to the neocortex? How about the memory in the neural networks of spiders that ensure species-specific, stereotypical webs built by different spider species? How about the memory of the monarch butterflies' migration route? What kinds of memories are these? Evolutionary selection was the only way to program these memories into the brain networks of the insects *during development*. Remember the Benzer mutations that could unpredictably, but selectably, meaningfully and heritably alter a certain memory and "reprogram" the network. This train of thought puts evolutionary principles at the base of the retrieval/storage process that is the fundamental modus operandi of the brain—and not only the neocortex. Self-selection, not programming. Is the evolutionary principle the same during development and function? We are facing two histories, the evolutionary history in our genes and the evolutionary history of the functioning brain in our memories. Back in 1978, Edelman felt that "no current theory attempts to relate the embryonic development of the brain to its mature higher functions in any consistent fashion."[28] Where we stand today will be the subject of the last two seminars, based on these key observations:

Both the developmental growth of a neural network and its function are based on algorithmic rules. There is no storage address or bit sequence for a memory, just as there is no such information in the genes that encode the growth of the neural network. The input wave and the neural network it rides all operate as autonomous agents, following their own rules, leading to decentralized storage and function with no master regulator or entity having oversight over the network. The rules themselves do not encode deterministic algorithmic growth, but principally probabilistic events that allow for flexibility and robustness. And all of this requires time and energy, step by step, as the information of who we are unfolds. We all are our history. Thank you.

3.2

Self-Assembly versus "Build First, Train Later"

The Ninth Discussion: On the Growth of Artificial Neural Networks

AKI (THE ROBOTICS ENGINEER): Now that you got me to accept that the brain is its developmental history, I still don't want to grow robots . . .

ALFRED (THE NEUROSCIENTIST): Still, decidedly hmmm. I can see that there are Minda's 'historical burdens of development' that influence function. That's fine, but that still doesn't mean that I want to know all about them. I still want to study function here and now.

MINDA (THE DEVELOPMENTAL GENETICIST): But you just heard of examples where the developmental history would mislead your quest for function. If you develop an entire research program to figure out the functional advantages of why the retina is the wrong way round, where would you be? Why should that not happen with connectomes?

ALFRED: If it's a remnant of development, I just want to recognize it as such and not have to worry about its history any longer. Anyway, I wonder how prevalent that really is.

PRAMESH (THE AI RESEARCHER): The argument is that evolutionary constraints can produce phenomena like that. This should at least make us suspicious. I think it is a typical biologist's intuition to look for a purpose just because something exists.

AKI: Yeah—you guys like to study X just because X exists . . .

ALFRED: Okay, I accept that. I am just not sure how much it really matters in the end. When people study the retina, they can still understand how it is wonderfully adapted, just based on the inverted design. This is how it is, this is how it functions, this is how we want to understand it.

MINDA: Except when wiring specificity would really be nonsensical, maybe not even serving a purpose, at least nothing straightforward. I like the adaption-exaptation differentiation.

ALFRED: I guess it doesn't hurt to be on the lookout, but it does hurt to get distracted on a tangent.

MINDA: Well, let me know when and how you decide what is tangential.

PRAMESH: I think there is no absolute criterion for what is tangential. It certainly depends on the question.

AKI: How helpful is that? How about some guidance on how to decide when development is important for function?

PRAMESH: Okay, let's start with the continuum from development to function argument. There is a fundamental principle here of algorithmic processes . . .

AKI: Wait, this is all again biology talk. We are successful with feeding information to ANNs to make them learn. As you know, the classic example of supervised learning is a dataset of thousands of handwritten numbers plus the label of what each number is. Train a rather simple net, and it will learn all right.

MINDA: . . . stuck in the pre-Sperry days . . .

PRAMESH: . . . and it really only works well for a specific task. To me, the idea of the 'function as continued development' argument is that new information continues to unfold as the algorithm continues to run. However, If you just want a network to learn number recognition, this may seem less important. Similarly, if Alfred just wants to understand a neural circuit underlying a specific visual task, like motion vision, then a circuit diagram may allow you to come close to understanding it.

ALFRED: Are you sure there is a parallel between the two?

PRAMESH: Neither of you would like to know anything about the growth of the network. And in both cases you look at an isolated subfunction of the way a brain works. Neither is artificial general intelligence.

ALFRED: There is a difference. The artificial neural net starts with random connectivity and then learns. But what I am studying—and I think the motion detection circuit is a pretty good example—is how the underlying circuit is wired up. That's not a circuit that learns much, it is simply wired in a very specific way to compute a specific sensory input, to serve a simple, fixed function.[1–3]

MINDA: Yes, Alfred's network got its information only through genetically encoded development and without learning, whereas artificial networks get their information only through learning and without genetically encoded development. The question remains how the two hang together. The analogy that the learning history is important for the artificial network seems to at least suggest a parallel.

PRAMESH: Exactly. In addition, the motion detection circuit is the outcome of evolutionary programming, ultimately through mutations in the genome. Maybe neural network function is the memory of its evolutionary programming.

AKI: This is getting a bit hard to swallow, my dear.

PRAMESH: Let's try this from a different angle. You want to know shortcuts, you want to know what you can leave out. Alfred wants to leave out the development of the neural circuit when studying function. You certainly want to leave out the development of specific circuitry in an ANN.

AKI: Yeeeeeeesss . . . ?

PRAMESH: How does information get into either system? In the biological neural circuit, information gets wired into the network through algorithmic growth. It literally unfolds in time. And there is no shortcut to that. In the ANN you get the information into the network through learning, one labeled bit of data at a time, and there is no shortcut, at least not if you use backpropagation. And in both cases, if you mess with the order of how information gets into the system over time, you'll change the endpoint information.

MINDA: It seems to me that training artificial networks with labeled data pairs one-by-one is already a step in the direction of algorithmic growth. So, how much more could you achieve if you would let network information unfold through algorithmic growth, instead of starting with a randomly prewired network?

PRAMESH: I think that is precisely the question. What do you think, Aki?

AKI: Whether there are artificial neural networks that grow and add more connections while learning? As you probably know better than me, there are comparably few people working on this. It's certainly not mainstream. It's not used in the AI systems that are used everywhere now. I also know of no task for which they would be better.

ALFRED: Maybe that depends on how narrowly you define the task . . .

MINDA: Right. In addition, obviously our brains grow while they learn. In fact, they grow by learning. There are dramatic developmental changes to the brain of a child as it matures and, well, develops an individual character.

AKI: True, yes. 'Only a child sees things with perfect clarity, because it hasn't developed all those filters which prevent us from seeing things we don't expect to see.'—Dirk Gently, one of my favorites.[4] Anyway, naïve, randomly wired ANNs before training have no preconceived notions either, you don't need growth for that. . . .

ALFRED: Our biases certainly seem to get worse with age. Usually the same biases we started out with.

MINDA: Are we really turning this into a discussion about our biases?

ALFRED: We could . . . maybe we should . . . we certainly know that it gets harder with age to change, and that correlates with changes in the brain. Actually, during most of our life, the number of synapses reduces. You have way fewer synapses than you had as a child. I think of this as part of sculpting a network that starts with incredible plasticity. This is the part you never have in an artificial network with fixed numbers of neurons and connections. Exactly the thing I am studying is what you are leaving out, Aki. Seems a big thing to leave out.

AKI: Learning in an ANN is also a sculpting process, just of the synaptic weights. Give me one good argument why I would need sculpting during a developmental growth process in addition. We can get a visual recognition network to work as good as we need it. Better than humans in fact.

PRAMESH: The sculpting process is clearly a similarity. But what else is your highly trained network good for? If you present the letter 'A' to a network that had been trained only with numbers, then the network will invariably think 'A' is some kind of number. ANNs are biased through their training history.

AKI: People are working on this and other approaches to artificial general intelligence, as you well know. What we do know is that nets that have been trained on one specific visual task typically are better at learning another visual task.

PRAMESH: Yes, I agree. But what if the tasks are as different as a PIN code and your grandmother. We are not particularly good with networks that use the same sub-networks for both . . .

AKI: How about this: amazing things may happen once you throw a bunch of nets together to generate a meganet with new properties. The things we currently get to work are cool enough as it is. We are going step by step.

MINDA: Sounds to me like you are going the first step to growing it step by step . . .

AKI: Ha!

MINDA: I think the closer you try to get to even artificial worm intelligence, the closer you will have to get to biological brains. And they do need to grow. Think about it: the whole neural network revolution of the last few years really is a big step towards coming back home to the biological template.

PRAMESH: There certainly are people who say that today's artificial neural nets have to become more brain-like to overcome current limitations. Geoffrey Hinton, the inventor of backpropagation, for one.[5]

AKI: Artificial general intelligence and artificial human intelligence are entirely different beasts. I don't think it has to become a worm or a fly or a human. I think an artificial intelligence can become as intelligent as a human without being like a human. How about artificial robot intelligence?

ALFRED: I think there are some features of whatever we call intelligence that any neural net needs to get right. For example, I've always felt that artificial neural networks are missing symbols—that's what we call representations of higher order phenomena.

PRAMESH: Not sure I understand. How higher-order properties are represented in the network is a big field of research. Maybe what you call symbols is nothing extra, but simply what happens to the network. I think of representations as distributed in the connections.

ALFRED: I don't hear you talk much about connectomes. All I hear in AI is feeding data to randomly connected initial networks.

AKI: Okay fine, I dare say you got me at least a little bit thoughtful on the idea of evolutionary programming and the development of connectivity. I think we can work on that.

MINDA: You've certainly not come closer to talking to biologists.

AKI: I just said that I started to like our little chats!

ALFRED: More than you like talking to your robots?

Seminar 9: From Algorithmic Growth to Artificial Intelligence

Self-organization is the bottom-up antithesis to top-down engineering. As we have seen in the fourth seminar on autonomous agents, the basic premise of self-organization is the emergence of order through local interactions of initially disordered components. With nobody in charge.

For living systems, self-organization is a dynamic process that seeks equilibrium, but never gets there. Such systems are characterized by feedback, a concept dating back to Norbert Wiener and the birth of cybernetics in the 1940s. Wiener defined cybernetics as "the scientific study of control and communication in the animal and the machine."[6] Remarkably, this definition implies the assumption, or at least the hope, that "control and communication" can be meaningfully compared between animals and machines. At its core, the idea of control and communication deals with information. Kolmogorov, one of the founders of algorithmic information theory, defined cybernetics as the "science concerned with the study of systems of any nature which are capable of receiving, storing and processing information so as to use it for control. "[7] Since its inception, cybernetics grappled with explanations for nonobvious, higher-level phenomena, often called "emergent phenomena." Autonomous agents, operating according to a set of rules, can create such phenomena through their interactions in space and time. Outputs can be complicated, even if the initial input information and rules are simple.

A typical feature of self-organizing systems is the tendency to move towards an equilibrium state defined as an attractor of that system. Attractors can be local or global minima in an energy landscape. This may sound more complicated than it is. When skiing down a slope, you are seeking a valley in such an energy landscape. Energy had to first be put into the system to get you on top of the hill (you probably paid for a ski lift to do that), and this energy is put to work when you effortlessly slide down. If you and your friends wanted to find the longest slope around without knowledge of the terrain, you could try all directions randomly to find out the shorter routes to local minima and finally the deepest valley, the global minimum. But how does any of this help us understand neural networks and the brain?

A key foundation for the study of the brain as a self-organizing system was laid by Ross Ashby (see fig. 4.2) in his seminal book *Design for a Brain—The Origin of Adaptive Behavior* in 1952. Donald Hebb had published his *The Organization of Behavior* in 1949, with the stated goal to develop a "theory of behavior that is based as far as possible on the physiology of the nervous system."[8] Ashby was a psychiatrist, but already in 1947 he had published a foundational paper cementing his more general interest in self-organizing dynamic systems.[9] In *Design for a Brain* Ashby followed this formal, cybernetics-oriented approach without mentioning or referencing Hebb.[10] He asked the simple question how a machine, obeying the laws of physics and described by mathematics, could produce adaptive behavior. His goal, very much in the spirit of cybernetics, was to understand what it would take to build an artificial brain. His key insight was that neurons must "self-co-ordinate" and that a designer of an artificial brain would face the same problem: "His problem is to get the parts properly co-ordinated. The brain does this automatically. What sort of a machine can be self-co-ordinating?"[10] As regards the "gene pattern," as he called the underlying genetic code, Ashby reasoned it could only control "a relatively small number of factors, and then these factors work actively to develop co-ordination in a much larger number of neurons. . . . Thus the work, if successful, will contain (at least by implication) a specification for building an artificial brain that will be similarly self-co-ordinating."[10]

How about ANNs, our fastest horses in the race for AI? As we have seen in retracing some of AI's turbulent history, today's ANNs are closer to an artificial brain than any approach previously taken in AI. However, self-organization has not been a major topic for much of the history of ANN research. When Rosenblatt started it all with his 1958 perceptron paper, he had fundamental questions in mind "to understand the capability of higher organisms for perceptual recognition, generalization, recall, and thinking"—in other words: biological brains.[11] Rosenblatt cited Ashby and von Neumann, but only to discuss how imperfection and randomness can lead to reliable output. There was no mention of self-organization.[11]

The perceptron is the archetype of the feedforward type of neural networks, which also include multilayer perceptrons and convolutional networks, amongst others.[12, 13] In supervised learning, such feedforward ANNs indeed do not obviously resemble self-organizing systems. The neurons in these neural networks do not compete or cooperate with each other to determine structure or function. "We just iterate a simple updating rule, which tells each unit which inputs it should receive, and this impetus cascades down the layers. Training really is more like a domino rally than a self-organizing system." Roland Fleming, who studies visual perception using deep learning at the University of Giessen in Germany, told me. But, adds Fleming: "deep learning is not just one thing"—and refers to a recent in-depth comparison of different neural network strategies for visual perception.[14] Things get more complicated (and lifelike) when neurons in the ANN not only connect to the next deeper layer, as in feedforward networks, but also connect within the same and preceding layers. Such networks are called recurrent neural networks and have a remarkable property: they can deal with sequences in time. The idea is that the state of the system at any point in time not only changes based on current input, but also incorporates information from previous time points, which are reflected in the recurrent connectivity.[13] While many visual perception tasks are well served with feedforward convolutional neural networks, speech and language processing in particular require the recognition of sequences and are better learned by recurrent network topologies. The world of deep learning has witnessed an explosion of network architectures that are suited particularly well for one task or another, all of which are introduced in detail elsewhere.[12, 13, 15] Here, I want to focus on the ways neural network topologies are inspired by biology and when they exhibit features of self-organizing systems. For example, feedforward convolutional networks are based on the ideas of receptive fields and the organization of the visual cortex. But the cortex is, of course, highly recurrent. Some network architectures are now modular, a key feature of biological brains; and modular architectures like the "committee machine" actually "vote" amongst collaborating neural networks.[15]

So, how much self-organization really is there in neural networks? As predicted by the early cyberneticists, feedback and noise are keys to self-organizing features. Even the simplest neural nets already exhibit the first steps towards self-organization based on the interactions and feedback amongst agents (neurons, nodes) of the network. For example, the aptly named and widely used method of gradient descent is a standard, algorithmic procedure essential for the training of ANNs. It is analogous to skiing down by repeatedly following a rule to turn towards the path with the steepest slope.[16] Standard feedforward ANNs utilize gradient descent for the backpropagation of an output error to change synaptic weights upstream in the network, a basis for feedback. The opportunities for feedback are greatly increased in implementations of multilayer recurrent neural networks.[17, 18] Also depending on the implementation, such networks, like many optimization functions, are easily caught in local minima without ever finding the best solution. Artificial neural network designers have done all kinds of things to get the network to search beyond local minima, including adding noise to synaptic changes.[19] Noise can help the ANN to avoid local minima, where it would otherwise get stuck the same way (deterministically) with every identically repeated run. Noise in neural nets can serve the same function as it does for the stochastic filopodial dynamics of a pathfinding growth cone: exploration of unknown terrain in search of the perfect target, the global minimum. Neither process is typically described in evolutionary terms, yet both are based on selection: random exploration of parameter space (a pool of variability) is required for robust selection of a solution. Von Foerster put it this way in 1960: "self-organizing systems do not only feed upon order, they will also find noise on the menu."[20] Add time and energy to a mix of agents, feedback and noise, and self-organization will not be far off. Self-amplification, competition and cooperation are features of self-organization that are now actively being discussed especially for unsupervised learning, another hallmark of our own brain's function.[15] Just because ANNs are not typically designed as self-organizing systems does not mean that they can avoid it.

How does self-organization relate to algorithmic growth? As we have seen in the first session, the ideas of noise and selection (the third seminar), and the local interactions of autonomous agents and local, rule-based interactions (the fourth seminar) are central to both self-organization and algorithmic growth. These concepts apply to individual steps and every iteration of the growth algorithm during the brain's development. And yet, studies on self-organizing systems are not typically concerned with developmental growth. Rather, the focus of self-organization is to understand how a fixed number of components (autonomous agents, players of the game) arrange themselves from a unordered state to a more ordered state, without the help of a master regulator.[9, 21] And if ANN developers do not typically think about their networks in terms of self-organizing systems, they think even less about growing them. ANNs from the 1958 perceptron to today's most complicated recurrent networks have this in common: somebody decides on a number of neurons, connects them randomly and then starts training. The brain's neural networks of course do not start with a fixed number of neurons nor are they randomly connected before they start to learn. This is quite a shortcut. But what exactly are we cutting short?

Algorithmic growth is characterized by a concatenation of steps, and each step may be subject to self-organization. For example, at some point in the growth of an animal organ, an unordered and often quite variable number of cells may engage in a game where every cell tries to tell every other cell to give up the attempt to achieve a certain fate. The rules of this particular game, as defined by molecular signaling mechanisms, are simple: the first cell to receive the "give up" signal will in turn stop telling others around it to give up. We have encountered this type of game in the third seminar: lateral inhibition based on Notch signaling is repeatedly used in growth programs of multicellular organisms to break symmetries and create patterns (see fig. 3.3).[22, 23] This is an example of a self-organizing process with players that transition from a less ordered to a more ordered state based on a mix of noise, feedback, time and energy. During development, the game is made possible by the output of preceding developmental steps that have put the players into the game in the first place. Once the players interact like the autonomous agents that they are, each newly self-organized pattern serves as input for the next step. In the example of lateral inhibition, such a next step could be programmed cell death for the loser and the determination of the growth program "neuron" for the winner. The subsequent neuronal program may include axon pathfinding, filopodial exploration and competitive synapse formation. Algorithmic growth unfolds information in the shape of a concatenation of self-organizing steps.

At different steps of the growth algorithm, different games will be played by different players following different rules. In the context of the developing brain, algorithmic growth is a framework for the integration of a series of continuously changing games of self-organization. I chose the term *self-assembly* to describe the course of such a series of self-organizing processes during an information-unfolding, time- and energy-consuming period of algorithmic growth. The definition and separation of these terms has changed over the decades and the meaning of self-assembly, as used here, differs from current uses in physics and other fields, but also provides an opportunity to compare notes between these fields.[24]

If brain development is a process of self-assembly, how does it relate to neural network function, which is typically considered neither in the context of growth nor self-organization? Few researchers are currently concerned with the developmental growth of ANNs, and they are far from practical applications compared to engineered network designs.[25–27] In the previous seminar, we already discussed function as an algorithmic extension of development. By contrast, current ANN technology is based on a clear separation of a top-down engineering period, followed by flipping the on switch and the start of a distinct functional process. In contrast to biological networks, information enters the ANN mostly during the functional period, and very little as part of its engineering. The idea of a random starting configuration is to ensure maximal entropy, meaning maximal disorder that ensures the absence of information content prior to training. Learning by feeding data or by self-play and reinforcement learning are the sources of information. The network topology and the learning rules do not change. There is no master regulator, but order does not solely emerge from player interactions. Rather, the players interact based on a continuous stream of new inputs. Information enters through the function of the network, not the development of the network.

Let's explore this idea of the separation of development and function a little more. We know algorithmic growth underlies the self-assembly of the biological brain, including the making of synaptic connections as well as their function (remember activity waves and the idea of "cells that fire together, wire together"). In other words, the basic functioning principles, the entire molecular machinery of transmission and plasticity, Hebb's synapse and memory, are all part of the growth algorithm during development. What changes when the brain starts functioning? Maybe function starts at the end of algorithmic growth, the last iteration, to continue for the rest of the brain's life as a single step with one set of rules. This seems to be a common assumption of fields like "learning and memory" and creators of ANNs alike, at least to the extent that they are not concerned with developmental history, only how it works once it works. A car has to be built during its "developmental period" in the factory, but once it is done, I just want to drive it. I may have to put in some maintenance to keep it from falling apart, but neither the car's maintenance nor its function are a continuation of its manufacturing process. Does the brain have a development period that simply stops when it is done, and once it is done, it is taken out of the factory and put to use (while slowly deteriorating and requiring occasional maintenance)?

There is little reason to think that algorithmic growth of biological networks ever stops (unless you are out of time or energy). Neurons continue to transition through new states, each producing output that serves as input for subsequent steps, changing the properties and therefore rules as the neuronal states change. Neurons live long lives, but they do age. Aging is often viewed as a process of slow functional deterioration due to continued use, much like in the analogy of a car. Alternatively, aging can be understood as an active process of progress down the road of continued development.[28] Clearly, the developmental events of cell type differentiation and axonal or dendritic growth mark a distinct developmental period of particularly dramatic progress, while subsequent function sees neurons remain, at least morphologically, much the same. Yet, the continuous process of feedback between the genome and its gene products keeps on running and continues to change the neuron. Gene products regulate both the current state of the neuron and change its future state by altering what new gene products will be made, step by step. These changes are partly cell-intrinsic, partly yet another level of feedback based on interactions with the neuron's surroundings. Aging is a continuation of development, because the algorithmic growth program keeps on running—provided time and energy.

We have begun to extend the scope of the concept of algorithmic growth. Fundamental to algorithmic growth is change in time, and an increase in order (local reduction of entropy) based on a continuous supply of energy. If the self-assembling brain and the aging brain represent a continuum, how about the functioning brain? During development, later time points may not be predictable from earlier time points and can only be reached by going through the entire sequence in time. As discussed in the last seminar, there is an interesting analogy to the way our memory functions: In contrast to the computer storage of information content at a specific address, many memories are only accessible as part of a sequence. If you want to be able to say the alphabet the wrong way round, you'll need to practice the reverse order as a fundamentally new sequence first. Try it. In addition to the analogy of a sequential processes, information storage (memory) in the brain is decentralized, flexible and robust in the manner of self-organizing systems. And self-organization itself has to occur in time, through a series of interactions that implement feedback, step by step. Even if the rules were to stay the same, i.e., the same learning rules and mechanisms apply during every iteration of a neural network's function, for every bit of new information that is simultaneously stored, changed and retrieved, learning is an algorithmic process in time. When we design artificial neural networks with random initial states, we simply shortcut the preceding growth process and flip the on switch for an algorithmic learning process.

Bear with me. We are slowly working through these analogies from development to function, and from biological to artificial neural networks, to arrive at a single, important question: *When does a neural network need to grow?* Clearly, for the developing brain the answer is easy, since we only know of brains that went through a growth process. For current artificial neural networks the answer seems equally easy, as almost all implementations of ANNs currently used to great fanfare and success of modern day AI are prefabricated, not grown. But then, there is still a considerable discrepancy between an image recognition AI and human intelligence. In fact, the discrepancy is very similar on many levels between the image recognition AI and, say, a fly. In 2017 a team including Geoffrey Hinton, one of the inventors of the backpropagation error correction and learning algorithm, published a deep convolutional neural network for image classification, with 650,000 neurons.[29] This artificial neural network has about five times more neurons than there are in the brain of the fruit fly *Drosophila*. And I am sure the artificial neural network is much better at image classification than a fly. Yet, the fly's brain is not bad at visual information processing—in addition to a number of other things. After all, the fly can fly. To do so, it processes motion faster than a human could, ensuring it doesn't hit anything midflight, counteracting wind speed and direction changes based on sensory input from multiple body parts, integrating smell, sound and light intensity and color, seeing and integrating polarized light that is invisible to humans, measuring and computing gravity, landing at the ceiling and flying away when you try to catch it. Tell an engineer to build an airplane that can do all that. And then build and train a neural network to control that plane. Of course, we want the neural network to also store and produce courtship singing and behavior, the control of walking, sleeping, metabolism and reproduction. As the list goes on, I get the impression that the difference between the intelligence of a fly and a human is rather little compared to the difference between the image classification ANN and the fly.

Of course, the comparison was never meant to be fair. The image classifying network was designed and trained to classify images. The fly's brain is an embodied entity that has evolved and developed to keep flies alive. But the comparison does highlight the sources of information and the shortcuts taken. The image classifier net is neither the product of evolution, nor the selection of algorithmically grown entities. As we have seen throughout the second session, a network like the fly's brain is evolutionarily programmed. Mutations in the genome may reveal nothing about how they will affect brain wiring, because the end point of brain wiring is information that unfolds through algorithmic growth. This is a form of learning: The neural network has learned to create a new behavior through (evolutionary) selection of a change that is encoded in the network through developmental growth. Neither the growth nor the selection process can be cut short in evolutionary programming of biological neural circuits.

In a field that is partly Artificial Life, and partly AI, some researchers therefore try to understand what it takes to evolve an AI. Arend Hintze and Chris Adami are amongst the pioneers who train ANNs not by feeding them large training datasets, but by evolving genomes that encode neural networks.[30, 31] The neural network determines the behavior of an agent that can be selected based on its behavioral fitness. For example, the agent may be an artificial rat in a maze and selected based on its performance to find its way through the maze. Thousands of iterations of behavioral selection of random mutations will yield a genome that encodes a high-performance behavior based on a specific, unpredictable distribution of synaptic weights in the ANN. In its simplest form, the genome contains one gene per synaptic weight. In effect, this "direct encoding" yet again does away with any kind of development. However, the evolutionary training period brings a kind of algorithmic learning process in through the back door. Efforts that include indirect and developmental encoding are much rarer in this field, but the pioneers are on it.[25, 32–36] There are now simulations of genomes that indirectly change developing gene regulatory networks, which in turn change synaptic weights of recurrent neural networks that drive agent behavior. This kind of work is in its infancy, but first experiments with indirect and developmental encoding already revealed remarkable effects on robustness and adaptability.[27, 37]

Evolved systems allow understanding of aspects of representation in networks that are beyond most current approaches, but they have not become part of mainstream applications. The game changer could be a realization of what evolutionarily programmed developing AIs can achieve that current technology does not. For now, Hintze still wonders how to explain what he does to his kids: "Other parents have normal job titles. But how do I describe what I do? Evolver of artificial intelligence?"

In our quest to answer the question "When does the network need to grow?," we have arrived at the difference between the "intelligence" of an image classifier and the "intelligence" of a fly. Not many people talk about artificial fly intelligence, but that's only because people are more interested in general problems and humans than in flies: the concepts of *artificial general intelligence* and *artificial human intelligence*. Maybe, we need to go back to the question of what intelligence and AI are supposed to be in the first place.

A classic definition of artificial intelligence is Minsky's "AI is the science of making machines do things that would require intelligence if done by men,"[38] echoed by Ray Kurzweil in 1990: "The art of creating machines that perform functions that require intelligence when performed by people."[39] By these definitions, the bar is pretty low. We are not talking about making human brains here, but only successfully mimicking a single ability of the human brain with a machine. The definitions did not say "making machines that do *simultaneously all* the things that require intelligence if done by humans." By Minsky's definition a car navigation system is most definitely intelligent, in fact, much more so than a poor human trying to understand an old-fashioned map (and without the use of a smart personal device). Minsky was of course very aware of the issues related to an AI that is better than a human at one thing, yet far from human intelligence (fig. 9.1). The problem of defining an intelligent system is exemplified in the famous Turing test: We could define artificial intelligence as the *ability to fool human intelligence*, but it turns out we humans are rather easily fooled.[40] More importantly: there is a lot out there that scientists consider intelligent, but there is no single, simple definition of intelligence that would cover it all.[41, 42] Comparisons help. I know what I mean when I talk about human or fly intelligence. But what is artificial *general* intelligence supposed to be, really?

Definitions of AI focus on function, just as efforts in cognitive psychology and neuroscience try to understand how information is processed during brain function. Just like self-organization and ANNs, it is much less common to find a discussion of AI in the context of brain development. This may seem curious given our discussion about how both scientific communities try to understand how to achieve or how to build the respective systems. The disconnect makes some sense if we

FIGURE 9.1. Marvin Minsky at the conference "AI@50" in 2006. Why is there no human-level AI yet?
Source: https://www.flickr.com/photos/steamtalks/483831102/.
Reproduced under the 'Attribution ShareAlike 2.0' License

think of being "built" as a question of engineering method and architecture, not growth. The history of AI is full of disagreements between schools of thoughts on such methods, including the logistic/symbolic versus connectionist, and the neats vs scruffies debates. There is an underlying assumption that once the best method is found, then AI will work like a brain. Or better. Or rather do things (better) that would require intelligence if done by humans, to follow Minsky's definition of AI.

Jeff Hawkins' idea of the memory-prediction system also followed this engineering tradition when trying to understand the method and architecture of the neocortex. The assumption is that once we really understand how the cortex works, we can feed it some information and see a truly intelligent simulation based on the real thing.[43] The memory-prediction system is conceptually based on network function, not growth.

If intelligence has its seat in the network topology plus the data that it is being fed, we may ask how important is topology versus data? As Fleming pointed out to me, in current implementations of deep learning, the network architecture embodies some insights about the world and the task to be performed, but the lion's share of the intelligence results from the learning process. In this seminar we are exploring the idea of the learning process itself as the tail end of the same algorithmic growth process that produced the network topology in the first place. By shortcutting developmental growth and starting with engineered network architectures, we have achieved today's AI. Studying the architecture and inner workings of the cortex certainly holds great promise to yield important insights into how the brain works and AI may have to work. But in the biological brain, wiring the cortex architecture is to grow its information content prior to as well as throughout the process of learning. Babies are not born with an engineered, adult cortex that just needs to be fed information by training it. The network self-assembles through processes that have much in common with those that define its adult function. We do not yet know at what level of detail a human cortex architecture needs to be, and can be, engineered in its adult form. The alternative remains not to start with the adult form as the basis for learning, but to make it the intermediate outcome of a continuous algorithmic process that started with something simpler.

In the connectionist view, information is encoded in the connectivity of the network components, which is in turn determined by the wiring diagram as well as the types and strengths of the connections. As the brain of the baby matures to become the brain of the toddler and the teenager, it is not just synaptic weights that change, but the growth of the individual's wiring patterns. In the case of human intelligence, the network architecture that demands our awe and attention is that of the human neocortex. In the case of insect intelligence, there is no neocortex, yet awe it demands nonetheless. There are structures in the insect brain that look anatomically completely different from any structure in the human brain. Yet, they have been implicated in learning and exhibit features that we have become familiar with in trying to understand both the human cortex as well as artificial neural networks. Parallel input channels, multilayered processing units, and recurrent wiring that allows for feedback are all to be found whenever learning and memory are involved.[44, 45]

Minsky and Papert delivered their blow to neural networks with their 1969 book *Perceptrons* based on limitations in the simplicity of the architecture, most critically the inability of a single layer perceptron to implement the simple XOR function.[46] The limitations were later shown to be far from fatal with the introduction of multilayered and recurrent networks that compute XOR just fine.[13] But what really makes today's artificial neural nets the academic and commercial success of AI is the feeding of big data. Of course the networks need to have certain properties, they need to have a minimal size, and different architectures have distinctive advantages. But in the end, they are really all enlargements and modulations of the original perceptron. The real game changer for the success of ANNs in AI was the amount of training data that could be fed in a reasonable amount of time. Most ANNs learn poorly on too little data. Supervised learning in feedforward ANNs requires a strong training set with individual examples presented to the network one by one, because of the nature of backpropagation error updating. The order in which the training data is presented matters, a reminder of the nature of an algorithmic process. Unsupervised learning does not even need to be told what to learn or what to look for to find salient features in data by itself, but it is still based on an iterative, algorithmic process. The key solution to problems many thought ANNs were incapable of solving simply was the amount of data available for training. In some cases, 1 million examples may yield little success, but suddenly, after 100 million data points, the network has learned. And thus "big data" and "deep learning" have become the names of the game.

Ten years after the publication of *On Intelligence,* Hawkins gave an interview with *Wired* magazine on his progress with large projects to realize the dream of deriving a new foundation for AI from an understanding of the cortex.[47] He had made great progress, but also learned to appreciate the limiting factor: time to learn. Hawkins: "Perhaps the biggest limits are training times. As you make a bigger brain, you still have to train it. We have to train humans and look how long it takes to train them. Our brains are learning continuously for years and years. This is the issue for us too. We spend most of our time training the system. Once you build a system, you've got to feed in the data. That's where most of the CPU cycles go, and memory and so on. A lot of the issue will be how we accelerate that."[47]

No kidding. Training is hard when it happens in the human brain, as connectivity matures, over years. Try to teach a 3-year-old empathy. Or a 6-year-old differential equations. And they do not learn after an on switch has been flipped for an adult network architecture. Imagine starting the process of learning like a baby with an adult, but "empty," brain. There are probably good reasons to assume that impressive results can be achieved even with that approach, but will it be human intelligence? We are still caught up in the idea that the computing substrate is an engineered entity and that training is a separate, subsequent process. The fact that the most successful AI applications are based on lengthy, rigorous learning is an indication for the importance of a time- and energy-dependent algorithmic process that is not easily cut short.

With the idea of the self-assembling brain in mind, we may wonder to what extent the network architecture can become the product, rather than the substrate, of learning. The history of AI started with fully engineered logical systems, even when they were knowledge-based expert systems of the '80s: vast amounts of information had to be obtained and fed into databases, but the process of getting the information into the system happened *prior* to flipping the on switch. The way ANNs learn is fundamentally different, but the system still has to be engineered first to have the property of being able to learn the way it does. How about the next step in this development: to replace the engineer even further, to extend learning into the development of the architecture itself. This is certainly what developing brains do. The question for AI is whether self-assembly rather than engineering is a necessary step to create a next level artificially intelligent learning system. AI's biggest successes to date came from leaving the learning to a scruffy self-organizing process rather than neat programming. Maybe this is enough. Or, maybe, this approach will reveal fundamental limitations as long as the learning system has not yet learned to learn by self-assembling itself. Especially if the principles of self-assembly during development and learning turn out to be the same.

These considerations bring us to another rather fundamental question in AI: *Do we actually want it to be anything like a biological brain?* From the perspective of evolutionary programming, all the biological messiness is necessary to produce the biological function. But AI has historically not been too appreciative of its biological inspiration, and evolutionary programming remains a fringe concept. John McCarthy, who gave AI its name, famously felt biology would impose unnecessary constraints: "AI does not have to confine itself to methods that are biologically observable."[48] In an interview in 2006 McCarthy was asked whether AI's goal was "not so much to make machines be like human, having human intellectual capabilities, but to have the equivalent of human intellectual capabilities" and he answered: "That's the way I see the problem. There are certainly other people who are interested in simulating human intelligence, even aspects of it that are not optimal. In particular, Allen Newell and Herbert Simon tended to look at it that way." About the brain he said "Certainly, we've gotten to know a lot about how the brain works. I don't think it's got very much that connects with artificial intelligence yet."[49] But this was 2006, and McCarthy died in 2011, just when the ANN revolution in AI really began.

By contrast, Geoffrey Hinton began a widely publicized lecture at MIT in 2014 with the words: "There's a lot of things wrong with the neural nets we are using. They've been quite successful at speech recognition and object recognition, particularly recently. But there are a lot of things very unlike the brain that I believe are making them work not as well as they could." What a difference eight years make! Hinton went further: "one thing wrong is that any complex set engineered system should have various levels of structure—neural nets have very few levels of structure. There's neurons, there's layers of neurons . . . which aren't at all like layers in cortex, and there's whole neural nets, and that's it. . . .

One thing that's missing in these neural nets is there is no explicit notion of an entity."[5] The absence of "levels of structure" sounds a lot like the absence of something that exists in the biological structure because of growth. The human brain is certainly based on levels of structure (fig. 9.2). Hinton's concern also reminds us of the levels problem we discussed in the seventh seminar and the shortcuts taken by current approaches. What Hinton wants is to reintroduce levels, from the encoding of the level of the neuron to subgroups of neurons encoding certain properties: *capsules* that do a lot of internal computation and then output a compact result. Hinton's capsules are inspired by bits of cortical columns, and he uses them to explain local representations akin to the famous *place cells*, neurons found in vertebrate brains that are active whenever the animal passes a certain location.[50, 51] Hinton is even challenging his own field by being critical about the very backpropagation method he once introduced to the field. There is nothing wrong with backpropagation for the applications it has solved. But if the goal is to do as brains do, we need to explore the limitations imposed by current shortcuts.

So, how many levels do we need? Let's start by assuming for a moment that all we need is a sufficiently well-designed, multilayer recurrent ANN that can be turned on and taught stuff. How long does the subsequent training take? Given that computers are vastly faster than neurons, the hope must be that training the network could be done orders of magnitudes faster than it would take an actual human brain. But wait: when we compare the speed of computers and brains, the question has to be: Faster at what? When we consider a neuron as a simulation in which we only have to simulate the strengthening and weakening of contacts with other neurons, then the simulation is certainly orders of magnitude faster than its counterpart in a biological brain. But is it enough to simulate synaptic weights to simulate an artificial human intelligence? The real neuron has all kinds of issues that one would like to ignore: it is a metabolic entity, taking care of its own nutrition balances and waste products, it responds to hormones and other substances floating around in the brain, and it has machinery that is only activated during development. It even may have machinery that

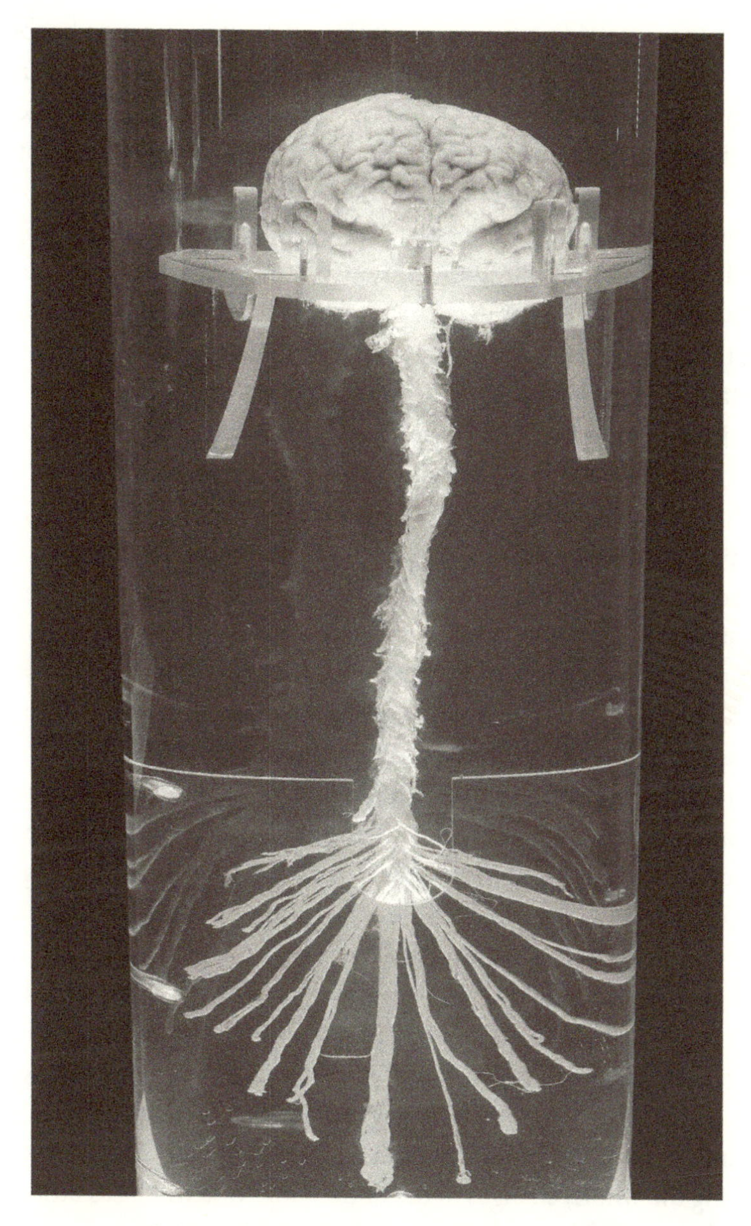

FIGURE 9.2. The human brain. Our brains are highly structured. The brain is not a random net at any point during its development or prior to learning. Photo by author; with permission of The Perot Museum of Nature and Science, Dallas, TX, USA

is only there as a side-effect of something else that its function has been selected for. A common idea in ANN design is that this is all part of hardware infrastructure, which can be congealed in a single parameter, like the synaptic weight. The key problem with this streamlined notion is this: evolutionary programming. The single parameter contains none of the parameter space necessary for evolutionary programming to modify an algorithmically growing network. As we have seen, the seemingly unimportant details may be the parameters that selection modifies and that will reprogram the network in a meaningful, selectable and heritable way. What is more, it is exactly these states of neurons and circuits, including metabolism and neuromodulation, that we currently learn to appreciate in circuit neuroscience as deeper means of "gating" the function of the whole. We come to the question of whether AI development could be any faster than development of a biological brain of the same level of complexity and capacities. The answer is linked to the question of what we want to generate: an AI for speech recognition and image classification, or fly, human or general AI? We may not need to and probably cannot simulate every molecule during function, let alone throughout algorithmic growth. We definitely have to pick our shortcuts wisely, as each will constrain the type of AI. And we may want to consider the possibility that function is not a state in itself, but just as much a progression through states as algorithmic growth during development. And with this, finally, we face the somewhat daunting idea that we may actually need to grow certain types of nets, in which case we'll need to program them by evolution. It will depend on what kind of intelligence we actually want. And similar to evolutionary programming, there is only one way to find out: by making the whole thing, or running the full simulation, assessing output performance and changing the underlying code for another run. When it comes to algorithmically growing the whole, changes in the underlying code will be as little predictable to its intelligence as genetic changes currently are for human or fly intelligence. Thank you.

3.3

Final Frontiers

BELOVED BELIEFS AND THE AI-BRAIN INTERFACE

The Tenth Discussion: On Connecting the Brain and AI

AKI (THE ROBOTICS ENGINEER): And here we all are, at the end of a long argument, and I guess you seriously want me now to go back and grow robots after all.

ALFRED (THE NEUROSCIENTIST): Are you sure about how biological you want them? You may need to buy them health insurance.

PRAMESH (THE AI RESEARCHER): It depends on the intelligence you want! We haven't really figured out what intelligence is, let alone artificial general intelligence. I always felt we made up that term just because we recognize the difference between a visual recognition task and the complex mix of tasks our brains really do. I like the idea of something like worm intelligence or fly intelligence. Clearly, a fly has a pretty small neural net that does a hundred things in parallel that we could call intelligent.

MINDA (THE DEVELOPMENTAL GENETICIST): . . . with a few more than 100 thousand neurons . . .

ALFRED: But that brings us back to connectomics. The fly brain is not a randomly wired network of 100 thousand neurons—it's really a composite of hundreds of interconnected sub-nets, all very specifically wired to do certain tasks. I liked that people are taking that idea to build modular artificial networks.

PRAMESH: The question is: To what extent does intelligence, or artificial general intelligence, depend on the particular connectomics of many different subnetworks?

MINDA: It's all part of it! As you said, it depends on what you want to simulate. I guess few people are interested in artificial fly intelligence after all.

PRAMESH: It is a fundamental problem, though. In the end, training a randomly connected network of 100 thousand neurons with a set of learning rules is obviously not the same as training the fly brain. And the complicated argument in the last seminar we heard about . . .

AKI: . . . suffered through . . .

PRAMESH: . . . was that the precise connectivity of the fly brain is already part of its developmental training. If algorithmic growth continues from the cell to connectivity to function to aging, then development is in some way part of learning.

MINDA: And evolutionary programming of this process is the reason why you cannot leave out any detail. You may still not know what to do with all the details we are studying in developmental biology. But in the end all the details will fit into the dynamic jigsaw puzzle.

AKI: Are you sure you are focusing on the rules when studying the details? Anyway, peace. Why not start with observing the execution in its details, as you do. The counterargument was that for many problems a simulation works just as well, and much faster, when leaving out irrelevant information. That'd be my way of finding the rules.

MINDA: There is no irrelevant information if you want the whole thing. With every bit you leave out, you restrict what kind of output you can get. Be careful what shortcuts you choose . . .

ALFRED: Come to think of it, choosing the right shortcuts is what we have to do every day. Only when you know what you can leave out do you understand what is important. But I am still bothered by the idea of evolutionary programming. I get that anything I leave out will change what can, well, 'unfold,' I guess. I might not have been able to predict it, but evolution can select it.

MINDA: And all of the molecular detail, and everything that happens during development, that's all part of this unfolding to have something to select.

PRAMESH: So, Alfred, how much can you really learn from studying the cell types and connectivity of a neocortex column, leaving out all the molecular detail and all of development?

ALFRED: Oh, a lot, for sure. A lot is known and more is actively being studied as we speak. But what are the limits of a simulation of the neocortex that hasn't learned throughout development, based on all the biological mess? I really don't know anymore.

AKI: Don't lose confidence in the cortex!

MINDA: Well, flies don't have one. They also have brain structures with many parallel fibers that are highly interconnected . . . recurrent, I guess? Anyway, their intelligence is largely genetically encoded and developmentally grown. They just continue to learn by training afterwards.

PRAMESH: And now imagine how information has grown into this network, first through connections and their molecular execution of rules and then through the continued modification of neuronal activity. It's all one long continuous process. The big flow of dynamic normality.

AKI: And it makes every one of us who we are, amen.

ALFRED: I like this picture. Whatever happens to you has to happen on top of everything that happened before. Every bit of information entering the network is different, really only has meaning, based on what was there before. It's a way of looking at things that I can relate to.'

MINDA: It's also a good way of describing why people don't change.

AKI: . . . and thus spoke the Seer on the Pole: 'You cannot see what I see because you see what you see. You cannot know what I know because you know what you know. What I see and what

I know cannot be added to what you see and what you know because they are not of the same kind. Neither can it replace what you see and what you know, because that would be to replace you yourself.'[1]

PRAMESH: Well, what I would love to learn from you, Aki, is how you would want to interface to the kind of information that gets wired into the brain through development. I know there are start-ups now that want to develop brain download and upload. I don't understand how that's going to work.

AKI: We have been building interfaces like this for many years. You can connect an artificial retina, a kind of bionic eye, to the visual cortex to restore some vision. And cortical interfaces are tested in paralysis patients in clinical trials. This is happening right now—and looks very promising.[2, 3]

ALFRED: I know. And Elon Musk is invested in this start-up that is trying to put in thousands of electrodes in the somatosensory and motor cortex. Never mind what cells they exactly record from. To them it's mostly about bandwidth. And merging with AI.

MINDA: Aren't we back to the 'engineered versus grown' discussion? Can you really just plug in and connect an extension that did not grow with the brain in the first place?

ALFRED: Well, we know that the cortex does actually have amazing plasticity, all going back to Lashley's equipotentiality. If there were a few more columns connected in the same way as the rest, I guess they would be integrated and taken over. But it would take time. And energy. The system would have to learn, bit by bit. . . .

AKI: That's exactly the idea. If you connect something with enough bandwidth, then the brain will learn to play with it, like, if you suddenly had a third arm. It would slowly become part of you. Come to think of it, I wouldn't like the battery to run out. Might feel awkward if something that has slowly become part of you is suddenly turned off.

PRAMESH: So how specific is the connectivity of these implants?

ALFRED: You can determine the region, but not the cell or any type of connectivity, really. You are just electrically stimulating hundreds or thousands of neurons in the vicinity. In the case of the visual implants that's better than nothing. It is actually very impressive that it works the way it does. The brain kinda does accommodate input—it's built to learn.

AKI: . . . and now imagine you can use this to connect to an AI.

MINDA: Honestly, this all sounds like science fiction. How did we get here? I mean, we started off with trying to figure out what we can learn from each other's approaches about neural networks. I think that point was well taken. The need to grow may yet haunt AI. And now you are talking about merging the two. Wouldn't they need to grow together?

PRAMESH: The idea was that they grow together by learning to work together. That may be just the bit that can't be cut short.

MINDA: After all I've heard, I would now have to say that learning together is only the later part of growing together. That's a big shortcut. And once you think about actually developmentally growing together, you cannot predict the outcome, only select the outcome at the population level. Evolutionary programming.

AKI: Oh, that's nothing. For years they have been talking about molecular assemblers in nanotechnology. The idea was that they can build themselves and anything really. Bottom-up engineering of cars, airplanes, anything. Now *that's* really out there. Or, well, there is rather a reason why it does *not* actually exist out there . . . [4, 5]

MINDA: Well, at least we know that growing stuff can be done: a tree does grow from a seed, bottom-up. A farmer can reproducibly grow a whole corn field every year from seeds. Your brain does develop from the information in an egg and a sperm. Growing stuff is fine. But growing your brain together with an AI?

ALFRED: I see issues if you'd have to program it by evolution, based on full simulations for each round of selection. Something reproducible will happen on average, like the average corn plant.

But what if part of the variability is your character—and you can only select based on outcomes. Awkward.

AKI: But again, all that only once you buy into the idea of 'developmentally growing together.' Let's first increase bandwidth and see how much people can learn and do with it.

ALFRED: There certainly is a fusion with biology coming. And people argue that we are already extending our brain function with smart phones, only there the connection is very slow . . . typing with your thumbs and all that . . . [6]

AKI: That's exactly what I meant, that's the bandwidth problem.

PRAMESH: But the fusion has to start with an understanding of the types of information we are dealing with. Yes, we are evolving neural networks to understand basic principles of, say, representations in networks. But we are still not much in touch with biologists over these.

MINDA: For me the idea was to learn something I would not normally hear about. I came here to see what I could pick up from the model-builders, as Gaze called them, when he contemplated about random connectivity and what it could and couldn't do.[7] And I think I got an idea of what type of information you want. Our molecular mechanisms are snapshots of a larger program. I've learned a lot about what you call algorithmic growth, some of which seems just different words for what developmental biologists have been studying all along. It doesn't help to just say, look, here is the original code as the basis of everything. That's what some people were thinking twenty years ago when the big genomes were deciphered. With the code in hand, is everything else just details? That's what our discussion of algorithmic growth has been all about. The model-builders want the rules of the program. We want the molecular mechanisms that unfold, if you will, during the run of the program.

ALFRED: Nice. I am still unclear though how much we have really embraced the idea of algorithmic growth in AI development? The random network design plus feeding big data can't be it.

PRAMESH: Yes, I am not sure any of us have embraced algorithmic growth the way we may need to. Nor the evolution of AI.

MINDA: I am not sure we have embraced how fundamental evolution really is. The principle is at work everywhere, it's not just a problem of fossils.

AKI: Well, what I have learned is that all my personal history and the algorithmic growth of my brain will make sure that I am by now so biased and cognitively stuck that I am not going to adopt much of this anymore. Like Douglas Adams's puddle of water that clings to the idea that the hole it's in must have been made to have it in it, just because it fits so well—while it is evaporating.[8] Clearly, if decades of my way of thinking have engrained themselves in my neocortex, then these few seminars won't have changed much. Isn't that how it works?

MINDA: Yes, paradigm shifts in science usually only happen when the last proponents of the old ideas die out . . .

ALFRED: This sounds like the brain does not like surprises. But then, it is only surprises you learn from. And the brain is built to learn. There is hope for you, Aki.

PRAMESH: As scientists, we may not be using our brains for what they are best at. Our history is both the basis and the bias of anything we try to figure out. To me the continuity of algorithmic growth is a bit of a new concept, and I think we all need to think about our shortcuts in light of that.

AKI: Depending where we want to end up. In our field, some people find it is more important for robots to be cuddly than what they actually say . . . who actually wants the super-AI?

PRAMESH: Maybe it's part of the continuity of algorithmic growth that history can't be stopped. I think your super-AI will come, whether you want it or not.

MINDA: We haven't touched on that discussion at all. From everything we discussed I didn't get the impression we were even close to what you call artificial general intelligence.

PRAMESH: Well, again, it depends on what you mean by it. I could say, if you can be beat at your game and fooled by a machine talking to you, then it is not only close but it is already here. On the other hand, if you mean a clone of yourself, I am not sure we will ever get there. But I shouldn't enter the predictions game. Like expectations, predictions are just extrapolations of our own historic biases. The faster things change, the more they will set us up for disappointments.

AKI: Well, I like to be surprised. Surprises are information!

Seminar 10: From Cognitive Bias to Whole Brain Emulation

Herbert Simon came to AI as a political scientist with a keen interest in bureaucracy. In 1948, while working as an administrator on the Marshall Plan following the Second World War, Simon published a seminal book entitled *Administrative Behavior*.[9, 10] At the time Shannon and Turing tried to teach computers chess, Simon asked questions about how human organizations got anything done at all. How do bureaucracies make decisions, and does that relate to how individuals make decisions? Simon's work very quickly became highly influential in psychology and economics alike. Models for economic decision making had largely been built on the idea of rational comparisons and the choice of the best option. Simon described in detail why and how people and their organizations do not work like that. Just to identify all possibilities and to compare them fairly for a given problem is usually not feasible and is itself an optimization problem. Consequently, Simon felt that people's choices were only partly rational, and partly an optimization of time investment, benefits, relative importance, etc.[10] By the time of the Dartmouth workshop in 1956, Simon was a widely recognized interdisciplinary scientist with credentials in cognitive psychology and decision making in politics and economy, but not quite as obviously in AI. And yet, he and his younger physics colleague Allen Newell were the only two attendants of the workshop who had already programmed a "reasoning program," called the *Logic Theorist*. Simon, who was rarely shy with his claims, said of Logic Theorist: "We have invented a computer program capable of thinking non-numerically, and thereby solved the venerable mind-body problem." A year after Dartmouth, he famously proclaimed: "It is not my aim to surprise or shock you—but the simplest way I can summarize is to say that there are now in the world machines that think, that learn and that create. Moreover, their ability to do things is going to increase rapidly until—in a visible future—the range of problems they can handle will be coextensive with the range to which human mind has been applied."[11] This was 1957.

From the get-go, AI was an interdisciplinary science, much more so than neurobiology. For decades the field remained a creative hodgepodge of computer enthusiasts, engineers, mathematicians, cyberneticists and psychologists, but not biologists. Of the participants of the Dartmouth workshop, McCulloch, a trained psychologist, was the only neurophysiologist. Simon clearly was interested in what we now call interdisciplinary work, and especially stated that he had always been interested in the nature of intelligence.[9] His study of human decision making had led him to recognize how people had to make decisions on incomplete information, using approximations and rules of thumb to find compromise solutions. The rule of thumb is often the easiest and quickest way to come to a solution and to be able to move on, which may be more important than finding the exhaustively optimal solution to a problem. This approach has become well-known as *heuristics,* both in behavioral sciences as well as in computer programming. In fact, Simon and Newell applied heuristics for their Logic Theorist. Hence, the earliest of reasoning programs, although based on formal logic, already utilized an essential, noisy process to achieve robust function. And function it did. Logic Theorist could independently prove several theorems of Russell and Whitehead's *Principia Mathematica* and is said to even have improved on one of the theorems.[11]

Heuristics play a major role in cognitive psychology as much as in our daily lives. Amos Tversky and Daniel Kahnemann spent the early 1970s in regular, intense discussion about questions and scenarios where our intuitions are likely to lead to incorrect judgements. As Kahnemann notes, this was a time when psychologists thought that human thinking was mostly rational unless disturbed by emotions.[12] In 1974 Tversky and

Kahneman published a paper entitled "Judgement and Uncertainty: Bias and Heuristics" in which they identified three heuristics, three shortcuts, the human brain takes to conclude something based on imperfect data.[12, 13] A classic example is "Steve the Librarian." Based on a short description of a tidy, introverted and detailed-oriented personality, most people would pick the job "librarian" out of a list of options as the most likely job for Steve. The problem is that librarians are much rarer than, say, engineers. But statistics are not immediately intuitive. The *judgement heuristic "representativeness"* is a functional shortcut—this is how the brain works, when it works quickly.[12] An educated guess is heuristic. Intuition in the absence of sufficient data is heuristic. Decisions based on what we lightheartedly call a "gut feeling" or "common sense" are heuristic. And they all are kinds of mental shortcuts. Note that a lack of common sense was a key factor Minsky identified when discussing why there still was no human AI at the celebration of "AI@50" ("Dartmouth Artificial Intelligence Conference: The Next Fifty Years") in 2006 (see fig. 9.1).

Some judgement heuristics appear universal, as does the need for slower thinking when it comes to statistics. But Tversky and Kahneman quickly realized that judgement heuristics that are the same in all people do not explain our differences in intuition. The educated guess is as good as the education it is based on. Education takes time and energy. The common sense and intuition on whether the family should buy a house are not the same for a 6-year-old and a math teacher. We all have our biases, as much as we are our history. An expert physician may recognize your symptoms before you describe them. And, as Kahneman points out "Most of us are pitch-perfect in detecting anger in the first words of a telephone call . . . and quickly react to subtle signs that the driver in the next lane is dangerous. Our everyday intuitive abilities are no less marvelous than the striking insights of an experienced firefighter or physician—only more common."[12] And, we may now add, these intuitive abilities are as different as people are. Which brings us close to the slippery slope of those disciplines that try to bridge levels as far apart as cognition and evolutionary or developmental biology. Evolutionary psychology is one of those. Typically, other psychologists are more annoyed with evolutionary psychology than any other type of scientist. It is a difficult debate and the critiques are profound: How do you experimentally test evolutionary psychological theories? Computer simulations are good, but where does that leave cognition? In the end, the levels problem always provides enough ammunition for a discussion-terminating blow: the higher level property cannot easily be reduced. But it is not one of my goals (or abilities) to enter the psychology debate. We got here because the levels problem is central to the idea of unfolding information through algorithmic growth. Whatever its debatable explanatory power, the brain's evolutionary and developmental history do affect its function. But how?

Let me start this discussion by making the maybe outrageous and definitively speculative statement that *cognitive bias is not a failure of brain function, but a feature of its algorithmic growth*. I am aware that the study of cognitive bias has a long and rich history. With respect to judgement heuristics, we have had a brief look at the foundational work by Tversky and Kahneman, which kickstarted decades of studies on the relation of brain function, judgement and decision making that is described elsewhere in detail.[12–14] Ideas of cognitive biases as features, not failures, based on evolutionary principles have been discussed in the field of evolutionary psychology and beyond for a long time.[15] Here, I want to focus on the potential role of algorithmic growth as an idea that, I think, fits neatly with current concepts of brain function and that merely echoes previous ideas common to social science and psychology from a different perspective.

To understand function as continued growth, we may start by revisiting typical features of the functioning brain at different developmental stages: there is something archetypical about the way a 3-year-old talks about the world that is very clearly different from a 6-year-old, a 10-year-old, a 16-year-old, a 25-year-old or a 60-year-old. Development never stops. Every child seems insatiably curious, everything is new. As the growing brain is programmed over years of development, representations of the world firmly establish themselves that will serve as a basis for the rest of the brain's life. As internal representations mature, new experiences become less new and enter the network as variations of previous experiences. Curiosity decreases with age in most people. The archetypical human adult has little need to be curious about the world. Been there, done that. The adult has a representation of the world built on experience, etched into the network through its individual growth process.

Brain function is not the computing of data (input) in comparison with memorized (stored) data based on logical operations, but instead a process of alignments and selection of patterns based on previous experiences. Consequently, every individual must have a logical, obvious, yet different perception of reality depending on personal history. Every input is either a familiar pattern or a surprise. For the child's brain everything is a surprise; the algorithm is in the middle of the growth process that builds the patterns with which new information can align and ultimately cease to be surprising. For the adult brain surprises mean extra work. The brain's job is to fit sensory input (including new ideas) into an existing brain with existing patterns that have grown out of past experience. This is a fundamental aspect of Hawkins' memory-prediction system: as long as predictions are met, everything functions smoothly and higher level processing is not required. You keep on walking without much mental effort until your next step does not touch the floor as expected. The unexpected is a potential emergency, a surprise that necessitates an alert in higher cortical layers. The better the predictions, the fewer and smaller the surprises. Roger Schank put it this way: "Information is surprises. We all expect the world to work out in certain ways, but when it does, we're bored. What makes something worth knowing is organized around the concept of expectation failure. . . . You learn something when things don't turn out the way you expected."[16]

The neurobiologist Randolf Menzel once told me: "The brain is a device that connects the past with the future." Menzel has spent decades figuring out the intelligence of bees.[17] Amongst insects, bees are particularly remarkable for their social behavior and abilities to learn and predict. As Menzel and his team have shown, bees explicitly learn to predict the future.[18] For example, when offered changing sugar concentrations in repeated trials at different locations (with increasing, decreasing or constant sugar concentrations), bees learn which location improved between past trials, such that in a subsequent trial they actually avoid the location that had a change to lower sugar concentrations in the last trials and preferred the location with the rising concentration.[18] The idea of sequences and our ability to predict the next item in a given sequence are classic elements of intelligence tests. And there is just no other way to learning the temporal sequence than by learning it, well, in time. It is a fundamentally algorithmic process, and there is no shortcut to learning by successive addition of the new experiences based on previous experiences.

I can train my biological or artificial neural network to learn (almost) anything. Then, when presented with something new, it will classify what it sees according to what it has learned. Imagine being trained on handwritten numbers and being presented with an elephant. We are all well-trained neural networks. And we are trained in time, with experience following experience. The same two experiences in reverse order have a different information content. Again Schank in 1995: "Everyone in the AI business, and everyone who is a viewer of AI, thinks there are going to be shortcuts. I call it the magic-bullet theory. . . . But we became intelligent entities by painstakingly learning what we know, grinding it out over time. . . . When you talk about how to get a machine to be intelligent, what it has to do is slowly accumulate information, and each new piece of information has to be lovingly handled in relation to the pieces already in there. Every step has to follow from every other step; everything has to be put in the right place, after the previous piece of information."[16] *How to Grow a Robot* by Mark Lee discusses similar ideas for the new field of Developmental Robotics.[19]

Information content in the brain has a history track. A new bit of information entering your brain is not simply up against a set of information that may or may not agree with the new content. Neural networks do not compare a single new incoming data point to existing data on an equal footing. The training history of the neural network leaves its traces. And this history is something that had to happen in time, with each new bit of data being learned by a network in its current state, algorithmically. The new bit of information is up against the entire history, cumulatively engrained patterns of bias. A typical ANN used in AI applications will learn differently if the same supervised training set is presented in a different order. It matters if an aberrant observation is made early or late. An AI-powered internet search engine that learned from a large biased dataset that nurses are more often associated with "her" and investors are more often associated with "him" in historical data will propagate the biased data as a biased prediction. Deductive logic dictates that a single observation of a white horse falsifies the hypothesis that horses have to be black, even if that hypothesis resulted from lifelong observations of thousands of black horses. If we had seen only black horses all our lives, then the new experience of a white horse is a violation, an aberration that, after all, is just a singular observation. If you have hitherto lived the good life on the assumption that horses are black, the single violation is hard to swallow. Maybe somebody just painted it white to try to mess with you? This is what makes the scientific method a difficult feat, one we seek to achieve *despite* the way our brains work. Our brains actually have a rather difficult time rationally comparing data. Evolution is not intuitive, because it is not something we experience every day. Algorithmic growth and information unfolding are not intuitive for the same reason.

What happens to your brain every day is this: new information arrives, it is sorted into your personal algorithmic timeline and becomes part of your history. You and I will learn something different from reading the same Wikipedia entry, with different ease, different associations. The same information will affect different memories in different people. When you learn something, you are prodding a system that tries to remain stable, and continues to make plausible predictions, based on the history of everything it ever learned so far. In this sense, the brain does not want to be surprised. It will align the new information as smoothly as possible with what is already there in the network. If your brain is already trained to perceive and interpret things in a certain way based on its history, it will strengthen pre-existing features over unexpected new ideas. This is why cognitive bias is a feature of algorithmic growth, even when it is a failure in its consequences.

If I live in a warped world, I will have warped experiences, and I need them to be warped just right to make the right predictions about the warped world I live in. Our experiences in communication between human brains are rich in these phenomena. Most people dislike change more and more the older they get, rather irrespective of what random situation life has placed them in. We are highly unlikely to change engrained beliefs in religion or political views, and we obviously feel we are right about our beliefs based on a rich history of experience. It doesn't matter how smart we are; smart people just have more elaborate means of feeding their cognitive and confirmation biases.[20] There are equally smart people on both sides of political divides, who each know they are right and the others rather stupid. We are continuously treating people and what they say as if they were rational data-processing input-output machines. Yet, nothing people say makes sense unless viewed through the lens of their history, their individual growth. Based on personal history alone, an individual's output is often surprisingly predictable. People who propose radically new ideas that do not align with what's in the network of most listeners are likely to meet resistance. The unknown is more likely to be bad than good. For something we have experienced before we have a low-surprise fit in our brain. The new, the unknown, the foreigner with, well "foreign" behavior, these things are potential dangers to the well-oiled, highly experienced, algorithmically grown network that wants to have things its own way. There are lots of places to go starting from this thought, which I will leave entirely to you.

I started this discussion with the thesis that algorithmic growth produces cognitive biases by the stepwise process of adding and aligning new bits of information to what is already there. My motivation was the parallel to algorithmic growth during development, which does exactly that. Of course there are entire fields in the psychological sciences dealing with the idiosyncrasies of our minds, none of which we touched on, and none of which are within the grasp of either the biologist or the AI researcher. My goal was not to provide a reductionist explanation for a complicated higher order phenomenon, but to discuss the predictions of algorithmic growth. This doesn't tell us much about psychology, but it allows us to identify patterns. And one of the patterns we encountered is the notion of the shortcut. It just won't do. Algorithmic growth needs to run its course to provide both the unfolding of information and the evolutionary programmability needed to understand the endpoint. When discussing cognition, we went far out on thin ice to try to understand such an endpoint, something we like to call complex. At the other end of the spectrum, scientists have been fascinated for decades with the idea of employing growth, starting from something very simple, to build something complex. How about growing a car, for example?

The ultimate in creating things by growing them from molecules was proposed in the 1980s by Eric Drexler, when he started the field of molecular nanotechnology.[4,21] Drexler envisioned molecular machines that could build things with precision, including themselves. He thereby envisioned that all you needed were such molecular assemblers, add some chemicals, and let the soup grow almost anything you wanted, from raw materials to cars. Key to Drexler's vision was the precision of a fundamentally deterministic process of growth. His vision inspired a generation of engineers, chemists and computer scientists to explore engineering at the nanoscale. In the beginning of the 2000s, a heated debate reached its peak on the feasibility of Drexler's version of nanotechnology. Richard Smalley, a chemist who had won the Nobel Prize in Chemistry for the nanoscale engineering of an astonishing macromolecular structure, said it couldn't be done. Trust in Drexler's vision faded, and while the debate is still ongoing to this day, the thrust of the field has shifted. Nanotechnology still is an immensely active research field of great economic importance, but the idea of growing cars bottom-up with molecular assemblers has dropped out of favor. The vision, the debate and the background of this story are all fascinating and summarized elsewhere.[5] We, who have just immersed ourselves into the intricacies of algorithmic growth, can ask a much simpler question, if simple is the right word. How do you design the code to grow the car?

Let's not beat around the bush. The code to grow a car cannot be *designed*, period. Drexler's nanotechnology idea of bottom-up growth is a good example of all the fallacies of information encoding we discussed in the context of the levels problem. You need time, energy—and algorithmic growth. The more complicated the output, the more likely algorithmic growth is unpredictable. The underlying code must

be programmed by evolution. The problem is not that such a code could not exist. There are codes out there to grow trees and human brains, so why not cars. Growth unfolds information and the unfolding is the only way the information comes to be. Of course, a fast simulation will get you there faster. But there is no shortcut to evolutionary programming. You have to try out code after code and select from the products. You can try to tweak the code a bit here or there, but the outcomes are more likely to be unpredictably different than tweaked in the way you intended. We know how it can be done, because we know how it *is* being done when we look at the code for an apple tree or the human brain. Of course, the precision of determinism is out the window. The best outcome for successfully growing a car, after some intensive breeding program, would be cars that are as different from each other as apple trees. And you might want to get health insurance for them, in case they get infected by a fungus.

Maybe we could start by not actually growing the car, but by simulating the process, much like the emulation of an artificial intelligence on a computer. As we have seen, when designing a simulation to grow anything of substantial complexity, we face the question of how much detail is dispensable. Do we need molecular diffusion simulations as discussed in the sixth and seventh seminars just to simulate the 302-neuron network of the worm *C. elegans*? If we wanted to simulate the growth of a car, we clearly would need to simulate the molecular assembler and its access to the substrate it assembles. As the structure grows, new levels of organization emerge. At some point there should be wheels and a clutch, whose higher level behavior could be simulated easily based on their higher order properties. Can we just stop simulating molecular dynamics at that point? These questions mirror those we previously asked for ANNs. There, we started at the other end, with the higher order properties, and then wondered how much lower level detail we really needed and how much could be cut short. In both cases we end up with the same problem: the information content that unfolds during the process of algorithmic growth and that can only be evolutionarily programmed cannot be shortcut by a higher order simulation. And this thought brings us to the synthesis of biological neural networks and artificial neural networks, between brain development and artificial intelligence.

In his influential book *Superintelligence*, Nick Bostrom provided one of the many versions of the fascinating idea of a whole brain emulation.[22] The basic idea is to scan a human brain such that it can be reproduced in a computer simulation. The consequences of this *brain upload*, the duplication of *you*, is a popular intellectual pastime in science-affine circles as well as for science fiction enthusiasts. As before, there are immensely important aspects to this discussion, not least societal, that are essential in other contexts. Here, we will completely ignore these aspects in favor of our focus on relevant information and information that unfolds during algorithmic growth. The question I would like to ask is this: *How much information do you need to scan in order to reproduce all relevant network information?* The question is closely related to questions we asked before: *How is information stored and retrieved? How much detail is needed for a given simulation? What shortcuts are possible?* The idea of reading out all *relevant* information from the brain is akin to the idea of Laplace's demon. In 1814, Pierre-Simon Laplace challenged the intelligentsia with the following thought experiment, translated into English in 1902: "Given for one instant an intelligence which could comprehend all the forces by which nature is animated and the respective situation of the beings who compose it—an intelligence sufficiently vast to submit these data to analysis—it would embrace in the same formula the movements of the greatest bodies of the universe and those of the lightest atom; for it, nothing would be uncertain and the future, as the past, would be present to its eyes."[23] To create a true whole brain emulation with no shortcuts, we need Laplace's demon as a brain scanner. We do not actually know whether the universe is ultimately deterministic, so we do not know whether this whole brain emulation would run a predictable future in every detail. But we do know that scanning of every detail with total precision is a theoretical and practical pipe dream, so maybe we are lucky the experiment cannot be done.

As before, we may hope for reasonable shortcuts. We *can* understand transmitter diffusion at a synapse without having to know the position

of every transmitter molecule. We *can* describe core properties in a McCulloch-Pitts neuron without knowing all the biologically messy stuff, given a defined context that can be safely assumed. And similarly, we *can* describe a molecular mechanism for a protein like netrin during axon pathfinding, without knowing all the exact tissue properties and other molecules floating around. We only get into trouble when the context of the details becomes relevant. And that information becomes relevant once we try to understand how evolutionarily programmed information unfolded in time and space. How much of that are we willing to give up in a simulation of a network that produces intelligent behavior? The molecular diffusion of transmitters and hormones? How much of that needs to be uploaded to preserve "you"? Or how much of "you" are you willing to give up, when backing up yourself for safe keeping?

In short, we are facing the same shortcut problem in biological neural networks and artificial neural networks, with brain development and artificial intelligence. The idea of the guidance cue explains an axon guidance problem on the basis of an important contribution to the system's algorithmic growth based on an intuitively understandable function. However, it is a shortcut, because it takes a contribution that, when knocked out, might produce an 80% penetrant phenotype and smooths over the context of metabolic state, tissue stiffness, etc. by calling these "lesser" mechanisms redundant or permissive. It is a shortcut because it takes steps out of the context of algorithmic growth. As in artificial neural networks, the shortcuts can be immensely useful. But neither in brain development, nor in the development of artificial general intelligence can we claim to have grasped the general principles without an appreciation for algorithmic growth.

Both systems can be meaningfully described at different levels, that is, different levels of shortcuts. There is a meaningful level of netrin function that can be relied on to be found in many different contexts. There is a meaningful description of neurotransmitter and neuromodulator diffusion that does not require single molecule dynamics and can be relied on to work in many different contexts. An instructive molecular mechanism during brain wiring or a computational change of a synaptic weight in an artificial neural net are workable shortcuts. We can leave out details for netrin function as long as it shows the function independent of the details of context. We can leave out the individual neurotransmitter molecules, as long as a diffusion formula captures all relevant function. And we can leave out biological messiness in an artificial neural network, as long as it does what we want from it.

In all cases, it would be well to be aware of what we give up with our shortcuts: In the worm nervous system with its 302 neurons, as in the human brain with its 86 billion neurons,[24] each neuron has its own history and its own place in the growth and function of the network. No two synapses are really the same. The hope of the shortcut is that an average synapse captures all properties, and it certainly gets us far. But there are properties, as seen already with the transmitter in the worm, that contribute to network function and that cannot be reduced below the level of the diffusion of the individual molecule. Let me make a bold prediction: a faithful whole brain emulation in the Bostrom sense of only the worm will require a full molecular dynamics simulation of all molecules in the worm. The stochastic dynamics of receptor movements and ion channel distributions and membrane lipids define the precise local electrical properties of bits of axons. We do identify those network functions that are sufficiently independent so we can describe them with a connectionist model of averaged excitatory or inhibitory synapses whose weights can be modulated. This is why we can look at successes of artificial neural networks in language processing or visual scene recognition. If we want real artificial worm intelligence, however, then we need the worm in all its molecular glory. If we want real artificial human intelligence, we need the human in all its molecular glory. And the only way to make either of them is, you guessed it, to run the entire algorithmic growth program, with no shortcuts.

Did I just say that there is no shortcut to producing artificial general intelligence other than developing the whole thing from a genetic code, physically or in a simulation, to the functioning brain? Well, it depends on what we mean with "general." If we mean artificial human intelligence in the sense of an intelligence that works like human individuals do, then yes, that's what I just said. But intelligence is not a simple

property that increases on a capacity scale from worms to humans. Worms have an intelligence that is entirely wormlike. Humans have an intelligence that is entirely humanlike. They are not just higher or lower on a scale, they are different based on the different ways the brains work (except that the 302 neurons of the worm do not really constitute a brain quite yet; take a fly then). An artificial general intelligence need not be humanlike, to be as smart as (or smarter than) a human. But if it didn't undergo the algorithmic growth of a human brain, it will not be human, even if we enjoy it better if it looks a bit like us (fig. 10.1). The artificial general intelligence will be its own history.

How far are we in modern-day cybernetics, from brain-machine interfaces to the elusive brain upload? There are lots of rumors in the tech world, and some of the world's leading engineers and entrepreneurs are on it. Elon Musk is such a one, and he is keen to extend the human brain. Musk is fascinated by the symbiotic relationship of your limbic system (that's the part of the brain responsible for emotions and drives) and the cortex, whose job, in Musk's words, it is "to make the limbic system happy."[25] In an interview in 2018 Musk laid out his vision that technology may be ready to integrate AI as a third layer, a digital extension of yourself, to do just that job. To Musk, phones are already such an extension, only with very slow data rates. In this view, the issue is mainly a problem of the *interface* and *bandwidth*.[25] This view was most clearly spelled out in the summer of 2019 at the launch event for a commercial endeavor to develop a high-bandwidth brain-machine interface: "The thing that will ultimately constrain our ability to be symbiotic with AI is bandwidth. . . . with a high bandwidth brain-machine interface, we can go along for the ride. We can effectively have the option of merging with AI."[6]

The advance of brain-machine interfaces into human lives is not a sci-fi scenario; it is happening right now (fig. 10.2). Input from the world is processed in the sensory cortex (e.g., visual, auditory or somatosensory), and output action in the world through the motor cortex. Consequently, Musk's company plans for implants of electrode arrays in primary motor cortex and somatosensory cortex.[6] An electrode array is a physical device with tiny fibers, in their case 1,000 on a 4 × 4 mm chip.

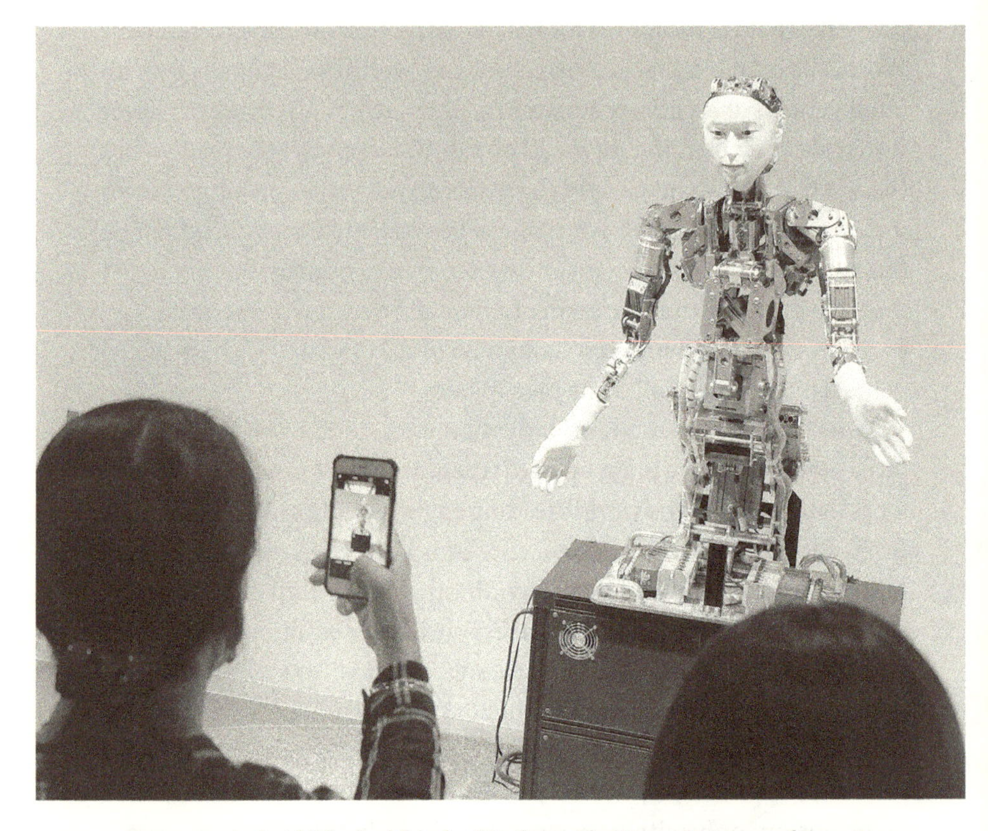

FIGURE 10.1. Android "Alter" exhibited at Miraikan—The National Museum of Emerging Science and Innovation, Tokyo, Japan. This android has been used in music performances and other impressive demonstrations. It represents something we can relate to, yet it is entirely engineered.
Photo by author, with permission of Miraikan, The National Museum of Emerging Science and Innovation, Tokyo, Japan

The electrodes need to get sufficiently close to neurons in the cortex to record action potentials, the bits of information single neurons send along their axons. What sounds like science fiction was in fact already done successfully in a human patient with paralysis in all four limbs (tetraplegia) in 2004.[26] This system, called *BrainGate,* was based on an early electrode array already developed in the mid-'90s[27] and allowed the first human patient to open and close a prosthetic hand.[26] The version implanted in the patient also was a 4 × 4 mm chip, only with 100

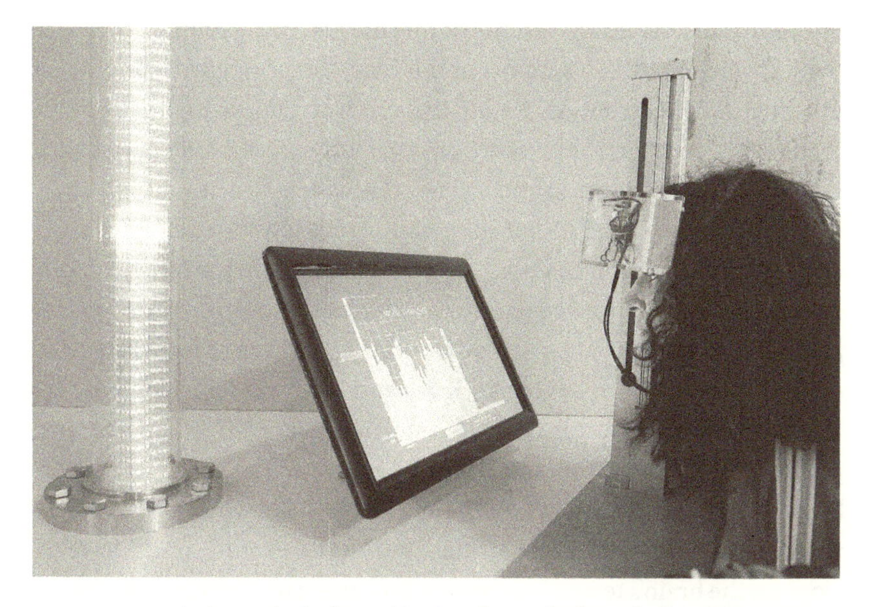

FIGURE 10.2. An interactive brain-machine interface on display at the Perot Museum in Dallas, Texas, USA. Electrical fields caused by neuronal activity can be picked up outside the cells—and in the case of this installation even outside the skull. Intense thinking (e.g., solving a math problem) leads to activity spikes, as seen on the screen. The installation is a "high striker" or "strongman game"—except that not the strength of the arm, but of the players thinking is measured. The closer the recording device to the actual neurons, the more information can be picked up by brain-machine interfaces.
Photo by author; with permission of The Perot Museum of Nature and Science, Dallas, TX, USA

electrodes. From 2009–2022 the BrainGate2 clinical trial has been ongoing to test feasibility in more patients with tetraplegia.[3] Based on these and subsequent successes by other teams,[28, 29] Musk's hope to have their 1,000 electrode chip in a human patient soon definitely does not sound like science fiction. In summer 2020 his team received "Breakthrough Device Designation" from the United States Food and Drug Administration.[30]

Musk's vision differs from current efforts to develop brain-machine interfaces when it comes to the idea of "merging with AI." The idea is that healthy, willing people may want to activate a high-bandwidth input-output extension to their cortex function not only to do things they otherwise could not, but to become symbiotic with an AI that resides in that extension. To Musk this means "mitigation of the existential threat of AI."[6] In support of this idea, the principle technologies of artificially intelligent ANNs and the ability to interface with the brain in the ways described above are both currently successfully developed. Having spent this entire seminar series discussing algorithmic information, we may now ask: What type of information transfer should we envision through such an interface? Max Hodak from the Musk-backed company effort acknowledged what we already know from existing implants of tetraplegic patients: "the first thing you have to learn is how to use it. Imagine you've never had arms, and then suddenly you have an arm, pick up a glass, this is not a cognitive task . . . it is like learning piano. . . . it's a long process."[6] This makes sense. The cortex is built to learn new things. I know of no theoretical reasons why a cortical network could not be extended artificially. But the extension will have to work on the brain's terms. The issue is this: there is no shortcut to learning how to play the piano. The cortex should be able to learn to adapt to any new input-output device that appropriately interacts within its network. If the new device is a prosthetic arm, the cortex will try to learn to use it in time. It will also miss it dearly if the thing is unplugged or the battery runs out. If the new device is itself an ANN with a mind of its own, bidirectional learning should ensue. Slowly, in time. Remember, if the cortex does not store bits of information, but decentralized patterns sufficient for the algorithmic unfolding of information (say your four-digit PIN), how is the interfaced AI going to read that output?

We may know soon enough where this and other remarkable adventures take us. If the past is any indication, then the future will be no linear extension of current technology. Engineering got us pretty far. In the case of AI, it got us as far as running simulations of well-designed ANNs on sufficiently fast computers to allow the feeding and processing of big data. But what the neural nets are actually doing has become some kind of happily ignored alchemy to most of those who use AI as a tool. When Musk declares that "if your biological self dies you can upload it to a new unit, literally,"[25] then we have reason to assume that engineers have a tractable vision of the path forward to solving issues like the bandwidth problem. But have they talked to neurobiologists who are honest enough to admit that they are still not sure how your four-digit PIN is actually stored in the brain? The information that defines each individual had to unfold and was programmed into a growth process by evolutionary principles. This seminar series was an attempt to understand this type of information, how it got into the brain and how it affects how the brain functions. The storage, retrieval and programming of algorithmic information is not an engineering problem, it is an evolutionary problem. Algorithmic information unfolds in a time- and energy-consuming process. The types of shortcuts we choose affect what intelligence we get. And for some things in life, there is just no shortcut. Thank you.

Epilogue

'Hi Alfred. It's Aki calling.'

'Aki! What a jolly surprise! How are your robots?'

'As happy as they've ever been.'

'Not happier?'

'No, not happier.'

'Are you okay, Aki? What's up?'

'All good, I guess. Pramesh sent me his new paper. Did you know about the work he did with Minda?'

'Nope. I don't really read his stuff. But funny if they did a thing together—it wasn't really Minda's cup of tea, I thought.'

'I only met her once briefly after the workshop. And there she sounded more positive. In fact, she kinda made it sound like none of this was really new to her . . . like, she had always thought that way, just with different words.'

'Of course. Nobody is going to define Minda's field for her! The three stages of progress: (1) This can't be. (2) This is trivial. (3) I've always said so. So, what did they do?'

'Honestly, it's a mixed bag. You know how Pramesh does all this computational evolution?'

'Yeah, of course. He doesn't design networks, but simulates neurons with some kind of genome to change synaptic weights in the network, and then selects the one that works best.'

'Kinda. But inclusion of development has remained a challenge. Their genomes usually just encode the network more or less directly.'

'I see, and now he teamed up with Minda to have the genome encode her type of developmental mechanisms for a network? Should keep a high-performance computing cluster busy . . .'

'Well, they did evolve Minda's pet system through some kind of developmental process, yes. It turns out there are multiple solutions, and now they are going back to find those in biology. I don't know, I'm not sure it's very general. That's the problem with our current AI as well—it's always specialized for one task, but never general.

'. . . and why does that suddenly interest you, Aki?'

'Well, the solutions themselves were pretty surprising, and pretty robust. There are probably solutions that we engineers would never have come up with. Come to think of it, the newest and most successful AI approaches are finally finding the best solutions only after learning from scratch, like DeepMind's AlphaZero when it learned chess by self-play. The next version, MuZero, doesn't even need to be told game rules anymore. The neural nets are still built first and trained later, kinda pre-Sperry. But the information content unfolds based on minimal input through reinforcement learning—isn't that algorithmic growth? In robotics the discussion is finally out about what types of intelligence would need to grow. The new field of Developmental Robotics![1–3]

'It all depends on the intelligence you want.'

'Well, I want that one.'

'Why did you call me then and not Pramesh?'

'Pramesh called me, actually. He wants to organize a meeting. But this time with sessions run by each of us, and not just listening to one neuro dude's narrative. We could invite him though, I guess. Anyway, I told Pramesh I'd talk to you about it.'

'I wondered when one of you would come up with that! Although I hesitate to add yet another meeting . . .'

'I know, it would be out of your comfort zone, out of your bias bubble . . .'

'. . . I would enjoy it, is what I was about to say. But thanks, Aki, that reminds me of what I've missed since we met . . .'

'Well, his idea is outside both our bubbles. He was very excited about making it all about evolution and entropy. If it wouldn't be for

Pramesh's polite ways, I'd say he was pretty out of control. I mean, evolution and entropy! What do you think?'

'Whatever that's supposed to mean. I always liked the nonintuitive idea that entropy, or disorder, can only increase in the universe as a whole—yet here we are. And I think I understand how evolution works. But I'm not sure I'd have much to contribute. What's the story?'

'Pramesh felt we only scratched the surface of what he calls the fundamental information question. You know, where brains come from: the sun shooting photons at a rock in space for a few billion years . . .'

'Oh dear. Okay, the sun could shine on a piece of rock for an eternity and do no more than keep it warm. But we are looking at energy powering the evolution of things . . . life, brains, engineers. And once you got engineers, they want to build better brains . . . it's always an algorithmic process.'

'But what's the difference from the rock that just keeps warm? Pramesh says evolution stores energy as information. For so much information locally, there must be a lot of disorder somewhere else in the universe, apparently.'

'But we are in the middle of it, as long as the sun keeps shining! I know a lot of people talking about runaway superintelligence, accelerated acceleration, all that stuff.'

'Do they talk about entropy?'

'Do they talk about evolution?'

'Stop it, Alfred! You know, there is a reason why my brain can only think the things it can think.'

'Yes, that's evolution.'

'I mean, you may love to think something is fundamental just because we don't know. Maybe the fundamental information question is just missing information.'

'Yes, that's information entropy.'

'Gee. Evolution and entropy then. I'm still an engineer. Give me a good old ANN to wrap my brain around and keep the universe in the drawer.'

'Yes, I know the feeling. A healthy sense of proportion . . .'

'Do you know Trin Tragula and the Total Perspective Vortex?'

'Oh no, is it again . . . ? . . . But please . . . tell me all about it . . .'

'He was a dreamer, a scientist I guess, except his wife thought he was an idiot. He spent too much time figuring out the nature of fairy cakes with enormous effort for very little obvious benefit. So she was very upset and insisted he should develop some sense of proportion. And so, the story goes: 'he built the Total Perspective Vortex—just to show her. And into one end he plugged the whole of reality as extrapolated from a piece of fairy cake, and into the other end he plugged his wife: so that when he turned it on she saw in one instant the whole infinity of creation and herself in relation to it. To Trin Tragula's horror, the shock completely annihilated her brain; but to his satisfaction he realized that he had proved conclusively that if life is going to exist in a Universe of this size, then the one thing it cannot afford to have is a sense of proportion.'[4]

'. . . and why exactly are you telling me this?'

'I'm just trying to be helpful. But it's your brain, of course. You can do with it what you want.'

GLOSSARY

activity-dependent process. In neurobiology, changes to a neural network based on electrical (spontaneous or evoked) neuronal activity that occurs as part of normal genetically-encoded development or following environmental influences.

adaptation. In neurobiology, a change of a neuron's development or function in response to a change in its surroundings. In evolutionary theory, a change of a phenotypic outcome based on one or more mutations that are selected in response to a change in the surroundings of the structure or organism exhibiting the phenotype.

adhesion molecule. In biology, a membrane protein that attaches to another molecule on a different surface (inside the same cell or outside the cell).

algorithm. A finite sequence of instructions based on one or more rules where the output of a preceding step in the sequence provides input for a subsequent step of the sequence. *See also* iteration.

algorithmic growth. In this book, a developmental or functional process that consumes energy and time while running an algorithm and leads to an endpoint that requires more information to describe than the initial state (reduced entropy). *See also* unfolding information.

algorithmic function. In this book, a functional process (especially in a neural network) that exhibits the properties of algorithmic growth.

algorithmic probability. A mathematical method to calculate the likelihood that a specific cause (or chain of events) may have led to a given (known) outcome. Also known as Solomonoff probability.

algorithmic information theory. A mathematical framework to calculate how and to what extent a given information content can be described by a smaller amount of information plus the time- and energy-consuming execution of a set of rules (an algorithm). *See also* randomness.

artificial general intelligence (AGI). A poorly defined term that expresses a goal of artificial intelligence research to develop a nonbiological system that performs many tasks (in parallel) in a manner that "would require intelligence if done by men" (*see also* artificial intelligence for part of the definition in quotes). It is unclear whether AGI would need to resemble human intelligence, yet any task-based AI that performs particularly well for a single task is typically described as "(above) human-level AI" for that single task.

artificial intelligence (AI). This book focuses on Marvin Minsky's original definition: "The science of making machines do things that would require intelligence if done by men." There is a continuous debate about the definition and direction of AI with no single agreed-upon definition.

artificial life (ALife). A wide-ranging field of studies that are devised to recreate natural "life-like" processes through software, hardware or chemically.

artificial neural network (ANN). A computational implementation of many interconnected computational units (neurons, typically based on the McCulloch-Pitts neuron) that can receive input, learn through changes of the connection strengths between neurons (synaptic weights) and produce an output. The first such implementation was the perceptron in 1958. ANNs of varying architectures have become the most successful programs in AI, especially since 2012. While the similarity to biological brains is largely inspirational, features learned from biology have been successfully implemented in ANNs, and, as in brains, the precise manner of how a specific information content is stored in an ANN becomes less clear with increasing network complexity.

association study. A correlation between two sets of observations, e.g., alcohol consumption and life span, or genome variations and phenotypic traits. It is difficult, and often impossible, to determine causality between the two parameters captured by the observations.

autapse. In neurobiology, a synapse formed between an axonal (presynaptic) and dendritic (postsynaptic) branch of the same neuron.

autonomous agent. An individual part or component of a larger system that operates according to its own rules in an unknown environment, to which it responds. Interactions of autonomous agents are a basis for self-organization.

axon. In neurobiology, the single cable-like (and sometimes branched) process that carries a neuron's electrical output signal. The axon connects a neuron to the input-receiving dendrites of other neurons.

axon guidance. In neurobiology, a field of study concerned with the directed developmental growth of axons.

backpropagation. A widely used algorithm for the training of artificial neural networks in supervised learning (through changing synaptic weights) based on the calculation and backward flow of error information through the layers of the network (using gradient descent).

bandwidth. In computation and engineering, the transfer rate (i.e., the amount of information that can be passed in a given time interval) between two agents (e.g., a human and a machine or two computer components).

behavioral trait. In biology and psychology, an umbrella term for phenotypic outcomes at the organismal level that encompasses tendencies (probabilities) to exhibit aggression, empathy, alcohol sensitivity, or any other measureable behavior.

blueprint. Originally a reproduction process for technical drawings that produces faithful copies of all the information contained in the original. Based on the algorithmic information approach applied in this book, a blueprint is, therefore, not a good metaphor for the relationship between the genome and a genetically encoded biological structure. *See also* unfolding information.

bottom-up technology. A description for approaches that contribute to the creation of an ordered system from smaller components through self-organization and without a central regulator or available endpoint information.

***Caenorhabditis elegans*.** A very small worm that lives in soil and has become a major genetic model organism to study animal development and behavior. The hermaphrodite (a worm with male and female characteristics) has a neural network of exactly 302 stereotypically arranged neurons.

cell-autonomous property. In biology, a cellular feature or process that does not require input from outside the cell.

cell surface receptor. *See* membrane receptor.

cellular automaton. A mathematical or computational model of (usually binary) cells in a (one- or higher-dimensional) lattice where the cells change states in time based on a set of rules that incorporates the states of the cell itself and its neighbors. Depending on the (usually very simple) rules, some cellular automata can display surprising life-like pattern dynamics and infinite complexity with infinite runtime. *See also* rule 110.

chemoaffinity theory. A theory that focuses on molecular interactions between pre- and postsynaptic neurons during the development of biological neural networks. Formulated at a time when non-molecular theories and learning were favored to explain wiring specificity, the original proposal by Roger Sperry envisioned strict and rigid molecular matching between synaptic partners in the brain (called "strict chemoaffinity" in this book). Modern developmental neuroscience often refers to chemoaffinity for a wider range of molecular interactions that contribute to neural network development (called "general chemoaffinity" in this book).

circadian clock. In biology, a molecular and cellular mechanism based on biochemical oscillation and feedback that cycles with a daily phase and is synchronized with solar time. The circadian clock in animals optimizes metabolic, neuronal and many other bodily functions based on a predictable day-night cycle.

cognitive bias. In psychology, a systematic deviation of an individual's perception of reality.

complexity. An overused term. In algorithmic information theory, (Kolmogorov) complexity is related to compressibility, e.g., a series of 1 million "0s" are easily compressed and not complex. However, by this definition a completely random series of "0s" and "1s" is maximally complex. More common and less well-defined applications of the idea of complexity therefore focus on non-random, highly ordered structures of unknown compressibility that are typically the outcome of algorithmic growth, e.g., the brain.

composite instruction. In this book, the set of components and parameters that must be in place and work together to ensure a specific decision made by an autonomous agent during algorithmic growth.

connectionism. In neuroscience, a range of ideas that focuses on the information content of neuronal connections to describe higher order properties of neural networks.

connectome. In neuroscience, a comprehensive description of neurons and their connections in a biological neural network. In its simplest form, the description only assigns (synaptic) connections based on anatomical observation without functional information. More complicated descriptions can include information about the nature and strength of individual synapses, as well as probabilities for their occurrence in different individuals.

convolutional neural network. A class of multi-layered (deep) artificial neural networks that are inspired by the biological visual cortex and that are particularly well-performing in visual recognition tasks.

cortex. In neurobiology, the cerebral cortex is the outermost and most recently evolved layer of the mammalian brain responsible for functions ranging for sensory integration to producing motor output (mostly in the six layers of the neocortex). The expansion of the cortex in primates correlates with levels of intelligent behavior.

cortical column. In neurobiology, a repetitive anatomical feature of the cortex characterized by shared input-specific response properties.

cybernetics. A historic, interdisciplinary field originally defined by Norbert Wiener in 1948 as "the scientific study of control and communication in the animal and the machine."

deep learning. An umbrella term for several machine learning approaches based on multilayered (deep) artificial neural networks.

dendritic tree. In neurobiology, the (often highly) branched structure that receives other neurons' electrical output signals. The dendritic tree receives and integrates the input information required for the neuron to decide on its own output signals.

deterministic process. An event in time that is 100% predictable and reproducible based on previous events. *See also* stochastic process.

differentiation. In biology, a process whereby a developing cell acquires a sufficiently new state to be classified as a new cell type, e.g., a specific subtype of neuron.

dissipative system. An energy-consuming (transforming) system that irreversibly transitions from one state to another state.

Drosophila melanogaster. The fruit fly is a genetic model organism since more than 100 years. It initially played key roles in the study of genetics, later developmental biology, and today almost any aspect of biomedical research. The connectome of the more than 100,000 neurons of the fly's brain is being assembled at the time of writing of this book, the first connectome of a brain on this scale.

Escherischia coli. A widely studied gut bacterium that is central to molecular biology-based research in laboratories across the world.

electrophysiology. In neuroscience, a range of methods to study the electrical properties of neurons. Electrophysiological experiments typically implicate recordings of voltage or current changes using electrodes inside or outside of neurons.

emergent property. A term that describes a feature of a system that was not a feature of any of the components that make up that system. Typically, an emergent property cannot be predicted based on the study of individual component properties, but arises from component interactions.

endpoint information. In this book, the information necessary and sufficient to describe (and fully depict) a phenotype, i.e., the outcome of a growth process, without consideration of how it came to be.

energy landscape. A graphical depiction and mapping of possible states of a system such that maxima and minima are represented as "hills" and "valleys."

entropy. The measurable property of disorder and uncertainty. Maximal entropy represents maximal disorder of a system. Entropy is a core concept of thermodynamics and information theory alike, linking energy and information. *See also* information entropy.

enzyme. In biology, a molecule (protein or RNA) that accelerates (catalyzes) a chemical reaction in cells and that often functions in a metabolic process required for cell function.

equipotentiality. A neurobiological theory formulated by Karl Lashley in the 1920s based on the idea that different parts of the brain (especially the cortex) can take over each other's roles when one part is damaged. While we now know that there are limits to this ability, the idea gave rise to an understanding of plasticity that is central to brain function.

evolutionary programming. In this book, the concept that a meaningful and heritable change to a biological neural network can only be achieved through random mutations in the genome and subsequent selection at the level of the phenotypic output after the entire developmental process is concluded. The concept is based on the hypothesis that the effect of a mutation is fundamentally unpredictable in the absence of any knowledge of previous outcomes due to the nature of algorithmic growth.

evolutionary selection. The fundamental idea of Charles Darwin that all naturally observed phenotypic outcomes came to be through selection based on a sufficiently large pool of variation and sufficient rounds of selection that allowed for gradual changes. Darwin distinguished natural selection in biology from intentional artificial selection by an experimenter.

exaptation. In evolutionary theory, the process that leads a trait or feature that happened to be available to serve a new selectable purpose. The exapted feature usually comes to be as a side-effect of another adaptation and is likely to be subject to its own adaptation after the initial repurposing event. *See also* spandrel.

excitability. In neurobiology, the property of neurons to very quickly generate electrical signals (action potentials) following above-threshold stimulation based on a positive feedback loop-controlled redistribution of charged molecules (ions) across the neuronal membrane.

feedback. In biology, engineering and computer science, the process of using (rerouting) output information of a cause-and-effect chain as input of the same chain, creating a loop. Feedback can be positive or negative, has been described as the basis of the field of cybernetics and is a key contributor in the generation of emergent properties.

feedforward neural network. An artificial neural network with unidirectional information flow through the layers from the input to the output layer (as opposed to bidirectional flow in a recurrent neural network), e.g., the original perceptron.

filopodium. A cellular membrane protrusion of finger-like appearance. Filopodia typically exhibit fast extension-retraction dynamics and have been implicated in the sensing of environmental cues, e.g., in the case of neuronal growth cone guidance.

floor plate. In biology, the ventral midline of the vertebrate spinal cord, which is central to the development of bilateral symmetry of the nervous system. The floor plate has been the focus of the field of axon guidance in the study of axonal midline crossing for many years.

genetic encoding. In biology, a term typically used to distinguish the genome as the source of information for the development or function of a biological structure as opposed to environmental information (learning). Stochastic processes and spontaneous neuronal activity can be part of a genetically encoded developmental program prior to the ability to sense or learn from environmental information. Hence, genetic encoding does not need to be deterministic.

gene map. In biology, a graphical depiction of the linear arrangement of genes and regulatory regions along the DNA (genome).

gene product. An RNA or protein that is the direct outcome of reading and transforming the genetic information of the DNA through transcription (DNA to RNA) and translation (RNA to protein).

genome. In biology, the genetic information in the DNA. The genome contains genes and regulatory regions, but does not depict the outcome of algorithmic growth processes, because such outcomes contain information resulting from continuous feedback processes between the genome and its gene products.

genome-wide association study (GWAS). *See* association study.

gradient descent. In computation and mathematics, an algorithm based on iterative optimization to find a local minimum that is representable in an energy landscape. *See also* backpropagation.

growth cone. In neurobiology, the distal end of a developing neuronal axon. The growth cone exhibits exploratory behavior and plays a central role in axon guidance.

growth factor. In biology, a chemical that promotes growth of a cell or an axonal growth cone. The precise timing and location of growth factor availability contains developmentally unfolded information that can play key roles for when and where cells mature or axons grow.

guidance cue. In neurobiology, the functional description of various secreted or surface-bound molecules that play roles in axon pathfinding by contributing to the guidance of the growth direction of a growth cone.

hard-wired neural network. In neurobiology, a neural network that is exclusively genetically encoded. Since spontaneous neuronal activity in the absence of environmental input can be part of genetic encoding, common usages of the term 'hard-wired' are often further restricted to exclude activity-dependent processes.

Hebb's rule. A basic concept formulated by Donald Hebb in 1949 describing how the strength of synaptic connections between neurons can change depending on stimulation. *See also* synaptic weight and plasticity.

heuristic. A problem-solving and decision-making approach based on incomplete information and guessing that does not try to find the optimal solution, but a sufficiently good solution while saving time and energy.

homophilic repulsion. In neurobiology, a functional property of cell surface protein interactions that mixes molecular behavior (homophilic, i.e., attractive same-type molecular binding) and cellular behavior (repulsion of filopodia or entire branches of axons or dendrites). The apparent oxymoron of "homophilic repulsion" is therefore resolved by (and an exemplification of) the levels problem.

housekeeping gene. In biology, an umbrella term for genes encoding gene products that serve widespread functions in many or all cells of an organism, e.g., a metabolic enzyme. The same gene product of a housekeeping gene may serve very different and highly specialized functions as part of composite instructions depending on the molecular and cellular context.

information entropy. In information theory, the amount of surprise or uncertainty associated with the outcome of a random process. The higher the predictability of the outcome, the lower the information entropy. Also called Shannon entropy.

information theory. A field of study concerned with the transmission and storage of information, founded by Claude Shannon. *See also* algorithmic information theory.

information unfolding. *See* unfolding information.

innate behavior. In biology and psychology, the instinct of an animal to exhibit a specific behavior. *See also* genetic encoding.

instructive mechanism. In biology, a molecular or cellular process that suffices to explain a specific (deterministic) molecular or cellular outcome, e.g., the instruction for a growth cone to grow in a specific direction. Every molecule's contribution to an instructive mechanism depends on molecular and cellular context, making the contribution a quantitative parameter on a gradient of what biologists label as "instructive" versus "permissive" with no agreed threshold to distinguish between the two. *See also* composite instruction and permissive mechanism.

intelligence. A poorly defined term to describe non-trivial information processing and output, commonly associated with biological or artificial neural network function. The term is further used to describe nontrivial information processing and output of other social (biological) and artificial (engineered) network structures of interacting autonomous agents.

invariant representation. In the context of biological and artificial neural network function, the idea that certain aspects of a memory appear to remain stable even though individual neurons or synapses change.

iteration. The repeated application of a rule using the output of the preceding step as the input of the subsequent step. A rule-based sequence of iterations constitutes an algorithm.

knock-out. In molecular genetics, the ablation of a gene through experimental genome manipulation for the purpose of studying development or function of a biological system in the absence of the gene and its products (loss-of-function study).

lateral inhibition. *See* Notch signaling.

levels problem. In this book, the phenomenon that the function of a gene or gene product at one level (e.g., its molecular mechanism) may not allow to predict its normal (wild type) or mutant effects on a process at a different level (e.g., behavior) due to unfolding information of algorithmic feedback processes.

ligand. In molecular biology, a molecule that binds to another molecule that functions as receptor.

lock-and-key mechanism. In molecular biology, the specific binding of two molecules, usually implying the exclusion of others. At the cellular level, the idea of molecular locks and keys has been proposed to explain neural network wiring specificity. *See also* chemoaffinity theory.

machine learning. An umbrella term for the algorithm-based improvement of computer performance through training (data input or self-play). Machine learning is commonly subdivided into supervised learning, unsupervised learning and reinforcement learning.

McCulloch-Pitts neuron. The basic concept and mathematical framework by Warren McCulloch and Walter Pitts from 1943 for a computational unit inspired by biological neurons that changes output strengths (synaptic weights) depending on inputs.

membrane receptor. In biology, a protein that resides in a cellular membrane and can bind to ligands. If the membrane receptor resides on the cell surface, binding to ligands outside the cell typically results in adhesion or intracellular signaling that changes cellular behaviors.

metabolism. In biology, the chemical reactions that sustain a living system and differentiate it from a nonliving system.

midline crossing. *See* floor plate.

molecular identification tag. In the context of the chemoaffinity theory, one or more molecules (usually membrane receptors on the cell surface) that uniquely define a neuron or synaptic partner for the purpose of synaptic matchmaking. More recently, the term has also been applied to membrane receptors that serve different roles of a unique (or close to unique) molecular identity, e.g., recognition of "self" versus "non-self" and self-avoidance.

monozygotic twins. In biology, two genetically identical individuals, resulting from the division of a single fertilized egg into two separately developing embryos.

multilayer perceptron. A class of feedforward artificial neural networks based on the addition of neuronal layers to the original single-layer perceptron designed by Frank Rosenblatt.

mutation rate. In biology, the frequency of changes of the DNA sequence of a gene or the genome of an organism.

nanotechnology. Engineering at the atomic scale. Molecular nanotechnology originally referred to the idea of controlled molecular assembly. *See also* bottom up technology.

neocortex. *See* cortex.

neural circuit. In neurobiology, a synaptically interconnected set of neurons that serves a defined function.

neuromodulator. In neurobiology, a chemical that acts on (usually many) neurons in a spatially and temporally defined manner. The function of a neuromodulator thereby depends on molecular diffusion and tissue properties to change network properties.

neuronal activity. In neurobiology, the electrical processes that lead to fast communication between neurons. *See also* activity-dependent process.

neuronal excitability. *See* excitability.

neuropil. In neurobiology, a brain region containing synaptic contacts, but no neuronal cell bodies.

neurotransmission. In neurobiology, the process of passing a signal from one (presynaptic) neuron to the a second (postsynaptic) neuron by converting an electrical signal of the presynaptic neuron into a chemical signal (diffusion of neurotransmitter molecules) and back into an electrical signal in the postsynaptic neuron.

Notch signaling. In developmental biology, a molecular mechanism based on membrane receptor interactions between cells and a negative feedback loop that ensures a single "winner" cell can never be in contact with another "winner" cell. This can lead to patterning of a two-dimensional tissue, a process also called lateral inhibition. More generally, Notch signaling enforces symmetry breaking by determining a winner and a loser; however, which individual cell wins or loses is a stochastic process.

optogenetics. In biology, a set of techniques that change cellular properties and function with light based on molecularly engineered, genetically encoded proteins that convert light energy into molecular action.

parameter space. In a mathematical model, the entirety of all possible contributors and their values (parameters) that could theoretically be explored.

penetrance. In genetics, the proportion of individuals that exhibit a specific phenotype based on the same genotype or mutation.

perceptron. The first computer implementation of an artificial neural network by Frank Rosenblatt in 1958, based on an input layer, an output layer and a single intermediate layer of neurons with random connectivity. Despite the perceptron's remarkable success, the inability to compute the logical XOR function (later solved with multilayer perceptrons) led to repercussions that slowed neural network research for many years.

permissive mechanism. In biology, a molecular or cellular process that "runs in the background" and does not determine a specific molecular or cellular outcome, e.g., metabolism that is required for the neuron to grow an axon, but contains little or no information about its growth direction. However, a unique composite of several permissive mechanisms in a specific cell at a specific time may constitute a highly specific instruction. *See also* composite instruction and instructive mechanism.

phenotype. In biology, the observable output state of a developmental or functional process based on genetic and environmental information.

plasticity. In biology, the ability of a developing or functional structure to adapt to changes in its surroundings. In neural network function, the ability of biological or artificial neural networks to learn and change depending on environmental input.

pre-specification. In this book, the concept that preceding algorithmic growth (the developmental history) has set up a state of the system that contains the information for a specific developmental decision (instruction), e.g., the pre-specification of two synaptic partner neurons that have been brought together at a specific place and time. *See also* post-specification

polygenicity. In genetics, the concept that a phenotype or the penetrance of its occurrence cannot be explained by a single gene or mutation. For example, there are many single gene mutations or single nucleotide polymorphisms (SNPs) that contribute to varying degrees to the phenotypic penetrance of common neurodevelopmental disorders, leading to a situation where the cause is partially unknown and labeled as "polygenic."

post-specification. In this book, the concept that a specific state of a system, e.g., specific synaptic connectivity, only occurs after the initial development of a less specific state of the system with less information content followed by a sculpting or pruning process that only leaves more specific (information containing) parts in place. *See also* pre-specification and pruning.

predisposition. In genetics, the concept that a sensitized background or developmental history (including environmental influences) increases the probability for the occurrence of a phenotype.

probability theory. In mathematics, a framework for the calculation of uncertainty and the likelihood of outcomes depending on knowledge about a (usually nondeterministic) system.

probabilistic biasing. In this book, the concept that a molecular or cellular mechanism quantifiably contributes to the likelihood of a developmental or functional outcome. For example, molecular interactions between neurons can increase the probability for synapse formation without making this genetic contribution deterministic.

programmed cell death. In biology, the well-characterized molecular and cellular mechanism that leads to the controlled removal of a cell (also called apoptosis). During most multicellular organisms' development, controlled programmed cell death is a common mechanism of the normal algorithmic growth program.

pruning. In neurobiology, the removal of synapses or axonal or dendritic branches during development or function. *See also* post-specificiation.

randomness. This book uses a definition based on algorithmic information theory: a process or system is random when the smallest amount of information that can describe the system is a full description of the system itself, i.e., the system is not compressible through a time- and energy-consuming process. *See also* stochastic process.

receptor protein. *See* membrane receptor.

recognition molecule. *See* molecular identification tag.

recurrent neural network. An artificial neural network with bidirectional information flow through the layers creating feedback loops (as opposed to unidirectional flow in a feed-forward neural network). The recurrent wiring creates memory, allows for processing of sequences, and has been particularly successful in learning sequential information, e.g., in speech recognition.

reductionism. A scientific and philosophical approach to understand a system based on its components. Sometimes reductionism is understood as implying that a system may be the simple sum of its parts, while other interpretations of reductionism focus on understanding components and rules that can give rise to emergent properties without trying to reduce the emergent properties themselves.

redundancy. In computer technology, the duplication of an information content, i.e., redundant identical hard drive backups. In genetics, the situation that the disruption of a single gene or mechanism does not (or only with low penetrance) causes a phenotype, because other gene functions and mechanisms can compensate for the disrupted function. In neural networks, the term is often used to describe the phenomenon that individual neurons or synaptic connections can be disrupted without majorly affecting network performance, including memory.

reinforcement learning. A form of training of artificial neural networks (or other learning algorithms) based on the concept of autonomous agents receiving a reward or benefit at the end of a process (e.g., a game of chess), i.e., learning through experience. *See also* machine learning.

resonance theory. In biology, a historic theory by Paul Weiss that focused on the role of plasticity and induced specificity in the development of neuronal connectivity.

reticular theory. In biology, a historic and incorrect speculation that the nervous system is a single continuous entity as opposed to being comprised of individual interconnected neurons.

retinal ganglion cell. The action potential firing output neuron of the vertebrate retina.

retinotectal projection. The axonal connections between the vertebrate eye and the brain established by retinal ganglion cells.

robustness. In biology, the quantifiable degree of invariance (stability) of a phenotype to a specific perturbation factor, e.g., environmental variability or genetic mutations.

rule. In this book, a set of one or more instructions sufficient to generate a specific behavior (output) of system components or autonomous agents given a specific input.

rule 110. A simple one-dimensional cellular automaton with proven Turing completeness, i.e., rule 110 is the simplest known universal Turing machine. *See also* unfolding information.

selection. *See* evolutionary selection.

self-organization. A process that increases the information content (reduces entropy) of a system of interacting components (autonomous agents) starting from an initially less ordered state of that system and requiring time and energy.

self-assembly. In this book, the series and interactions of self-organizing processes that constitute algorithmic growth, characterized by unfolding information (entropy reduction) and the consumption of energy in time. More commonly, self-assembly describes the spontaneous interactions of components leading to a more ordered state of a system comprised of the components.

self-avoidance. In neurobiology, a developmental process whereby stochastically growing axonal or dendritic branches of the same neuron molecularly recognize each other (but not non-self branches from other neurons), leading to spreading of the branched structure.

sensitized background. In genetics, the concept that a specific genotype or mutation increases the probability for a phenotype only if other mutations or environmental factors push the system above a threshold for the occurrence of that phenotype with a given penetrance. *See also* predisposition.

sequencing. In biology, a set of techniques that allow to read the primary sequence of the DNA, RNA or the amino acid sequence of proteins. Much of biology focuses on relating the information content of these primary sequences to observations in living systems.

Shannon entropy. *See* information entropy.

single-nucleotide polymorphism (SNP). In biology, a naturally occurring variation of a single base pair in the DNA (genome).

spandrel. In evolutionary biology, a feature or trait that occurs as the byproduct of the evolutionary adaptation of another feature or trait. *See also* exaptation.

splicing. In biology, the molecular process of producing different proteins from the same gene locus on the DNA through the generation of different (alternative) intermediate messenger RNA molecules.

spontaneous activity. In neurobiology, the property of neurons and their networks to produce and transmit electrical activity without external stimulus. *See also* excitability.

stigmergy. In biology and the study of self-organizing systems more generally, self-organization based on indirect interactions of autonomous agents through traces in the environment, rather than direct agent interactions.

stochastic process. In probability theory, an event in time that is not reproducible or predictable based on knowledge of previous events, e.g., rolling dice. *See also* deterministic process and randomness.

supervised learning. A form of training of artificial neural networks (or other learning algorithms) based on examples, e.g., a large collection of handwritten numbers, each with a correct digital representation as 0..9. An ANN trained this way will classify everything as 0..9, also the picture of an elephant.

symbol-processing artificial intelligence. The processing of representations (e.g., alphanumerical symbols) and syntax based on formal logic to produce "intelligent" computational behavior. Symbol-processing AI was the leading type of AI until the deep learning revolution around 2012.

symmetry-breaking mechanism. *See* Notch signaling.

synapse. In neurobiology and artificial neural networks, the contact point where information is passed between two (and sometimes more) neurons.

synaptic matchmaking. In developmental neurobiology, the concept that two neurons form a synaptic contact based on a molecular key-and-lock mechanism that is specific to this pair of neurons under exclusion of others.

synaptic weight. The connection strength (information transmission strength) at the contact point between two biological or artificial neurons. *See also* Hebb's rule and McCulloch-Pitts neuron.

synaptotropic growth. In developmental neurobiology, a simple rule set that leads to the growth of neuronal branches in the direction where most synaptic contacts are found. Synaptotropic growth is based on stochastic axonal or dendritic branching and selective stabilization and creation of new branch origins upon contact.

top-down technology. A description for approaches to create or understand an ordered system based on available endpoint information, including through break-down into subsystems or reverse engineering.

transcript. In biology, the RNA molecule generated from reading the DNA sequence of a gene.

transcription factor. In biology, a gene-encoded protein that itself binds to regulatory regions of the DNA (usually of other genes, but sometimes also the gene encoding for itself) to positively or negatively regulate expression of genes under control of that regulatory region. A cascade of transcription factors that regulate each other's expression is a prime example for unfolding information based on feedback between the genome and its own products.

transcriptome. In biology, the set of all RNAs, i.e., the molecules read (transcribed) from DNA (the genome).

translation. In biology, the process of generating a protein from a messenger RNA.

Turing completeness. In computational theory, a system that can simulate any Turing machine, i.e., a system capable of calculating any computation based on a limited set of states, a transition (computation) function and infinite memory storage.

undecidability. In computational theory, a problem for which it is proven that the answer (or outcome) cannot be determined (formally calculated) with certainty.

unfolding information. In this book, the concept that more information is required to describe the endpoint of an algorithmic growth process than the information required to describe the initial state from which the system developed (reduced entropy). The missing information between the initial state and the outcome originates in a time- and energy-consuming process.

unsupervised learning. A form of training of artificial neural networks (or other learning algorithms) devised to find unknown patterns in data without being given examples of these patterns. *See also* machine learning.

REFERENCES

Section 1 (Introduction, First Discussion, The Historical Seminar)

1. Sperry, R. W. Embryogenesis of behavioral nerve nets, in *Organogenesis.* (eds. R. L. Dehaan & H. Ursprung) (New York: Holt, Rinehart and Winston, 1965).
2. Reppert, S. M. & de Roode, J. C. Demystifying monarch butterfly migration. *Curr Biol* **28**, R1009–R1022 (2018).
3. Mouritsen, H. et al. An experimental displacement and over 50 years of tag-recoveries show that monarch butterflies are not true navigators. *Proc Natl Acad Sci* **110**, 7348–7353 (2013).
4. Clancy, K. Is the brain a useful model for artificial intelligence? *Wired* **28.06**, https://www.wired.com/story/brain-model-artificial-intelligence/ (2020).
5. Crevier, D. *AI: The Tumultuous History of the Search for Artificial Intelligence.* (New York: Basic Books, 1993).
6. Wolfram, S. *A New Kind of Science.* (Champaign, IL: Wolfram Media, 2002).
7. LeCun, Y., Bengio, Y. & Hinton, G. Deep learning. *Nature* **521**, 436–444 (2015).
8. Alpaydin, E. *Machine Learning.* (Cambridge, MA: MIT Press, 2016).
9. Zador, A. M. A critique of pure learning and what artificial neural networks can learn from animal brains. *Nat Commun* **10**, 3770 (2019).
10. Hassabis, D., Kumaran, D., Summerfield, C. & Botvinick, M. Neuroscience-inspired artificial intelligence. *Neuron* **95**, 245–258 (2017).
11. Hinton, G., in *Brain & Cognitive Science—Fall Colloquium Series.* https://www.youtube.com/watch?v=rTawFwUvnLE (Cambridge, MA: MIT, 2014).
12. Tonegawa, S., Liu, X., Ramirez, S. & Redondo, R. Memory engram cells have come of age. *Neuron* **87**, 918–931 (2015).
13. Marder, E. Neuromodulation of neuronal circuits: Back to the future. *Neuron* **76**, 1–11 (2012).
14. Dayan, P. Twenty-five lessons from computational neuromodulation. *Neuron* **76**, 240–256 (2012).
15. Adams, D. N. *The Hitchhiker's Guide to the Galaxy.* (London: Pan Books, 1979).
16. Feynman, R. On his blackboard at time of death in 1988; as quoted in the *Universe in a Nutshell* by Stephen Hawking (1988).
17. von Waldeyer-Hartz, W. Über einige neuere Forschungen im Gebiete der Anatomie des Zentralnervensystems. *Deutsche medizinische Wochenschrift* **17** (1891).
18. von Waldeyer-Hartz, W. *Über einige neuere Forschungen im Gebiete der Anatomie des Zentralnervensystems.* (Schleswig-Holstein, Germany: Hansebooks, 2018).

19. Finger, S. *Origins of Neuroscience: A History of Explorations into Brain Function.* (Oxford: Oxford University Press, 1994).
20. Golgi, C. The neuron doctrine—Theory and facts. *Nobel Lecture December 11, 1906* (1906).
21. von Gerlach, J. Von dem Rückenmark, in *Handbuch der Lehre von den Geweben des Menschen und der Thiere.* (ed. S. Stricker) 663–693 (Leipzig, Germany: Engelmann, 1871).
22. López-Muñoz, F., Jesús, B., & Alamo, C. Neuron theory, the cornerstone of neuroscience, on the centenary of the Nobel Prize award to Santiago Ramón y Cajal. *Brain Research Bulletin* **70**, 391–405 (2006).
23. Sabbatini, R. Neurons and synapses: The history of its discovery. *Brain & Mind Mag* (2003).
24. Harrison, R. G. The outgrowth of the nerve fiber as a mode of protoplasmic movement. *J Exp Zool* **9**, 787–846 (1910).
25. Abercrombie, M. Obituary—Ross Granville Harrison, 1870–1959. *Biograph Mem Fellows Roy Soc* (1961).
26. Overton, J. Paul Alfred Weiss. *Biograph Mem* **72**, 373–386. (Washington, DC: National Academy of Sciences, 1997).
27. Weiss, P. Selectivity controlling the central-peripheral relations in the nervous system. *Biol Rev*, **1**, 494–531. (1936).
28. Weiss, P. *Principles of Development: A Test in Experimental Embryology.* (New York: Henry Holt and Company, 1939).
29. Sperry, R. W. The effect of crossing nerves to antagonistic muscles in the hind limb of the rat. *J Comp Neurol* **75** (1941).
30. Shepherd, G. M. *Creating Modern Neuroscience: The Revolutionary 1950s.* (Oxford: Oxford University Press, 2009).
31. Mitchell, K. J. *Innate.* (Princeton, NJ: Princeton University Press, 2018).
32. Sperry, R. W. Functional results of muscle transplantation in the hind limb of the albino rat. *The Anatomical Record* **75**, 51 (1939).
33. Sperry, R. W. Reestablishment of visuomotor coordinations by optic nerve regeneration. *The Anatomical Record* **84**, 470 (1942).
34. Sperry, R. W. Visuomotor coordination in the newt (*Triturus viridescens*) after regeneration of the optic nerve. *J Comp Neurol* **79**, 33–55 (1943).
35. Sperry, R. W. Chemoaffinity in the orderly growth of nerve fiber patterns and connections. *Proceedings of the National Academy of Sciences* **50**, 703–710 (1963).
36. Hunt, R. K. & Cowan, W. M. The chemoaffinity hypothesis: An appreciation of Roger W. Sperry's contribution to developmental biology, in *Brain Circuits and Functions of the Mind. Essays in Honor of Roger W. Sperry* (ed. C. Trevarthen) 19–70 (Cambridge: Cambridge University Press, 1990).
37. Sperry, R. W. Regulative factors in the orderly growth of neural circuits. *Growth* **10**, 63–87 (1951).
38. Sperry, R. W. Effect of 180 degree rotation of the retinal field on visuomotor coordination. *J Comp Neurol* **92** (1943).
39. Gefter, A. The Man Who Tried to Redeem the World with Logic. *Nautilus* **21** (2015).
40. McCulloch, W. S., Pitts, W. A logical calculus of the ideas immanent in nervous activity. *Bull of Math Biophys* **5**, 115–133 (1943).

41. Hebb, D. O. *The Organization of Behavior: A Neuropsychological Theory*. (Oxford: Wiley, 1949).
42. Shannon, C. E. A mathematical theory of communication. *Bell System Tech J* **27**, 379–423 (1948).
43. Solomonoff, R. J. A formal theory of inductive inference. Part I. *Inform Contr* **7**, 1–22 (1964).
44. Solomonoff, R. J. A formal theory of inductive inference. Part II. *Inform Contr* **7**, 224–254 (1964).
45. Solomonoff, R. *A Preliminary Report on a General Theory of Inductive Inference*. AFOSR TN-60–1459, **V-131** (US Air Force Office of Technical Research; Cambridge: Zator Company, 1960).
46. Bernstein, J. A. I. Interview with Marvin Minsky. *New Yorker* (1981).
47. Pitts, W., McCulloch, W. S. How we know universals the perception of auditory and visual forms. *Bull Math Biophys* **9**, 127–147 (1947).
48. Russell, S. J., Norvig, P. *Artificial Intelligence. A Modern Approach*. (Englewood Cliffs, NJ: Prentice Hall, 1994).
49. Solomonoff, R. & Rapoport, A. Connectivity of random nets. *Bull Math Biophys* **13**, 1–10 (1951).
50. Solomonoff, R. The discovery of algorithmic probability. *J Comp Sys Sci* **55**, 73–88 (1997).
51. Sperry, R. W. Review of: The Formation of Nerve Connections by R. M. Gaze. *Quarterly Rev Biol* **46**, 198 (1971).
52. von Neumann, J. Probabilistic logics and the synthesis of reliable organisms from unreliable components, in *Automata Studies, in Ann Math Stud* **34**. (ed. C. E. Shannon & McCarthy, J.) (Princeton, NJ: Princeton University Press, 1956).
53. Winograd, S., Cowan, J. *Reliable Computation in the Presence of Noise*. (Cambridge, MA: MIT Press, 1963).
54. Arbib, M. A. *Brain, Machines, and Mathematics*. (Maidenhead, UK: McGraw-Hill, 1964).
55. Gaze, R. M. *The Formation of Nerve Connections*. (London: Academic Press, 1970).
56. Sperry, R. W. Physiological plasticity and brain circuit theory, in *Biological and Biochemical Bases of Behavior*. (ed. H. F. Harlow, Woolsey, C. N.) 401–424 (Madison, University of Wisconsin Press, 1958).
57. Shatz, C. J. Emergence of order in visual system development. *Proc Natl Acad Sci* **93**, 602–608 (1996).
58. Katz, L. C. & Shatz, C. J. Synaptic activity and the construction of cortical circuits. *Science* **274**, 1133–1138 (1996).
59. Hassan, B. A. & Hiesinger, P. R. Beyond molecular codes: Simple rules to wire complex brains. *Cell* **163**, 285–291 (2015).
60. Lashley, K. S. In search of the engram. *Symp Soc Exp Biol* **4**, 454–483 (1950).
61. Lashley, K. S. *Brain Mechanisms and Intelligence*. (New York: Hafner, 1929).
62. Cowan, J. D. The engineering approach to the problem of biological integration, in *Symposium on Cybernetics of the Nervous System*. (ed. N.Wiener, & Schadé, J. P.) (New York: Elsevier, 1962).
63. Minsky, M. Logical versus analogical or symbolic versus connectionist or neat versus scruffy. *AI Mag* **12**, 34–51 (1991).
64. Friedberg, R. M. A learning machine, Part I. *IBM J Res Devel* **2**, 2–13 (1958).

65. Friedberg, R. M., Dunham, B. and North, T. A learning machine: Part II. *IBM J Res Devel* **3**, 282–287 (1959).
66. Holland, J. A universal computer capable of executing an arbitrary number of subprograms simultaneously. *Proc Eastern Joint Comp Conf*, 108–112 (1959).
67. Rosenblatt, F. The perceptron: A probabilistic model for information storage and organization in the brain. *Psychol Rev* **65**, 386–408 (1958).
68. Rosenblatt, F. *Principles of Neurodynamics: Perceptrons and the Theory of Brain Mechanisms.* (Washington, DC: Spartan Books, 1961).
69. Minsky, M. & Papert, S. *Perceptrons: An Introduction to Computational Geometry.* (Cambridge, MA: MIT Press, 1969).
70. Hawkins, J., Blakeslee, Sandra *On Intelligence: How a New Understanding of the Brain Will Lead to the Creation of Truly Intelligent Machines.* (New York: St. Martin's Griffin, 2005).
71. Attardi, D. G. & Sperry, R. W. Preferential Selection of Central Pathways by Regenerating Optic Fibers. *Exp Neurol* **7**, 46–64 (1963).
72. Gaze, R. M. & Sharma, S. C. Axial differences in the reinnervation of the goldfish optic tectum by regenerating optic nerve fibres. *Exp Brain Res* **10**, 171–181 (1970).
73. Gaze, R. M. & Keating, M. J. The visual system and "neuronal specificity." *Nature* **237**, 375–378 (1972).
74. Sperry, R. W. Models, new and old, for growth of retinotectal connections, in *From Theoretical Physics to Biology.* (ed. M. Marois) 191–215 (Amsterdam: North-Holland Publishing, 1975).
75. Meyer, R. L. & Sperry, R. W. Retinotectal specificity: Chemoaffinity theory, in *Neural and Behavioral Specificity. Studies on the Development and Behavior and the Nervous System* **3** 111–149 (1976).
76. Murray, M. Regeneration of retinal axons into the goldfish optic tectum. *J Comp Neurol* **168**, 175–195 (1976).
77. Schmidt, J. T. Retinal fibers alter tectal positional markers during the expansion of the retinal projection in goldfish. *J Comp Neurol* **177**, 279–295 (1978).
78. Sharma, S. C. & Romeskie, M. Immediate "compression" of the goldfish retinal projection to a tectum devoid of degenerating debris. *Brain Res* **133**, 367–370 (1977).
79. Gierer, A. Model for the retino-tectal projection. *Proc R Soc Lond B Biol Sci* **218**, 77–93 (1983).
80. Yoon, M. G. Neural reconnection between the eye and the brain in goldfish, in *Brain Circuits and Functions of the Mind. Essays in Honor of Roger W. Sperry.* (ed. C. Trevarthen) 86–99 (Cambridge: Cambridge University Press, 1990).
81. Yoon, M. G. Retention of the original topographic polarity by the 180 degrees rotated tectal reimplant in young adult goldfish. *J Physiol* **233**, 575–588 (1973).
82. Yoon, M. G. Readjustment of retinotectal projection following reimplantation of a rotated or inverted tectal tissue in adult goldfish. *J Physiol* **252**, 137–158 (1975).
83. Cheng, H. J., Nakamoto, M., Bergemann, A. D. & Flanagan, J. G. Complementary gradients in expression and binding of ELF-1 and Mek4 in development of the topographic retinotectal projection map. *Cell* **82**, 371–381 (1995).
84. Drescher, U. et al. In vitro guidance of retinal ganglion cell axons by RAGS, a 25 kDa tectal protein related to ligands for Eph receptor tyrosine kinases. *Cell* **82**, 359–370 (1995).
85. Sharma, S. C. Retinal projection in a non-visual area after bilateral tectal ablation in goldfish. *Nature* **291**, 66–67 (1981).

86. Willshaw, D. J. & von der Malsburg, C. How patterned neural connections can be set up by self-organization. *Proc R Soc Lond B Biol Sci* **194**, 431–445 (1976).
87. Hope, R. A., Hammond, B. J. & Gaze, R. M. The arrow model: Retinotectal specificity and map formation in the goldfish visual system. *Proc R Soc Lond B Biol Sci* **194**, 447–466 (1976).
88. Laverack, M. S. Book Review of The Formation of Nerve Connections. *Science Progress* **59**, 605–606 (1971).
89. Hubel, D. H. & Wiesel, T. N. Aberrant visual projections in the Siamese cat. *J Physiol* **218**, 33–62 (1971).
90. Dreyfus, H. L. *Alchemy and Artificial Intelligence*. (Santa Monica, CA: Rand Corp, 1965).
91. Dreyfus, H. L. *What Computers Can't Do*. (New York: Harper & Row, 1972).
92. Taube, M. *Computers and Common Sense: The Myth of Thinking Machines*. (New York: Columbia University Press, 1961).
93. Lytinen, S. L. A unification-based integrated natural language processing system. *Comp Math Applic* **23**, 51–73 (1992).
94. Brockman, J. *The Third Culture*. (New York: Simon and Schuster, 1996).
95. Schank, R. Chapter 9: Information is Surprises, in *The Third Culture*. (ed. J. Brockman) (New York: Simon and Schuster, 1996).
96. Sterling, P. & Laughlin, S. *Principles of Neural Design*. (Cambridge, MA: MIT Press, 2015).
97. Austin, J. L. *Sense and Sensibilia*. (Oxford: Oxford University Press, 1962).
98. McCorduck, P. *Machines Who Think*, 2nd ed. (Natick, MA: A. K. Peters, Ltd., 2004).
99. Hutson, M. Has artificial intelligence become alchemy? *Science* **360**, 478 (2018).

Section 2 (Second Discussion and Seminar)

1. LeCun, Y., Bengio, Y. & Hinton, G. Deep learning. *Nature* **521**, 436–444 (2015).
2. Wolfram, S. *A New Kind of Science*. (Champaign, IL: Wolfram Media, 2002).
3. Adams, D. N. *So Long, and Thanks for All the Fish*. (London: Pan Books, 1984).
4. Berlekamp, E. R., Conway, J. H. & Guy, R. K. *Winning Ways for your Mathematical Plays*. (London: Academic Press, 1982).
5. Wolfram, S. Undecidability and intractability in theoretical physics. *Phys Rev Lett* **54**, 735–738 (1985).
6. Packard, N. H. & Wolfram, S. Two-dimensional cellular automata. *J Stat Phys* **38**, 901–946 (1985).
7. Prusinkiewicz, P. & Lindenmayer, A. *The Algorithmic Beauty of Plants*. (New York: Springer-Verlag, 1990).
8. Kolmogorov, A. N. Three approaches to the quantitative definition of information. *Internatl J Comp Math* **2**, 157–168 (1968).
9. Solomonoff, R. J. A formal theory of inductive inference. Part I. *Info Contr* **7**, 1–22 (1964).
10. Hiesinger, P. R. & Hassan, B. A. The evolution of variability and robustness in neural development. *Trends Neurosci* **41**, 577–586 (2018).
11. von Neumann, J. The general and logical theory of automata—Lecture at the Hixon Symposium, in *John von Neumann Collected Works* **V**. (ed. A. H. Taub) (Oxford: Pergamon Press, 1948).
12. von Neumann, J. *Theory of self-reproducing automata*. (Urbana: University of Illinois Press, 1967).

13. Gardner, M. Mathematical Games. The fantastic combinations of John Conway's new solitaire game "life." *Sci Am* **223**, 120–123 (1970).
14. Aron, J. First replicating creature spawned in life simulator. *New Sci* (2010).
15. Casti, J. L. & Wolfram, S. A new kind of science. *Nature* **417**, 381–382 (2002).
16. Cook, M. Universality in elementary cellular automata. *Compl Sys* **15**, 1–40 (2004).
17. Solomonoff, R. J. A formal theory of inductive inference. Part II. *Info Cont* **7**, 224–254 (1964).
18. Chaitin, G. J. On the simplicity and speed of programs for computing infinite sets of natural numbers. *JACM* **16**, 407–422 (1969).
19. Bennett, C. H. Logical depth and physical complexity, in *The Universal Turing Machine—a Half-Century Survey*. (ed. R. Herken) 227–257 (Oxford: Oxford University Press, 1988).
20. Shannon, C. E. A mathematical theory of communication. *Bell System Tech J* **27**, 379–423 (1948).
21. Avery, J. S. *Information Theory and Evolution*, 2nd ed. (London: World Scientific Publishing, 2012).
22. Maxwell, J. C. *Theory of Heat*. (London: Longmans, Green, and Co., 1871).
23. Vaquerizas, J. M., Kummerfeld, S. K., Teichmann, S. A. & Luscombe, N. M. A census of human transcription factors: function, expression and evolution. *Nat Rev Genet* **10**, 252–263 (2009).
24. Hamburger, V. The history of the discovery of the nerve growth factor. *J Neurobiol* **24**, 893–897 (1993).
25. Levi-Montalcini, R. Ontogenesis or neuronal nets: The chemoaffinity theory, 1963–1983, in *Brain Circuits and Functions of the Mind. Essays in Honor of Roger W. Sperry*. (ed. C. Trevarthen) 3–16 (Cambridge: Cambridge University Press, 1990).
26. Willshaw, D. J. & von der Malsburg, C. How patterned neural connections can be set up by self-organization. *Proc R Soc Lond B Biol Sci* **194**, 431–445 (1976).

Section 3 (Third Discussion and Seminar)

1. Leder, P. The genetics of antibody diversity. *Sci Am* **246**, 102–115 (1982).
2. Tonegawa, S. Somatic generation of antibody diversity. *Nature* **302**, 575–581 (1983).
3. Kise, Y. & Schmucker, D. Role of self-avoidance in neuronal wiring. *Curr Opin Neurobiol* **23**, 983–989 (2013).
4. Zipursky, S. L. & Sanes, J. R. Chemoaffinity revisited: Dscams, protocadherins, and neural circuit assembly. *Cell* **143**, 343–353 (2010).
5. Schmucker, D. et al. *Drosophila* Dscam is an axon guidance receptor exhibiting extraordinary molecular diversity. *Cell* **101**, 671–684 (2000).
6. Watson, F. L. et al. Extensive diversity of Ig-superfamily proteins in the immune system of insects. *Science* **309**, 1874–1878 (2005).
7. Schmucker, D. & Chen, B. Dscam and DSCAM: Complex genes in simple animals, complex animals yet simple genes. *Genes Dev* **23**, 147–156 (2009).
8. Matthews, B. J. et al. Dendrite self-avoidance is controlled by Dscam. *Cell* **129**, 593–604 (2007).

9. Hughes, M. E. et al. Homophilic Dscam interactions control complex dendrite morphogenesis. *Neuron* **54**, 417–427 (2007).
10. Sanes, J. R. & Zipursky, S. L. Synaptic specificity, recognition molecules, and assembly of neural circuits. *Cell* **181**, 536–556 (2020).
11. de Wit, J. & Ghosh, A. Specification of synaptic connectivity by cell surface interactions. *Nat Rev Neurosci* **17**, 22–35 (2016).
12. Adams, D. N. *The Restaurant at the End of the Universe*. (London: Pan Books, 1980).
13. Shatz, C. J. Emergence of order in visual system development. *Proc Natl Acad Sci* **93**, 602–608 (1996).
14. Hassan, B. A. & Hiesinger, P. R. Beyond molecular codes: Simple rules to wire complex brains. *Cell* **163**, 285–291 (2015).
15. Hazelbauer, G. L. Bacterial chemotaxis: The early years of molecular studies. *Annu Rev Microbiol* **66**, 285–303 (2012).
16. Crevier, D. *AI: The Tumultuous History of the Search for Artificial Intelligence*. (New York: Basic Books, 1993).
17. Barkai, N. & Leibler, S. Robustness in simple biochemical networks. *Nature* **387**, 913–917 (1997).
18. Alon, U., Surette, M. G., Barkai, N. & Leibler, S. Robustness in bacterial chemotaxis. *Nature* **397**, 168–171 (1999).
19. Tindall, M. J., Porter, S. L., Maini, P. K., Gaglia, G. & Armitage, J. P. Overview of mathematical approaches used to model bacterial chemotaxis I: The single cell. *Bull Math Biol* **70**, 1525–1569 (2008).
20. Hassan, B. A. et al. atonal regulates neurite arborization but does not act as a proneural gene in the *Drosophila* brain. *Neuron* **25**, 549–561 (2000).
21. Langen, M. et al. Mutual inhibition among postmitotic neurons regulates robustness of brain wiring in *Drosophila*. *Elife* **2**, e00337 (2013).
22. Linneweber, G. A. et al. A neurodevelopmental origin of behavioral individuality in the *Drosophila* visual system. *Science* **367**, 1112–1119 (2020).
23. Bray, S. J. Notch signalling in context. *Nat Rev Mol Cell Biol* **17**, 722–735 (2016).
24. Collier, J. R., Monk, N. A., Maini, P. K. & Lewis, J. H. Pattern formation by lateral inhibition with feedback: A mathematical model of delta-notch intercellular signalling. *J Theor Biol* **183**, 429–446 (1996).
25. Agi, E., Kulkarni, A. & Hiesinger, P. R. Neuronal strategies for meeting the right partner during brain wiring. *Curr Opin Neurobiol* **63**, 1–8 (2020).
26. Meijers, R. et al. Structural basis of Dscam isoform specificity. *Nature* **449**, 487–491 (2007).
27. Sharma, S. C. Retinal projection in a non-visual area after bilateral tectal ablation in goldfish. *Nature* **291**, 66–67 (1981).
28. Clements, J., Lu, Z., Gehring, W. J., Meinertzhagen, I. A. & Callaerts, P. Central projections of photoreceptor axons originating from ectopic eyes in *Drosophila*. *Proc Natl Acad Sci* **105**, 8968–8973 (2008).
29. Van der Loos, H. & Glaser, E. M. Autapses in neocortex cerebri: Synapses between a pyramidal cell's axon and its own dendrites. *Brain Res* **48**, 355–360 (1972).
30. Bekkers, J. M. & Stevens, C. F. Excitatory and inhibitory autaptic currents in isolated hippocampal neurons maintained in cell culture. *Proc Natl Acad Sci* **88**, 7834–7838 (1991).

31. White, J. G. Neuronal Connectivity in *Caenorhabditis elegans*. *Trends Neurosci* **8**, 277–283 (1985).
32. Krishnaswamy, A., Yamagata, M., Duan, X., Hong, Y. K. & Sanes, J. R. Sidekick 2 directs formation of a retinal circuit that detects differential motion. *Nature* **524**, 466–470 (2015).
33. Katz, L. C. & Shatz, C. J. Synaptic activity and the construction of cortical circuits. *Science* **274**, 1133–1138 (1996).
34. Kolodkin, A. L. & Hiesinger, P. R. Wiring visual systems: Common and divergent mechanisms and principles. *Curr Opin Neurobiol* **42**, 128–135 (2017).
35. Schmidt, J. T. *Self-Organizing Neural Maps: The Retinotectal Map and Mechanisms of Neural Development*. (London: Academic Press, 2020).
36. Kirschfeld, K. Die Projektion der optischen Umwelt auf das Raster der Rhabdomere im Komplexauge von Musca. *Exp Brain Res* **3**, 248–270 (1967).
37. Braitenberg, V. Patterns of projection in the visual system of the fly. I. Retina-lamina projections. *Exp Brain Res* **3**, 271–298 (1967).
38. Wiener, N. *Cybernetics: Or Control and Communication in the Animal and the Machine*. (Cambridge, MA: MIT Press, 1948).
39. Agi, E. et al. The evolution and development of neural superposition. *J Neurogenet* **28**, 216–232 (2014).
40. Clandinin, T. R. & Zipursky, S. L. Making connections in the fly visual system. *Neuron* **35**, 827–841 (2002).
41. Clandinin, T. R. & Zipursky, S. L. Afferent growth cone interactions control synaptic specificity in the *Drosophila* visual system. *Neuron* **28**, 427–436 (2000).
42. Hiesinger, P. R. et al. Activity-independent prespecification of synaptic partners in the visual map of *Drosophila*. *Curr Biol* **16**, 1835–1843 (2006).
43. Edwards, T. N. & Meinertzhagen, I. A. Photoreceptor neurons find new synaptic targets when misdirected by overexpressing runt in *Drosophila*. *J Neurosci* **29**, 828–841 (2009).
44. Langen, M. et al. The developmental rules of neural superposition in *Drosophila*. *Cell* **162**, 120–133 (2015).
45. Xu, C. et al. Control of synaptic specificity by limiting promiscuous synapse formation. *bioRxiv*, 415695 (2018).
46. He, H. H. et al. Cell-intrinsic requirement of Dscam1 isoform diversity for axon collateral formation. *Science* **344**, 1182–1186 (2014).
47. Lashley, K. S. Structural variation in the nervous system in relation to behavior. *Psychol Rev* **54**, 325–334 (1947).
48. Ayroles, J. F. et al. Behavioral idiosyncrasy reveals genetic control of phenotypic variability. *Proc Natl Acad Sci* **112**, 6706–6711 (2015).
49. Hiesinger, P. R. & Hassan, B. A. The evolution of variability and robustness in neural development. *Trends Neurosci* **41**, 577–586 (2018).
50. Kain, J. S. et al. Variability in thermal and phototactic preferences in *Drosophila* may reflect an adaptive bet-hedging strategy. *Evolution* **69**, 3171–3185 (2015).
51. Hong, W., Mosca, T. J. & Luo, L. Teneurins instruct synaptic partner matching in an olfactory map. *Nature* **484**, 201–207 (2012).
52. Finch, C. E. & Kirkwood, T. B. L. *Chance, Development and Aging*. (Oxford: Oxford University Press, 2000).
53. Mitchell, K. J. *Innate*. (Princeton, NJ: Princeton University Press, 2018).

54. Clarke, P. G. The limits of brain determinacy. *Proc Biol Sci* **279**, 1665–1674 (2012).
55. Lewontin, R. C. Random factors: An interview with 2017 Thomas Hunt Morgan Medal recipient Richard C. Lewontin. *Genetics* **207**, 1213–1214 (2017).
56. Reed, T., Sprague, F. R., Kang, K. W., Nance, W. E. & Christian, J. C. Genetic analysis of dermatoglyphic patterns in twins. *Hum Hered* **25**, 263–275 (1975).
57. Skandalis, A. Estimation of the minimum mRNA splicing error rate in vertebrates. *Mutat Res-Fund Mol M* **784**, 34–38 (2016).
58. Imamura, A. et al. Genetic and environmental factors of schizophrenia and autism spectrum disorder: Insights from twin studies. *J Neural Transm* (Vienna) (2020).

Section 4 (Fourth Discussion and Seminar)

1. Edelman, G. M. Group selection and phasic reentrant signaling: A theory of higher brain function, in *The Mindful Brain* (Cambridge, MA: MIT Press, 1978).
2. Dawkins, R. *The Blind Watchmaker*. (New York: Norton & Company, 1986).
3. Adams, D. N. *Dirk Gently's Holistic Detective Agency*. (London: William Heinemann, 1987).
4. Drexler, K. E. *Engines of creation. The coming era of nanotechnology*. (New York: Doubleday, 1986).
5. White, J. G., Southgate, E., Thomson, J. N. & Brenner, S. The structure of the nervous system of the nematode *Caenorhabditis elegans*. *Philos Trans R Soc Lond B Biol Sci* **314**, 1–340 (1986).
6. Sulston, J. E. Post-embryonic development in the ventral cord of *Caenorhabditis elegans*. *Philos Trans R Soc Lond B Biol Sci* **275**, 287–297 (1976).
7. Weltfussball.de.
8. Kise, Y. & Schmucker, D. Role of self-avoidance in neuronal wiring. *Curr Opin Neurobiol* **23**, 983–989 (2013).
9. Kim, J. H., Wang, X., Coolon, R. & Ye, B. Dscam expression levels determine presynaptic arbor sizes in drosophila sensory neurons. *Neuron* **78**, 827–838 (2013).
10. He, H. H. et al. Cell-intrinsic requirement of Dscam1 isoform diversity for axon collateral formation. *Science* **344**, 1182–1186 (2014).
11. Dascenco, D. et al. Slit and receptor tyrosine phosphatase 69D confer spatial specificity to axon branching via Dscam1. *Cell* **162**, 1140–1154 (2015).
12. Vaughn, J. E., Henrikson, C. K. & Grieshaber, J. A. A quantitative study of synapses on motor neuron dendritic growth cones in developing mouse spinal cord. *J Cell Biol* **60**, 664–672 (1974).
13. Vaughn, J. E. Fine structure of synaptogenesis in the vertebrate central nervous system. *Synapse* **3**, 255–285 (1989).
14. Niell, C. M., Meyer, M. P. & Smith, S. J. In vivo imaging of synapse formation on a growing dendritic arbor. *Nat Neurosci* **7**, 254–260 (2004).
15. Minsky, M. *Society of Mind*. (New York: Simon and Schuster, 1988).
16. von Foerster, H. On self-organizing systems and their environments, in *Self-Organizing Systems* 31–50 (University of Illinois at Urbana-Champaign, 1960).
17. Grassé, P. P. La reconstruction du nid et les coordinations interindividuelles chez *Bellicositermes natalensis* et *Cubitermes sp.* la théorie de la stigmergie: Essai d'interprétation du comportement des termites constructeurs. *Insectes Sociaux* **6**, 41–80 (1959).

18. Grassé, P. P. The automatic regulations of collective behavior of social insect and "stigmergy." *J Psychol Norm Pathol (Paris)* **57**, 1–10 (1960).
19. Heylighen, F. Stigmergy as a generic mechanism for coordination: Definition, varieties and aspects. *ECCO Working Paper* (2011–12).
20. Ashby, W. R. Principles of the self-organizing dynamic system. *J Gen Psychol* **37**, 125–128 (1947).
21. Wiener, N. *Cybernetics: Or Control and Communication in the Animal and the Machine.* (Cambridge, MA: MIT Press, 1948).
22. Ashby, W. R. *Design for a Brain.* (London: Chapman & Hall, 1952).

Section 5 (Fifth Discussion and Seminar)

1. Maturana, H. R. & Varela, F. J. *Autopoiesis and Cognition: the Realization of the Living,* 2nd ed. (Dordrecht, Holland: D. Reidel Publishing, 1980).
2. Adams, D. N. *The Hitchhiker's Guide to the Galaxy.* (London: Pan Books, 1979).
3. Ashby, W. R. *Design for a Brain.* (London: Chapman & Hall, 1952).
4. Weiner, J. *Time, Love, Memory.* (New York: Random House, 1999).
5. Ranganathan, R. Putting evolution to work. *Cell* **175**, 1449–1451 (2018).
6. Hughes, D. P. et al. From So simple a beginning: The evolution of behavioral manipulation by fungi. *Adv Genet* **94**, 437–469 (2016).
7. Fredericksen, M. A. et al. Three-dimensional visualization and a deep-learning model reveal complex fungal parasite networks in behaviorally manipulated ants. *Proc Natl Acad Sci* **114**, 12590–12595 (2017).
8. Elya, C. et al. Robust manipulation of the behavior of *Drosophila melanogaster* by a fungal pathogen in the laboratory. *Elife* **7** (2018).
9. Takemura, S. Y. et al. A connectome of a learning and memory center in the adult *Drosophila* brain. *Elife* **6** (2017).
10. Takemura, S. Y. et al. The comprehensive connectome of a neural substrate for "ON" motion detection in *Drosophila. Elife* **6** (2017).
11. Scheffer, L. K. et al. A connectome and analysis of the adult *Drosophila* central brain. *Elife* **9** (2020).
12. Mouritsen, H. et al. An experimental displacement and over 50 years of tag-recoveries show that monarch butterflies are not true navigators. *Proc Natl Acad Sci* **110**, 7348–7353 (2013).
13. Hershey, A. D. & Chase, M. Independent functions of viral protein and nucleic acid in growth of bacteriophage. *J Gen Physiol* **36**, 39–56 (1952).
14. Watson, J. D. & Crick, F. H. Molecular structure of nucleic acids: A structure for deoxyribose nucleic acid. *Nature* **171**, 737–738 (1953).
15. Sturtevant, A. H. A third group of linked genes in *Drosophila ampelophila. Science* **37**, 990–992 (1913).
16. Morgan, T. H. Chromosomes and associative inheritance. *Science* **34**, 636–638 (1911).
17. Morgan, T. H. Sex limited inheritance in *Drosophila. Science* **32**, 120–122 (1910).
18. Sperry, R. W. Regulative factors in the orderly growth of neural circuits. *Growth* **10**, 63–87 (1951).

19. Benzer, S. From the gene to behavior. *JAMA* **218**, 1015–1022 (1971).
20. Benzer, S. Behavioral mutants of *Drosophila* isolated by countercurrent distribution. *Proc Natl Acad Sci* **58**, 1112–1119 (1967).
21. Konopka, R. J. & Benzer, S. Clock mutants of *Drosophila melanogaster*. *Proc Natl Acad Sci* **68**, 2112–2116 (1971).
22. Siwicki, K. K., Hardin, P. E. & Price, J. L. Reflections on contributing to "big discoveries" about the fly clock: Our fortunate paths as post-docs with 2017 Nobel laureates Jeff Hall, Michael Rosbash, and Mike Young. *Neurobiol Sleep Circadian Rhythms* **5**, 58–67 (2018).
23. Rosbash, M. Molecular control of circadian rhythms. *Curr Opin Genet Dev* **5**, 662–668 (1995).
24. Sehgal, A. et al. Rhythmic expression of timeless: A basis for promoting circadian cycles in period gene autoregulation. *Science* **270**, 808–810 (1995).
25. Kaneko, M., Helfrich-Forster, C. & Hall, J. C. Spatial and temporal expression of the period and timeless genes in the developing nervous system of *Drosophila*: Newly identified pacemaker candidates and novel features of clock gene product cycling. *J Neurosci* **17**, 6745–6760 (1997).
26. Kobayashi, Y., Ye, Z. & Hensch, T. K. Clock genes control cortical critical period timing. *Neuron* **86**, 264–275 (2015).
27. Gill, K. S. A mutation causing abnormal mating behavior. *Drosophila Information Service* **38**, 33 (1963).
28. Gailey, D. A. & Hall, J. C. Behavior and cytogenetics of fruitless in *Drosophila melanogaster*: Different courtship defects caused by separate, closely linked lesions. *Genetics* **121**, 773–785 (1989).
29. Demir, E. & Dickson, B. J. fruitless splicing specifies male courtship behavior in *Drosophila*. *Cell* **121**, 785–794 (2005).
30. Manoli, D. S. et al. Male-specific fruitless specifies the neural substrates of *Drosophila* courtship behaviour. *Nature* **436**, 395–400 (2005).
31. Halder, G., Callaerts, P. & Gehring, W. J. Induction of ectopic eyes by targeted expression of the eyeless gene in *Drosophila*. *Science* **267**, 1788–1792 (1995).
32. Vernes, S. C. Genome wide identification of fruitless targets suggests a role in upregulating genes important for neural circuit formation. *Sci Rep* **4**, 4412 (2014).
33. Nüsslein-Volhard, C. *Coming to Life*. (Carlsbad, CA: Kales Press, 2006).
34. Wolfram, S. *A New Kind of Science*. (Champaign, IL: Wolfram Media, 2002).
35. Hiesinger, P. R. & Hassan, B. A. Genetics in the age of systems biology. *Cell* **123**, 1173–1174 (2005).
36. Morgan, T. H. The origin of five mutations in eye color in *Drosophila* and their modes of inheritance. *Science* **33**, 534–537 (1911).
37. Hiesinger, P. R. et al. Activity-independent prespecification of synaptic partners in the visual map of *Drosophila*. *Curr Biol* **16**, 1835–1843 (2006).
38. Zhai, R. G. et al. Mapping *Drosophila* mutations with molecularly defined P element insertions. *Proc Natl Acad Sci* **100**, 10860–10865 (2003).
39. Brose, N., Petrenko, A. G., Sudhof, T. C. & Jahn, R. Synaptotagmin: A calcium sensor on the synaptic vesicle surface. *Science* **256**, 1021–1025 (1992).

40. Petrenko, A. G. et al. Binding of synaptotagmin to the alpha-latrotoxin receptor implicates both in synaptic vesicle exocytosis. *Nature* **353**, 65–68 (1991).
41. Littleton, J. T., Stern, M., Schulze, K., Perin, M. & Bellen, H. J. Mutational analysis of *Drosophila* synaptotagmin demonstrates its essential role in Ca(2+)-activated neurotransmitter release. *Cell* **74**, 1125–1134 (1993).
42. Rothenfluh, A. et al. Distinct behavioral responses to ethanol are regulated by alternate RhoGAP18B isoforms. *Cell* **127**, 199–211 (2006).
43. Corl, A. B. et al. Happyhour, a Ste20 family kinase, implicates EGFR signaling in ethanol-induced behaviors. *Cell* **137**, 949–960 (2009).

Section 6 (Sixth Discussion and Seminar)

1. Wolfram, S. *A New Kind of Science.* (Champaign, IL: Wolfram Media, 2002).
2. Sperry, R. W. Chemoaffinity in the orderly growth of nerve fiber patterns and connections. *Proceedings of the National Academy of Sciences* **50**, 703–710 (1963).
3. Meyer, R. L. & Sperry, R. W. Retinotectal specificity: Chemoaffinity theory, in *Neural and Behavioral Specificity. Studies on the Development and Behavior and the Nervous System.* **3** 111–149 (London: Academic Press, 1976).
4. Gaze, R. M. & Keating, M. J. The visual system and "neuronal specificity." *Nature* **237**, 375–378 (1972).
5. Adams, D. N. *So Long, and Thanks for All the Fish.* (London: Pan Books, 1984).
6. Bonhoeffer, F. & Huf, J. Position-dependent properties of retinal axons and their growth cones. *Nature* **315**, 409–410 (1985).
7. Bonhoeffer, F. & Huf, J. Recognition of cell types by axonal growth cones in vitro. *Nature* **288**, 162–164 (1980).
8. Henke-Fahle, S. & Bonhoeffer, F. Inhibition of axonal growth by a monoclonal antibody. *Nature* **303**, 65–67 (1983).
9. Rathjen, F. G., Wolff, J. M., Chang, S., Bonhoeffer, F. & Raper, J. A. Neurofascin: A novel chick cell-surface glycoprotein involved in neurite-neurite interactions. *Cell* **51**, 841–849 (1987).
10. Drescher, U. et al. In vitro guidance of retinal ganglion cell axons by RAGS, a 25 kDa tectal protein related to ligands for Eph receptor tyrosine kinases. *Cell* **82**, 359–370 (1995).
11. Cheng, H. J., Nakamoto, M., Bergemann, A. D. & Flanagan, J. G. Complementary gradients in expression and binding of ELF-1 and Mek4 in development of the topographic retinotectal projection map. *Cell* **82**, 371–381 (1995).
12. Cheng, H. J. & Flanagan, J. G. Identification and cloning of ELF-1, a developmentally expressed ligand for the Mek4 and Sek receptor tyrosine kinases. *Cell* **79**, 157–168 (1994).
13. Walter, J., Muller, B. & Bonhoeffer, F. Axonal guidance by an avoidance mechanism. *J Physiol (Paris)* **84**, 104–110 (1990).
14. Kania, A. & Klein, R. Mechanisms of ephrin-Eph signalling in development, physiology and disease. *Nat Rev Mol Cell Biol* **17**, 240–256 (2016).
15. Harris, W. A. & Holt, C. E. Dedication to Friedrich Bonhoeffer. *J. Neurobiol.* **56**, 1–2 (2004).
16. Harris, W. A. & Holt, C. E. From tags to RAGS: Chemoaffinity finally has receptors and ligands. *Neuron* **15**, 241–244 (1995).

17. Sperry, R. W. Review of: The Formation of Nerve Connections by R. M. Gaze. *Quarter Rev Biol* **46**, 198 (1971).
18. Yoon, M. G. Neural reconnection between the eye and the brain in goldfish, in *Brain Circuits and Functions of the Mind. Essays in Honor of Roger W. Sperry*. (ed. C. Trevarthen) 86–99 (Cambridge: Cambridge University Press, 1990).
19. Gaze, R. M. *The Formation of Nerve Connections*. (London: Academic Press, 1970).
20. Sperry, R. W. Embryogenesis of behavioral nerve nets, in *Organogenesis*. (eds. R. L. Dehaan & H. Ursprung) (New York: Holt, Rinehart and Winston, 1965).
21. Hassan, B. A. & Hiesinger, P. R. Beyond molecular codes: Simple rules to wire complex brains. *Cell* **163**, 285–291 (2015).
22. Hope, R. A., Hammond, B. J. & Gaze, R. M. The arrow model: Retinotectal specificity and map formation in the goldfish visual system. *Proc R Soc Lond B Biol Sci* **194**, 447–466 (1976).
23. Gaze, R. M. & Sharma, S. C. Axial differences in the reinnervation of the goldfish optic tectum by regenerating optic nerve fibres. *Exp Brain Res* **10**, 171–181 (1970).
24. Schmidt, J. T. *Self-Organizing Neural Maps: The Retinotectal Map and Mechanisms of Neural Development*. (London: Academic Press, 2020).
25. Sudhof, T. C. Synaptic neurexin complexes: A molecular code for the logic of neural circuits. *Cell* **171**, 745–769 (2017).
26. Sanes, J. R. & Yamagata, M. Many paths to synaptic specificity. *Annu Rev Cell Dev Biol* **25**, 161–195 (2009).
27. Goldstein, B. Sydney Brenner on the genetics of *Caenorhabditis elegans*. *Genetics* **204**, 1–2 (2016).
28. Brenner, S., Stretton, A. O. & Kaplan, S. Genetic code: The "nonsense" triplets for chain termination and their suppression. *Nature* **206**, 994–998 (1965).
29. Weiner, J. *Time, Love, Memory*. (New York: Random House, 1999).
30. Friedberg, E. C. *Sydney Brenner—A Biography*. (Cold Spring Harbor, NY: Cold Spring Harbor Laboratory Press, 2010).
31. Brown, A. *In the beginning was the worm. Finding the secrets of life in a tiny hermaphrodite*. (New York: Columbia University Press, 2003).
32. White, J. G. Neuronal connectivity in *Caenorhabditis elegans*. *Trends Neurosci* **8**, 277–283 (1985).
33. White, J. G., Southgate, E., Thomson, J. N. & Brenner, S. The structure of the nervous system of the nematode *Caenorhabditis elegans*. *Philos Trans R Soc Lond B Biol Sci* **314**, 1–340 (1986).
34. Agi, E., Kulkarni, A. & Hiesinger, P. R. Neuronal strategies for meeting the right partner during brain wiring. *Curr Opin Neurobiol* **63**, 1–8 (2020).
35. openworm.org.
36. Sarma, G. P. et al. OpenWorm: Overview and recent advances in integrative biological simulation of *Caenorhabditis elegans*. *Philos Trans R Soc Lond B Biol Sci* **373** (2018).
37. Szigeti, B. et al. OpenWorm: An open-science approach to modeling *Caenorhabditis elegans*. *Front Comput Neurosci* **8**, 137 (2014).
38. Brenner, S. The genetics of *Caenorhabditis elegans*. *Genetics* **77**, 71–94 (1974).
39. Serafini, T. et al. The netrins define a family of axon outgrowth-promoting proteins homologous to *C. elegans* UNC-6. *Cell* **78**, 409–424 (1994).

40. Hedgecock, E. M., Culotti, J. G. & Hall, D. H. The *unc-5*, *unc-6*, and *unc-40* genes guide circumferential migrations of pioneer axons and mesodermal cells on the epidermis in *C. elegans*. *Neuron* **4**, 61–85 (1990).
41. Tessier-Lavigne, M., Placzek, M., Lumsden, A. G., Dodd, J. & Jessell, T. M. Chemotropic guidance of developing axons in the mammalian central nervous system. *Nature* **336**, 775–778 (1988).
42. Tessier-Lavigne, M. Wiring the brain: The logic and molecular mechanisms of axon guidance and regeneration. *Harvey Lect* **98**, 103–143 (2002).
43. Serafini, T. et al. Netrin-1 is required for commissural axon guidance in the developing vertebrate nervous system. *Cell* **87**, 1001–1014 (1996).
44. Brankatschk, M. & Dickson, B. J. Netrins guide *Drosophila* commissural axons at short range. *Nat Neurosci* **9**, 188–194 (2006).
45. Dominici, C. et al. Floor-plate-derived netrin-1 is dispensable for commissural axon guidance. *Nature* **545**, 350–354 (2017).
46. Varadarajan, S. G. et al. Netrin1 produced by neural progenitors, not floor plate cells, is required for axon guidance in the spinal cord. *Neuron* **94**, 790–799 e793 (2017).
47. Wu, Z. et al. Long-range guidance of spinal commissural axons by netrin1 and sonic hedgehog from midline floor plate cells. *Neuron* **101**, 635–647 e634 (2019).
48. Charron, F., Stein, E., Jeong, J., McMahon, A. P. & Tessier-Lavigne, M. The morphogen sonic hedgehog is an axonal chemoattractant that collaborates with netrin-1 in midline axon guidance. *Cell* **113**, 11–23 (2003).
49. Jaworski, A. et al. Operational redundancy in axon guidance through the multifunctional receptor Robo3 and its ligand NELL2. *Science* **350**, 961–965 (2015).
50. de Wit, J. & Ghosh, A. Specification of synaptic connectivity by cell surface interactions. *Nat Rev Neurosci* **17**, 22–35 (2016).

Section 7 (Seventh Discussion and Seminar)

1. Edelman, G. M. *Topobiology. An Introduction to Molecular Embryology*. (Basic Books, New York; 1988).
2. Adams, D. N. *Life, the Universe and Everything*. (London: Pan Books, 1982).
3. Pfeiffer, B. D. et al. Tools for neuroanatomy and neurogenetics in *Drosophila*. *Proc Natl Acad Sci* **105**, 9715–9720 (2008).
4. Friedberg, E. C. *Sydney Brenner—A Biography*. (Cold Spring Harbor, NY: Cold Spring Harbor Laboratory Press, 2010).
5. Bernstein, J. G., Garrity, P. A. & Boyden, E. S. Optogenetics and thermogenetics: Technologies for controlling the activity of targeted cells within intact neural circuits. *Curr Opin Neurobiol* **22**, 61–71 (2012).
6. Ernst, O. P. et al. Photoactivation of channelrhodopsin. *J Biol Chem* **283**, 1637–1643 (2008).
7. Gradinaru, V., Mogri, M., Thompson, K. R., Henderson, J. M. & Deisseroth, K. Optical deconstruction of parkinsonian neural circuitry. *Science* **324**, 354–359 (2009).
8. Han, X., Qian, X., Stern, P., Chuong, A. S. & Boyden, E. S. Informational lesions: Optical perturbation of spike timing and neural synchrony via microbial opsin gene fusions. *Front Mol Neurosci* **2**, 12 (2009).
9. Miesenbock, G. Lighting up the brain. *Sci Am* **299**, 52–59 (2008).

10. Borst, A. & Helmstaedter, M. Common circuit design in fly and mammalian motion vision. *Nat Neurosci* **18**, 1067–1076 (2015).
11. Ashby, W. R. *Design for a Brain*. (London: Chapman & Hall, 1952).
12. von Neumann, J. The general and logical theory of automata—Lecture at the Hixon Symposium, in *John von Neumann Collected Works* **5**. (ed. A. H. Taub) (Oxford: Pergamon Press, 1948).
13. Deisseroth, K. & Hegemann, P. The form and function of channelrhodopsin. *Science* **357** (2017).
14. Krueger, D. et al. Principles and applications of optogenetics in developmental biology. *Devel* **146** (2019).
15. Hartmann, J., Krueger, D. & De Renzis, S. Using optogenetics to tackle systems-level questions of multicellular morphogenesis. *Curr Opin Cell Biol* **66**, 19–27 (2020).
16. Agi, E., Kulkarni, A. & Hiesinger, P. R. Neuronal strategies for meeting the right partner during brain wiring. *Curr Opin Neurobiol* **63**, 1–8 (2020).
17. Adesnik, H. & Naka, A. Cracking the function of layers in the sensory cortex. *Neuron* **100**, 1028–1043 (2018).
18. Kolodkin, A. L. & Hiesinger, P. R. Wiring visual systems: Common and divergent mechanisms and principles. *Curr Opin Neurobiol* **42**, 128–135 (2017).
19. Fischbach, K. F. & Hiesinger, P. R. Optic lobe development. *Adv Exp Med Biol* **628**, 115–136 (2008).
20. Ngo, K. T., Andrade, I. & Hartenstein, V. Spatio-temporal pattern of neuronal differentiation in the *Drosophila* visual system: A user's guide to the dynamic morphology of the developing optic lobe. *Dev Biol* **428**, 1–24 (2017).
21. Schrödinger, E. *What Is Life? The Physical Aspect of the Living Cell*. (Cambridge: Cambridge University Press, 1944).
22. Wilson, E. O. *Naturalist*. (Washington, DC: Island Press, 1994).
23. Weiner, J. *Time, Love, Memory*. (New York: Random House, 1999).
24. Agarwala, K. L., Nakamura, S., Tsutsumi, Y. & Yamakawa, K. Down syndrome cell adhesion molecule Dscam mediates homophilic intercellular adhesion. *Brain Res Mol Brain Res* **79**, 118–126 (2000).
25. Kise, Y. & Schmucker, D. Role of self-avoidance in neuronal wiring. *Curr Opin Neurobiol* **23**, 983–989 (2013).
26. Zipursky, S. L. & Sanes, J. R. Chemoaffinity revisited: Dscams, protocadherins, and neural circuit assembly. *Cell* **143**, 343–353 (2010).
27. He, H. H. et al. Cell-intrinsic requirement of Dscam1 isoform diversity for axon collateral formation. *Science* **344**, 1182–1186 (2014).
28. Kim, J. H., Wang, X., Coolon, R. & Ye, B. Dscam expression levels determine presynaptic arbor sizes in *Drosophila* sensory neurons. *Neuron* **78**, 827–838 (2013).
29. Vigen, T. *Spurious Correlations*. (New York: Hachette Books, 2015).
30. Bush, W. S. & Moore, J. H. Chapter 11: Genome-wide association studies. *PLoS Comput Biol* **8**, e1002822 (2012).
31. Pearson, T. A. & Manolio, T. A. How to interpret a genome-wide association study. *JAMA* **299**, 1335–1344 (2008).
32. Dorgaleleh, S., Naghipoor, K., Barahouie, A., Dastaviz, F. & Oladnabi, M. Molecular and biochemical mechanisms of human iris color: A comprehensive review. *J Cell Physiol* (2020).

33. Ganna, A. et al. Large-scale GWAS reveals insights into the genetic architecture of same-sex sexual behavior. *Science* **365** (2019).
34. Mills, M. C. How do genes affect same-sex behavior? *Science* **365**, 869–870 (2019).
35. Zhan, S. et al. The genetics of monarch butterfly migration and warning colouration. *Nature* **514**, 317–321 (2014).
36. Oberhauser, K. S. Concerns that captive breeding affects the ability of monarch butterflies to migrate. *Nature* **573**, 501–502 (2019).
37. Tenger-Trolander, A., Lu, W., Noyes, M. & Kronforst, M. R. Contemporary loss of migration in monarch butterflies. *Proc Natl Acad Sci* **116**, 14671–14676 (2019).
38. Wu, Z. et al. Long-range guidance of spinal commissural axons by netrin1 and sonic hedgehog from midline floor plate cells. *Neuron* **101**, 635–647 e634 (2019).
39. Kiral, F. R. et al. Autophagy-dependent filopodial kinetics restrict synaptic partner choice during *Drosophila* brain wiring. *Nat Commun* **11**, 1325 (2020).
40. Shatz, C. J. Emergence of order in visual system development. *Proc Natl Acad Sci* **93**, 602–608 (1996).

Section 8 (Eighth Discussion and Seminar)

1. Sarma, G. P. et al. OpenWorm: Overview and recent advances in integrative biological simulation of *Caenorhabditis elegans*. *Philos Trans R Soc Lond B Biol Sci* **373** (2018).
2. Adams, D. N. *The Restaurant at the End of the Universe*. (London: Pan Books, 1980).
3. Solomonoff, R. The discovery of algorithmic probability. *J Comp Sys Sci* **55**, 73–88 (1997).
4. Schwab, I. R. *Evolution's Witness: How Eyes Evolved*. (Oxford: Oxford University Press, 2012).
5. Dawkins, R. *The Blind Watchmaker*. (New York: Norton & Company, 1986).
6. Gould, S. J. *Ontogeny and Phylogeny*. (Cambridge, MA: Belknap Press, 1985).
7. Sanes, J. R. & Zipursky, S. L. Design principles of insect and vertebrate visual systems. *Neuron* **66**, 15–36 (2010).
8. Gould, S. J. & Lewontin, R. C. The spandrels of San Marco and the Panglossian paradigm: A critique of the adaptationist programme. *Proc R Soc Lond B Biol Sci* **205**, 581–598 (1979).
9. Gould, S. J. The exaptive excellence of spandrels as a term and prototype. *Proc Natl Acad Sci* **94**, 10750–10755 (1997).
10. Buss, D. M., Haselton, M. G., Shackelford, T. K., Bleske, A. L. & Wakefield, J. C. Adaptations, exaptations, and spandrels. *Am Psychol* **53**, 533–548 (1998).
11. Kirschfeld, K. Die Projektion der optischen Umwelt auf das Raster der Rhabdomere im Komplexauge von Musca. *Exper Brain Res* **3**, 248–270 (1967).
12. Langen, M. et al. The developmental rules of neural superposition in *Drosophila*. *Cell* **162**, 120–133 (2015).
13. Sterling, P. & Laughlin, S. *Principles of Neural Design*. (Cambridge, MA: MIT Press, 2015).
14. Mountcastle, V. B. Modality and topographic properties of single neurons of cat's somatic sensory cortex. *J Neurophysiol* **20**, 408–434 (1957).
15. Mountcastle, V. B. An organizing principle for cerebral function: The unit model and the distributed system, in *The Mindful Brain*. (eds. G. M. Edelman & V. B. Mountcastle) (Cambridge, MA: MIT Press, 1978).

16. Hubel, D. H. & Wiesel, T. N. Receptive fields of single neurones in the cat's striate cortex. *J Physiol* **148**, 574–591 (1959).
17. Hubel, D. H. & Wiesel, T. N. Receptive fields, binocular interaction and functional architecture in the cat's visual cortex. *J Physiol* **160**, 106–154 (1962).
18. Wiesel, T. N. & Hubel, D. H. Single-cell responses in striate cortex of kittens deprived of vision in one eye. *J Neurophysiol* **26**, 1003–1017 (1963).
19. Crowley, J. C. & Katz, L. C. Development of ocular dominance columns in the absence of retinal input. *Nat Neurosci* **2**, 1125–1130 (1999).
20. Horton, J. C. & Adams, D. L. The cortical column: A structure without a function. *Philos Trans R Soc Lond B Biol Sci* **360**, 837–862 (2005).
21. Nikolaou, N. & Meyer, M. P. Lamination speeds the functional development of visual circuits. *Neuron* **88**, 999–1013 (2015).
22. Hiesinger, P. R. & Hassan, B. A. The evolution of variability and robustness in neural development. *Trends Neurosci* **41**, 577–586 (2018).
23. Hawkins, J., Blakeslee, S. *On Intelligence: How a New Understanding of the Brain Will Lead to the Creation of Truly Intelligent Machines.* (New York: St. Martin's Griffin, 2005).
24. Tonegawa, S., Liu, X., Ramirez, S. & Redondo, R. Memory engram cells have come of age. *Neuron* **87**, 918–931 (2015).
25. Poo, M. M. et al. What is memory? The present state of the engram. *BMC Biol* **14**, 40 (2016).
26. Wang, S. From birth onward, our experience of the world is dominated by the brain's continual conversation with itself, in *Think Tank*. (ed. D. J. Linden) 34–39 (New Haven, CT: Yale University Press, 2018).
27. Freiwald, W., Duchaine, B. & Yovel, G. Face processing systems: From neurons to real-world social perception. *Annu Rev Neurosci* **39**, 325–346 (2016).
28. Edelman, G. M. Group selection and phasic reentrant signaling: A theory of higher brain function, in *The Mindful Brain* (Cambridge, MA: MIT Press, 1978).

Section 9 (Ninth Discussion and Seminar)

1. Silies, M., Gohl, D. M. & Clandinin, T. R. Motion-detecting circuits in flies: Coming into view. *Annu Rev Neurosci* **37**, 307–327 (2014).
2. Borst, A. Neural circuits for motion vision in the fly. *Cold Spring Harb Symp Quant Biol* **79**, 131–139 (2014).
3. Borst, A. & Helmstaedter, M. Common circuit design in fly and mammalian motion vision. *Nat Neurosci* **18**, 1067–1076 (2015).
4. Adams, D. N. *Dirk Gently's Holistic Detective Agency.* (London: William Heinemann, 1987).
5. Hinton, G. in *Brain & Cognitive Science—Fall Colloquium Series.* https://www.youtube.com/watch?v=rTawFwUvnLE (Cambridge, MA: MIT, 2014).
6. Wiener, N. *Cybernetics: Or Control and Communication in the Animal and the Machine.* (Cambridge, MA: MIT Press, 1948).
7. Umpleby, S. Definitions of cybernetics. *The Larry Richards Reader 1997–2007*, 9–11 (2008).
8. Hebb, D. O. *The Organization of Behavior: A Neuropsychological Theory.* (Oxford: Wiley, 1949).

9. Ashby, W. R. Principles of the self-organizing dynamic system. *J Gen Psychol* **37**, 125–128 (1947).
10. Ashby, W. R. *Design for a Brain*. (London: Chapman & Hall, 1952).
11. Rosenblatt, F. The perceptron: A probabilistic model for information storage and organization in the brain. *Psychol Rev* **65**, 386–408 (1958).
12. Aggarwal, C. C. *Neural Networks and Deep Learning: A Textbook*. (New York: Springer, 2018).
13. Alpaydin, E. *Machine Learning*. (Cambridge, MA: MIT Press, 2016).
14. Fleming, R. W. & Storrs, K. R. Learning to see stuff. *Curr Opin Behav Sci* **30**, 100–108 (2019).
15. Haykin, S. *Neural networks and learning machines*, 3rd ed. (Upper Saddle River, NJ: Pearson Prentice Hill, 2008).
16. Curry, H. B. The method of steepest descent for non-linear minimization problems. *Quart Appl Math* **2**, 258–261 (1944).
17. Schmidhuber, J. Deep learning in neural networks: An overview. *Neural Networks* **61**, 85–117 (2015).
18. Mandic, D. P. & Chambers, J. A. *Recurrent Neural Networks for Prediction: Learning Algorithms, Architectures and Stability*. (New York: John Wiley & Sons, 2001).
19. Goodfellow, I., Bengio, Y. & Courville, A. *Deep Learning*. (Cambridge, MA: MIT Press, 2016).
20. von Foerster, H. On self-organizing systems and their environments, in *Self-Organizing Systems* 31–50 (University of Illinois at Urbana-Champaign, 1960).
21. Camazine, S. et al. *Self-Organization in Biological Systems*. (Princeton, NJ: Princeton University Press, 2001).
22. Bray, S. J. Notch signalling in context. *Nat Rev Mol Cell Biol* **17**, 722–735 (2016).
23. Collier, J. R., Monk, N. A., Maini, P. K. & Lewis, J. H. Pattern formation by lateral inhibition with feedback: A mathematical model of Delta-Notch intercellular signalling. *J Theor Biol* **183**, 429–446 (1996).
24. Halley, J. D. & Winkler, D. A. Consistent concepts of self-organization and self-assembly. *Complexity* **14**, 10–17 (2008).
25. Khan, G. M., Miller, J. F. & Halliday, D. M. Breaking the synaptic dogma: Evolving a neuro-inspired developmental network, in *Simulated Evolution and Learning, 7th International Conference, SEAL 2008* (ed. X. Li), *LNCS* **5361**, 11–20 (Berlin: Springer-Verlag, 2013).
26. Khan, M. M., Ahmad, A. M., Khan, G. M. & Miller, J. F. Fast learning neural networks using Cartesian genetic programming. *Neurocomputing* **121**, 274–289 (2013).
27. Hintze, A., Hiesinger, P. R. & Schossau, J. Developmental neuronal networks as models to study the evolution of biological intelligence, in Artificial Life Conference, 2020 Workshop on Developmental Neural Networks. https://www.irit.fr/devonn/2020/07/13/hintze.html (2020).
28. McDonald, R. B. *Biology of Aging*, 2nd ed. (Philadelphia, PA: Taylor & Francis, 2019).
29. Krizhevsky, A., Sutskever, I. & Hinton, G. E. ImageNet classification with deep convolutional neural networks. *Comm ACM* **60**, 84–90 (2017).
30. Schossau, J., Adami, C. & Hintze, A. Information-theoretic neuro-correlates boost evolution of cognitive systems. *Entropy* **18** (2016).
31. Sheneman, L. & Hintze, A. Evolving autonomous learning in cognitive networks. *Sci Rep* **7**, 16712 (2017).

32. Miller, G. F., Todd, P. M. & Hegde, S. U. Designing neural networks using genetic algorithms. *Proc Third Internatl Conf Gene Algor*, 379–384 (1989).
33. Harding, S. & Banzhaf, W. Artificial development, in *Organic Computing* 201–219 (Spring, Berlin, Heidelberg; 2009).
34. Miller, J. F., Wilson, D. G. & Cussat-Blanc, S. Evolving programs to build artificial neural networks. *Emergence Complex Comp* **35**, 23–71 (2020).
35. Balaam, A. Developmental neural networks for agents. *Lect Notes Artif Int* **2801**, 154–163 (2003).
36. Cangelosi, A., Parisi, D. & Nolfi, S. Cell-division and migration in a genotype for neural networks. *Network-Comp Neural* **5**, 497–515 (1994).
37. Miller, J. F. Evolving developmental neural networks to solve multiple problems, in *Artificial Life Conference Proceedings*, 473–482 (Cambridge, MA: MIT Press, 2020).
38. Crevier, D. *AI: The Tumultuous History of the Search for Artificial Intelligence*. (Basic Books, New York, 1993).
39. Kurzweil, R. *The Age of Intelligent Machines*. (Cambridge, MA: MIT Press, 1990).
40. Legg, S. & Hutter, M. Tests of machine intelligence. *50 Years of Artificial Intelligence* **4850**, 232–242 (2007).
41. Gottfredson, L. S. Mainstream science on intelligence: An editorial with 52 signatories, history, and bibliography (Reprinted from *The Wall Street Journal*, 1994). *Intelligence* **24**, 13–23 (1997).
42. Legg, S. & Hutter, M. Universal intelligence: A definition of machine intelligence. *Mind Mach* **17**, 391–444 (2007).
43. Hawkins, J., Blakeslee, Sandra *On Intelligence: How a New Understanding of the Brain Will Lead to the Creation of Truly Intelligent Machines*. (New York: St. Martin's Griffin, 2005).
44. Arena, P., Cali, M., Patane, L., Portera, A. & Strauss, R. Modelling the insect mushroom bodies: Application to sequence learning. *Neural Netw* **67**, 37–53 (2015).
45. Cognigni, P., Felsenberg, J. & Waddell, S. Do the right thing: Neural network mechanisms of memory formation, expression and update in *Drosophila*. *Curr Opin Neurobiol* **49**, 51–58 (2018).
46. Minsky, M. & Papert, S. *Perceptrons: an introduction to computational geometry*. (Cambridge, MA: MIT Press, 1969).
47. Garling, C. in *Wired* (2015).
48. Rajamaran, V. John McCarthy—Father of Artificial Intelligence. *Resonance* **19**, 198–207 (2014).
49. McCarthy, J. in *cnet* interview. (ed. J. Skillings) (2006).
50. Moser, M. B., Rowland, D. C. & Moser, E. I. Place cells, grid cells, and memory. *Cold Spring Harb Perspect Biol* **7**, a021808 (2015).
51. O'Keefe, J. A review of the hippocampal place cells. *Prog Neurobiol* **13**, 419–439 (1979).

Section 10 (Tenth Discussion and Seminar)

1. Adams, D. N. *Mostly Harmless*. (London: William Heinemann, 1992).
2. Second-Sight-Medical-Products. https://clinicaltrials.gov/ct2/show/NCT03344848 (ed. C.g. I. NCT03344848) (2017).

3. Hochberg, L. R. in https://clinicaltrials.gov/ct2/show/NCT00912041. (ed. C.g. I. NCT00912041) (2009–2022).
4. Drexler, K. E. *Engines of creation. The coming era of nanotechnology*. (New York: Doubleday, 1986).
5. Regis, E. in *Wired* (2004).
6. Musk, E. Neuralink Launch Event. https://www.youtube.com/watch?v=r-vbh3t7WVI (2019).
7. Gaze, R. M. *The Formation of Nerve Connections*. (London: Academic Press, 1970).
8. Adams, D. N. *The Salmon of Doubt*. (London: William Heinemann, 2002).
9. Crevier, D. *AI: The Tumultuous History of the Search for Artificial Intelligence*. (New York; Basic Books, 1993).
10. Simon, H. *Administrative Behavior. A Study of Decision-Making Processes in Administrative Organizations*. (New York: Macmillan, 1947).
11. Russell, S. J. N., Peter *Artificial Intelligence. A Modern Approach*. (Englewood Cliffs, NJ: Prentice Hall, 1994).
12. Kahneman, D. *Thinking Fast and Slow*. (New York: Farrar, Straus and Giroux, 2011).
13. Tversky, A. & Kahneman, D. Judgment under uncertainty: Heuristics and biases. *Science* **185**, 1124–1131 (1974).
14. Tversky, A. & Kahneman, D. The framing of decisions and the psychology of choice. *Science* **211**, 453–458 (1981).
15. Haselton, M. G., Nettle, D. & Andrews, P. W. The evolution of cognitive bias, in *The Handbook of Evolutionary Psychology*. (ed. D. M. Buss) 724–746 (New York: John Wiley & Sons, 2005).
16. Schank, R. Chapter 9: Information is surprises, in *The Third Culture*. (ed. J. Brockman) (New York: Simon and Schuster, 1996).
17. Menzel, R. & Eckholdt, M. *Die Intelligenz der Bienen*. (Munich: Knaus Verlag, 2016).
18. Gil, M., De Marco, R. J. & Menzel, R. Learning reward expectations in honeybees. *Learn Mem* **14**, 491–496 (2007).
19. Lee, M. H. *How to Grow a Robot*. (MIT Press, 2020).
20. Shermer, M. *Why people believe weird things*. (New York: Henry Holt and Company, 1997).
21. Drexler, K. E. Molecular engineering: An approach to the development of general capabilities for molecular manipulation. *Proc Natl Acad Sci* **78**, 5275–5278 (1981).
22. Bostrom, N. *Superintelligence. Paths, Dangers, Strategies*. (Oxford: Oxford University Press, 2014).
23. Laplace, P. S. *A Philosophical Essay on Probabilities*, trans. from French 6th ed. (New York: John Wiley & Sons, 1902).
24. Azevedo, F. A. et al. Equal numbers of neuronal and nonneuronal cells make the human brain an isometrically scaled-up primate brain. *J Comp Neurol* **513**, 532–541 (2009).
25. Musk, E. The Joe Rogan Experience #1169. https://thejoeroganexperience.net/episode-1169-elon-musk/ (2018).
26. Hochberg, L. R. et al. Neuronal ensemble control of prosthetic devices by a human with tetraplegia. *Nature* **442**, 164–171 (2006).

27. Maynard, E. M., Nordhausen, C. T. & Normann, R. A. The Utah intracortical electrode array: a recording structure for potential brain-computer interfaces. *Electroencephalogr Clin Neurophysiol* **102**, 228–239 (1997).
28. Gilja, V. et al. Clinical translation of a high-performance neural prosthesis. *Nat Med* **21**, 1142–1145 (2015).
29. Nuyujukian, P. et al. Cortical control of a tablet computer by people with paralysis. *PLoS One* **13**, e0204566 (2018).
30. Musk, E. Neuralink progress report 2020. https://www.youtube.com/watch?v=iOWFXqT5MZ4 (2020).

Epilogue

1. Silver, D., et al. A general reinforcement learning algorithm that masters chess, shogi, and Go through self-play. *Science* 362, 1140–1144 (2018).
2. Schrittwieser, J., et al. Mastering Atari, Go, chess and shogi by planning with a learned model. *Nature* 588, 604–609 (2020).
3. Lee, M. H. *How to Grow a Robot.* (MIT Press, 2020).
4. Adams, D. N. *The Restaurant at the End of the Universe.* (London: Pan Books, 1980).

INDEX

Page numbers in *italics* refer to figures.

ablation: of cells or tissues, 70, 188, 239; of target, 68–70, 188
actin cytoskeleton, 183–85
action potential, 225
activity-dependent refinement, 119, *129–30*
activity-dependent plasticity, 57, 119
Adami, Chris, 227
Adams, Douglas, 35, 86, 117, 141, 163, 190, 216, 242, 265, 289, 293, 314
adaptation: evolutionary, 249–51, 260; neuronal, 40–43
address code: and ID tags, 115, 196, 206; and relative positioning, 191; rigidity of, 70, 194–97; synaptic, 70, 188, 194–97, 206
adhesion molecule, 117, 127–28, 134. *See also* cell surface receptor
Administrative Behavior (book title), 294. *See also* Simon, Herbert
AGI. *See* artificial general intelligence
aggression, gene for, 174, 226
aging, 274
AI. *See* artificial intelligence
alchemy problem, 78, 310
alcoholism, gene for, 226
alcohol sensitivity, 185
algorithm, definition of, 6
algorithmic function, 185, 257
algorithmic growth, 6, 92. *See also* developmental growth
algorithmic information, 6, *50–51*, 98, 180, 311; and brain-machine-interfaces, 310–11; in contrast to endpoint information, 7–8, 98; and memory storage, 256; and synaptic specification, 212
algorithmic information theory, 6, *50–54*, 87. *See also* algorithmic information
algorithmic probability, 54. *See also* Solomonoff, Ray
Alife. *See* artificial life
AlphaZero, 313
ANN. *See* artificial neural network
alternative splicing. *See* splicing
Amara's law, 35
analogical tradition, 60
apple tree: and algorithmic growth, 87–88, 114–17, *123*; genetic encoding of, 86–87, 101, *123*; and variability, 101, 114, 144
Arbib, Michael, 55
artificial general intelligence, 10, 264–67, 278, 287; and ANN architecture, 79, 288; compared to biological intelligence, 27–32, 264–67, 278; and developmental growth, 305–7
artificial human intelligence, 27–28, 267, 278, 284, 306. *See also* human intelligence
artificial intelligence: and brain development, 304–5; definition of, 120, 278; evolution of, 277; history of, 47, 61–75; human-level, 76, 267; general (*see* artificial general intelligence)

artificial life, 6, 63, 92, 153, 277
artificial neural network: in AI applications, 12, 57, 236; comparison to biological neural network, 30–34, 48, 72, 83, 242, 281, 305–6; training of, 48–51, 243–44, 271–75, 299; first, 52 (*see also* Perceptron; SNARC); growing of, 146, 265
artificial retina, 290
artificial worm intelligence, 266, 306
Ashby, W. Ross, 156, *157*, 268
association study. *See* genome-wide association studies
attractive signal, 90, 120, 194, *204*. *See also* guidance cue
attractor: in self-organizing systems, 268; molecular long range, 234
Austin, J. L., 77
autapses, 14, 118, 127
autism, 138
auto-associative memory, 256
autonomous agent, 16, 140–44, 148–59
autopoiesis, 163
Avery, John Scales, 50
axon: branches of, 150–52; choices of (*see* axon guidance); innervation of, 40–42, 68–71, 145, *152*, 193, 196, *204*
axonal competition, 194
axonal growth, 108, 126, *204*, 220–23
axonal growth cone. *See* growth cone
axon-axon interaction, *132*, 194
axon guidance, 41, 107–8, 158, *204–9*, 220–23. *See also* chemoaffinity theory; debris guidance
axon pathfinding, *105*, 158, 188, 193–96, *204–9*, 220–23. *See also* axon targeting
axon patterning, 126–27, *132*, 151–52. *See also* axon-axon interaction
axon regeneration, 46, 68–70
axon sorting. *See* axon patterning
axon targeting, 148, *205*, 223

backpropagation, 27, 260, 271, 281–84
bacteria, 120
bacterial motor protein complex, 120
Baker, Bruce, 178
bandwidth, 290, 307
bee, 298
behavioral mutant, 18, 175, 177
behavioral trait, 8, 20, 174, 226
Bellen, Hugo, 183
Bennett, Charles, 51, 98
Benzer, Seymour, 167, *172*, *173*, *200*, *201*
Benzer paradox, 176
biased random walk, 121–22. *See also* chemotaxis
big bang, 112, 181
big data, 281
biological neural network: growth of, 45, 59, 68, 85; in a brain, 9, 13, 68; learning in, 27 (*see also* learning); synthesis with ANN, 304–5
bionic eye, 290
binary digit, 21, 49
bit. *See* binary digit
blind spot, 249
Blind Watchmaker, The (book title), 245
blueprint: in engineering, 4, 88–89 (*see also* endpoint information); and the genetic code, 5, 89
BMI. *See* brain-machine interface
Boltzmann's constant, 99
Bonhoeffer, Friedrich, 58, 70, 193
Bostrom, Nick, 304
bottom-up approach, 10, 267. *See also* bottom-up technology
bottom-up technology, 142, 291. *See also* bottom-up approach
brain determinacy, 136; 4, 79, 119 (*see also* development; developmental growth; neuronal development); analogy to rule, 110, 187
brain development: and artificial intelligence, 304–5; continuity with brain function, 273; random processes in, 134–36; and self-organization, 278
brain upload: 304–11

BrainGate, *308*. *See also* brain-machine interface
brain-machine interface, 307–9
Braitenberg, Valentino, 130, 133
branching pattern, 114, 144, *152*
Brenner, Sydney, *173*, 199–*200*, *201–204*, 217
burden of development, 21, 243
bureaucracy, 294
Butler, Samantha, 206

Caenorhabditis elegans, 128, *200–203*
Cajal, Ramon y, 37–42, *100*
calcium sensor, 184–85
capsules (in ANN development), 284
Carnap, Rudolf, 47
Carnot, Nicolas, 99
cartridge (in the fly brain), *251*
CCD sensor, 246
cell adhesion molecule, 117, 127–134. *See also* cell surface receptor
cell autonomous program, of a developing neuron, 155–158, 195
cell division, *103*
cell non-autonomous program, of a developing neuron, 155. *See also* cell autonomous program
cell surface receptor, 106–7, 114–19, 124, 150, 189–191, 212, 235, 254
cells that fire together, wire together, 48, 57–58, 119, *129–30*, 273–74
cell types (differentiation of), *103–4*, 274
cellular automaton, 5–7, 13–14, 94–97, 148
Chaitin, Gregory, 51, 98
chaos, deterministic. *See* deterministic chaos
Chedotal, Alain, 206
chemical tag, *172*, 192–97. *See also* chemoaffine tag
chemical transmitter, 183–84
chemoaffine molecules, 46, 194. *See also* chemoaffinity theory
chemoaffine tag, 68–69, 106, 126. *See also* chemical tag; chemoaffine molecules; chemoaffinity theory
chemoaffinity. *See* chemoaffinity theory
chemoaffinity theory, 46, 68–73, 107, 130, 188–90, 195–99
chemoattractant, 19, 107, *205*
chemoreceptor, 120–21
chemotaxis, 120–22, 241
chess, 51–53, 313
choices: human rational, 294; of a neuron during development, 34, 126, 140–146, *205*, 213
Chomsky, Noam, 75
circadian clock, 177–178. *See also* daily rhythm
circadian rhythm. *See* circadian clock
circuit properties, 226
Clandinin, Thomas R., *132*
Clausius, Rudolf, 99
code: algorithmic, 88–89, 109; genetic (*see* genetic code); molecular address (*see* address code); for a PIN, 29, 258, 266; for a target in brain wiring, *69–70* (*see also* target code)
cofactor, 178
cognitive bias, 77, 297–300
cognitive psychology, 278, 294–95
Cohen, Stanley, 107
columns, in the brain, 253. *See also* cortical column
commissural axon, *205*
common sense, 296
compensation, in neural networks, 218–19
competition: axonal, 194; in self-organization, 271
complexity, 13, 87, *95–98*; irreducible, 245; Kolmogorov (*see* Kolmogorov complexity); of neural networks, 197–98; unpredictable, 181–82
composite instruction, 20, 107–9, 117, *209*, 214–16, *229*
composite property, *209–10*. *See also* composite instruction
compression (data): data, 6, *50–51*; neural map, 68–70, 188

computation: based on von Neumann architecture, 255; in neural network, 255, 284; Turing's theory of, 53
computational evolution, 84, 277–78, 312
computational model, 72, *198*
computational neuroscience, 55
computation time, 51, 98
connectionist view, 60–61, 192, 218, 279–80
connectome, 217. *See also* connectomics
connectomics, *200*, 252. *See also* connectome
contact guidance theory, 41
context: dependence on, 146–148, 150, 169, 206–10, 212–15, 222–29; developmental, in vivo, 158–159; genetic, 178, 185, 228–29, 232–34; of instructive signals, 113, 148, 206–10, 212–15 (*see also* composite instruction)
contrast enhancement (in vision), 246
convergent evolution, 250
convolutional network, 269–270
Conway, John, 86, 92–93, 141
Cook, Matthew, 94
cooperation (in self-organization), 271
cordyceps, 171
correlated neuronal activity, 20, *129*
cortex, 253–57, 259, 270, 280–82, 290, 307–10
cortical column, 253–54, 289
courtship behavior, 178
Cowan, Jack D., 55
Crick, Francis, *173*, 199
crystallins, 250
curiosity, 298
cybernetics, *49*, 63, 130–32, 156, 164, 268–70
cytoskeleton, 185

daily rhythm, 176. *See also* circadian clock
Dartmouth workshop, 51, 61–62, *97*, 294–96
data compression, 6, *50*–51
Dawkins, Richard, 89, 140, 245
de Bivort, Benjamin, *135*–38
debris guidance (of neuronal axons), 70
decentralizated information storage, 29, 257–61, 310
decision-making process, 207, 294–97
deductive logic, 300
deep learning, 270, 280–81. *See also* machine learning
DeepMind, 313
default behavior (of neurons), 158–59
Delbrück, Max, *173*, 224
dendritic self-avoidance, *123*–26
dendritic tree, *100*–101, 116–17, *123*–26, 144, *152*
dependency on context. *See* context
deterministic brain. *See* brain determinacy
deterministic chaos, 85–86
deterministic system, 13, 85–86, *95*
development: burden of, 21, 243; precision of, 34, 131–32; stochastic, 54, 136–38
developmental constraint, 250–52
developmental growth, 178, 212–14, 254–61, 273, 280. *See also* algorithmic growth
developmental history, 21, 243–45, 249–254, 297. *See also* evolutionary history
developmental neurobiology, 8, 195
developmental robotics, 299, 313. *See also* robot
developmental timing, 70, 188
dichotomy: formal logic versus probability, 54, 61; nature versus nature (*see* nature versus nurture dichotomy); neat versus scruffy (*see* scruffy versus neat dich otomy); necessity versus sufficiency, 213; precision versus imprecision, 54; specificity versus plasticity, 46; tidy-looking, 54, 74, 77
Dickson, Barry, 178
differentiation (of cells). *See* cell types
disorder, neurodevelopmental. *See* neurodevelopmental disorder
disspative system, 112
DNA-binding protein, 178. *See also* transcription factor
Dougherty, Ellsworth, 199
Drexler, Eric, 142, 302
Dreyfus, Hubert, 74, 78
Drosophila, 124, 171–74, 182, 217, 233

Dscam (gene name), 115–17, 127, 134, 150, 224–25
dynamic normality, 155, 163

E. coli, 120
Edelman, Gerald, 139, 211, 260–61
Edmonds, Dean, 52–53. *See also* SNARC
educated guess, 296. *See also* heuristics
electrical wiring, 4, 88
electrode, 290, 307–9
electrophysiological recording, *69*, 73, 183
embodied entity, 276
embryonic development, 45, 197, 246–47, 260
emergent phenomena, 268
emergent property, 32
endpoint information, 5–7, 89, 98–101, 106–9, 252
energy landscape, 268
engineering: of ANNs, 273, *279–82*; bottom up, 169, 187, 291, 302 (*see also* nanotechnology); electrical, 202; evolutionary approach, 170; genome, 185; interface with biology, 63–64, 310–11 (*see also* brain-machine interface); of proteins, 170; perspective versus biology, 245, 255, *279–82*; reverse approach, 10; versus self-organization, 267, *279–82*
entropy, *49–54*, 99, 112, 273, 313–14
environmental information: contribution to brain wiring, 4, 36, 85–88, 119; as part of the genetic program, 15, 36, 59, 108, 119, *137*; and the resonant theory, *41–42* (*see also* Resonance Theory); and spontaneous activity, *129*–30 (*see also* activity-dependent fine-tuning); through learning (*see* learning); versus genetic information, 36, 48, 54, 85–88, 136, *137–38* (*see also* nature versus nurture dichotomy)
Eph receptors (gene name), 194–96
ephrins (gene name), 194–96
equilibrium, 112, 268
equipotentiality (Karl Lashley), 44, 60, 253
ethanol sensitivity, 185
evolution, *137*, 171, *247–52* (*see also* evolutionary programming; evolutionary selection); of a behavioral trait, 231–34; predictability of, 151, 181–82, 186
evolutionary algorithms, 63
evolutionary arms race, 171
evolutionary history, burden of, 246–249, 255, 261
evolutionary principle, *123*, *137*, 143, 187, 254, 260–61
evolutionary programming, 171, 181–85, 260–61
evolutionary psychology, 296
evolutionary selection, 101, 151, 245. *See also* evolution; evolutionary principle
exaptation, 250–52
expectation failure, 298
experience (in neural networks), 23, 258, 297–301
expression pattern, genetic, 212
eye-brain mapping, 130–32. *See also* retinotectal projections
eyeless/Pax6 (gene name), 178–79, 234

face recognition, 259–60
factual explosion, 179–81
feedback: in ANNs, 268–75; in artificial life research, 63; between autonomous agents, 146, 155, 222, 270; and compensation, 218; in the context of cybernetics and Norbert Wiener, 63, 164, 268–75; in developmental patterning, *125*, 222; in a feedback loop, 125, 131, 177; between the genome and its products, 5, *103*, 230; and molecular synaptic specification, 192, 197–99, 208–9; and unpredictability of outcomes, 168, 255
feedback loop, *125*, 177
feedforward neural network, 269–71
Feynman, Richard, 37

filopodia: as autonomous agents, 149–55, 221–22, 235; dynamics of, 150, 233–35; first observation of, 41; random exploration of, 139, 151–52, 215–16, 271; selection of, 149, 235; in synaptic partner selection, 215–16, 233–35
fingerprint, 136–37
flagellum, 120–22
Flanagan, John, 193
Fleming, Roland, 270, 280
flexibility: developmental, 68–70, 199, 202; of neural networks, 59–60, 257
floor plate, *204–5*
fly room, 174, 224
formal logic, 9, 47, 54, 61–63, 295
forward genetic screen, 18, 166–68, 182, 203
Franklin, Rosalind, *173*
Friedberg, Errol, *200*
Friedberg, Richard, 63
fruitless (gene name), 178–82, 228, 234
fruity (gene name), 178
functional plasticity, 54, 57, 73
functionalist view, 45

game of life, 86, 92–94
gating, (neural circuit), 186
gay gene, 228–31
Gaze, Raymond Michael, 55–60, 68, 71–77, 130, 191–92, 195–98
gene map, 174,
gene product: developmental function of, 8, *103*, 169, 223, 274; feedback on genome of, *103*, 274
general chemoaffinity, 196
generality, 180, 207
general principle, 180, 207
genetic algorithm, 63
genetic basis: of aggression, 174–76, 226–27; of alcohol sensitivity, 185, 226–31; of empathy, 174–76, 226–31; of intelligence, 226–27; of sexual orientation, 174, 178, 226–31
genetic code, 5–7, 88–90, 108–9, 140; feedback on, *103*; outcome predictability of, 5–7, 230; replication errors of, *137*
genetic determinism, 84, 136–37
genetics, polygenic. *See* polygenic genetics
genetic sensitization. *See* sensitized background
genetic screen. *See* forward genetic screen
genetic sufficiency, 178, 213
genome: in ANN evolution, 277; and behavior, *172*, 228 (*see also* genetic basis); as a blueprint, 89; and encoding of stochastic processes, *137*; feedback with its products, *103*, 274 (*see also* feedback; transcription factor cascade); information content of, 48, 109, 221–22; what is encoded by, 59, 72, 89, 106, 221 (*see also* gene product)
genome-wide association studies, 227–31
Gill, Kubir, 177
Golgi, Camillo, 37–41
Gould, Stephen Jay, 249–51
GPS (global positioning system), 4
gradient: from instructive to permissive, *209*, 213–14i; n neurodevelopment, 71, 189, 194–96
gradient descent, 270–71
grandmother neuron, 258
Grassé, Pierre-Paul, 156
growth cone, 106 (*see also* axon pathfinding; filopodia); and attraction of repulsion, 194, *204* (*see also* axon pathfinding); as autonomous agent, 139–41, 148–52, 154
growth factor: as part of composite instructions, 108, 206; as permissive or instructive signal, 19, 107, 158–59, 206 (*see also* nerve growth factor)
guidance cue, 90–91, 187–88; cell surface molecules as, 190–91; as permissive of instructive signal, 107, 234; roles during brain wiring, 190–91, 234, 305

guidance molecule, 88. *See also* guidance cue
guidance receptor. *See* guidance cue
gut feeling. *See* heuristics
GWAS (genome-wide association study). *See* genome-wide association studies

Haeckel, Ernst, 246–47
half-tectum, in classic regeneration experiments, 68–70
Hall, Jeff, 176, 178
Hamburger, Viktor, 44
hardware infrastructure, and role for neural network implementations, 286
hard-wired neural network, 36, 133
Harrison, Ross, 38–39
Hassan, Bassem, 124
Hawkins, Jeff, *67–68*, 255–56
Hebb, Donald O., 47–48, 52, 268
Hebbian learning rule, 48
Hebb's law. *See* Hebbian learning rule
Hebb synapse, 48, *129*
Heberlein, Ulrike, 185
heterosexuality, genetics of, 228–30. *See also* homosexuality
heuristics, 295–97
higher-order property, *157*, 236
Hinton, Geoffrey, 267, 276, 283–84
Hintze, Arend, 277
Hitchhiker's Guide to the Galaxy, 35, 163
Hodgkin and Huxley, model for action potential, 241
Holland, John, 63
homologous response, as defined by Paul Weiss, 40
homophilic repulsion, 225. *See also* Dscam
homosexuality, gene for, 226, 230. *See also* heterosexuality
honey bee, 298
Hope, Tony, 72
housekeeping genes, 166, 183
Hubel, David, 74, 253
human intelligence, 267, 275–76

identification tag. *See* molecular identification tag
image classification, 276, 286
immune system, comparison with nervous system function and development, 139, 260
immune system, function of, 114–16, 119, *123*
imprecision: in science, 54; and selection, 114
incomplete information, in decision making, 295
information: algorithmic, 5–7, *49–50*, 95–98; endpoint, 98–*102*, *157*; and energy, 99–*100*; environmental, *129*, *137*; genetic, *103*, 108–9, *137*, 175, 226; incomplete (*see* incomplete information); irrelevant, 101–2, 207; missing (*see* information entropy) storage, 29–30, 257 (*see also* memory); unfolding, *95–98*,*103*, 179–81;
information entropy, 50, 99
information problem, 1–3, 38, 68–71, 175, 197
information theory, 49–51
innate behavior, 45, 136
instructive mechanism, 19, 107, 117, 193, 208–10. *See also* composite instruction
instructive signal. *See* instructive mechanism
intelligence, 72, 153; artificial (*see* artificial intelligence); definition of, 278–79; gene for, 174–76, 226; human (*see* human intelligence)
intelligent machines, 304
interface, of brain and machine, 290, 307–10
invariant representation, 256–59
irreducible complexity, 245
iteration, 6–7, *96*

Jacob, François , *173*
Jessell, Thomas M., *204*
judgement heuristic, 296–97

Kahneman, Daniel, 295
Keating, Michael, *198*

Kirschfeld, Kuno, 130–33
Kirschner, Marc, 179
knock-out, *105*, 150, 221, 233
Kolmogorov, Andrey, 51, 98, 268
Kolmogorov complexity, 6, 98
Konopka, Ron, 176–77
Kurzweil, Ray, 278

language processing, 270, 306
Laplace, Pierre-Simon, 304
Laplace's demon, 304
Lashley, Karl S., 44–47, 59–60, 134. *See also* equipotentiality
lateral inhibition, 124–25, 272
layers: in artificial neural networks, 63, 269–71, 281; in the brain, 221–23, *248*, 254
learning: and algorithmic function, 243, 275; of ANNs, 45, *65*, 273, 282; with backpropagation, 27, 281; of biological brains, 48, *137*, 299; of brain-machine interface, 310; deep (*see* deep learning); as continued development, 30, 79, 119, *137*, 265, 280–83; and memory, 48, 274; supervised, 27, 263, 269, 281; unsupervised, 30, 271
learning history, 264
learning mechanism, 48, 275
learning rules, 48. *See also* Hebb, Donald O.
Lee, Mark H., 299
levels problem, 20, 111, 151–53, 165, 215, 222–28
Levi-Montalcini, Rita, 107
Lewis, Edward B., 174, 224
Lewontin, Richard C., 136, 250
Lighthill report, 74
light-sensing cell, 131, 246–49
limbic system, 307
Lindenmayer, Aristid, 87, 101–2
linguistic theory, 75
little King Kong, 170–71
local decision, 140–44, 155
local interaction, 148, 267–71
lock-and-key mechanism, 114, 127
logical depth, 98
Logic Theorist, *62*, 294–95
long-range attraction, 207–8
L-system, 87, 101–12

machine evolution, 63
machine learning, 109. *See also* deep learning
mass action (Karl Lashley), 44
master regulator, 261, 272–73. *See also* eyeless/Pax6
matchmaking, synaptic, 117–18, 126–27, 196, 208
Maturana, Humberto, 163
Maxwell, James Clerk, 99
Maxwell's demon, 99
McCarthy, John, 61–62, 76, 283
McCorduck, Pamela, 78
McCulloch, Warren S., 46–47, 295
McCulloch-Pitts neuron, 47, 53, 60–64, 305
mechanism: general, 207; instructive (*see* instructive mechanism); molecular (*see* molecular mechanism); permissive (*see* permissive mechanism)
medulla (fly brain structure), 221–23
membrane receptor, 179. *See also* cell surface receptor
memory, 242, 256–61 (*see also* learning); computer, 91; invariant, 256; retrieval, 256; storage, 28–30, 256
memory-prediction system, 255, 279–80, 298. *See also* Hawkins, Jeff
Menzel, Randolf, 298
metabolic enzyme, *103*, 166–69, 190–91, 232
metabolism, role for brain wiring, 113, 183, 286
midline crossing, *205*
migration, of monarch butterfly, 2, 31, 172, 230, 260
Miller, George, 52
mind-body problem, 295
minimum: global, 268, 271; local, 271
Minsky, Marvin, 52–55, 59–*62*, 67, 95, 153, 279
missing information, *49–50*, 89, 99, 314

Mitchell, Kevin J., 136
molecular address code. *See* address code
molecular assembler, 291, 302–3
molecular clock, 167, 177
molecular code, 199, 214
molecular dynamics simulation, 303, 306
molecular identification tag, 72, 115–16, 206. *See also* chemoaffinity theory; Sperry molecule
molecular machine, 142, 177, 302
molecular mechanism, 149–50, 195 (*see also* instructive mechanism); and the levels problem, 216, 219–20; relevance of for phenotypic outcome, 164–68, 177–79, 184, 192, 204
molecular nanotechnology. *See* nanotechnology
molecular target code. *See* target code
monarch butterfly migration, 2–3, 31, 172, 230–31
Monod, Jacques, 173
monozygotic twins, 136–38
Morgan, Thomas Hunt, 174, 182
morphological rigidity, 57, 76
motion detection, 264
motion vision. *See* motion detection
motor cortex, 290, 307
motor protein, in bacterial chemotaxis, 120
Mountcastle, Vernon, 253
multilayer perceptron, 269
Musk, Elon, 307–10
mutagen, to induce mutations for genetic screens, 18, 166, 175–78, 203
mutation, 8, 18, 105, 113–15, 167 (*see also* evolutionary programming; single nucleotide polymorphism); and predictability of phenotypic outcome, 85, 167–69, 174–76; random, 132 (*see also* mutagen)
mutation rate, 166, 203
MuZero, 313

nanomachine, 142–43
nanoscale, 302
nanotechnology, 142, 291, 302
Nash, John, 61
natural language, 74–75
nature versus nurture dichotomy, 42, 54, 136, 137
necessity and sufficiency (in genetics), 178, 213, 234
negative feedback loop, 177. *See also* feedback
Nell2 (gene name), 206
nematode, 199. See also *Caenorhabditis elegans*
neocortex. *See* cortex
neocortex column. *See* cortical column
NGF. *See* nerve growth factor
nerve growth factor, 19, 107–8, 158–59, 206
netrin (gene name), 204–9, 212, 234, 305–6
network topology, 36, 252, 273, 280
neural circuit, 216, 217–19, 225
neural Darwinism, 139, 260
neural group selection, 139, 260
neural network: artificial (*see* artificial neural network); biological (*see* biological neural network); compensation in (*see* compensation); complexity of, (*see* complexity); experience in (*see* experience); feedforward (*see* feedforward neural network); flexibility of (*see* flexibility); hardware of (*see* hardware infrastructure); hard-wired (*see* hard-wired neural network); random architecture of (*see* random architecture); random connectivity of (*see* random connectivity); recurrent (*see* recurrent neural network); redundancy in (*see* redundancy); relevance of growth for (*see* relevance); robustness of (*see* robustness); self-organization in (*see* self-organization)
neural superposition, 131–33, 251
neurodevelopment. *See* neuronal development
neurodevelopmental disorder, 138

neurogenetics, 128, *172*, 217
neuromodulator, 24, 33, 240–41
neuron: artificial, 47, 60, *102*, 284 (*see also* McCulloch-Pitts neuron); biological, 1, 12, 37, *100*, 108, 274; in culture, 158
neuronal aging, 274–75
neuronal activity: 15, 59, 128–29, *309*; genetically encoded, 57–59, *129*, *137*; under optogenetic control, 218–19; spontaneous, *129*
neuronal circuit. *See* neural circuit
neuronal connectivity, 13, 195, *248*
neuronal development, 104, *123*–28, 145, 150–59, 195, 221–23, 274 (*see also* algorithmic growth; brain development; branching pattern; development); genetic contribution to, 4, 8, 108, 179; molecular mechanisms in, 104–5, 195
neuronal differentiation: *103*, 146
neuronal excitability, *129*, 226
neuronal identity, 115
neuronal group selection. *See* neural group selection
neuronal growth, 108, *123*–28, 141, 150–59, 221–23. *See also* algorithmic growth; neuronal development
neuronal properties, 226, 236
neuron doctrine, 37–40
neuron types, 42, 197, *248*. *See also* neuronal differentiation
neuropil (synaptic brain region), 221–23
neurotransmitter, 183, 305
Newell, Allen, *62*, 283
noise: in ANNs, 271; as a pool of variation, 14–15, 115, *123*–25, *129*, 134; and randomness, *95*, 114, 136
nonadditive genetic effect, 230
Notch signaling, 124–25, 272
Nüsslein-Volhard, Christiane, 179

one-to-one mapping, 71–72, 89, 196, 212, 227
ontogeny, 246
open system, in thermodynamics, 112
OpenWorm project, 203, 236, 241
optimization function, 271
optimization problem, 294
optogenetics, 218–21
order from noise, 156
output error, in ANNs, 271. *See also* backpropagation

Papert, Seymour, *65*–*67*
parameterization, 225
parameter space: in ANN function, 271; for evolutionary programming, 23, 286
penetrance: and genetic context, *229*–34; phenotypic, 108, 114, 176, 213
Perceptron (first ANN), 63–66, 269, 272. *See also* Rosenblatt, Frank
Perceptrons (book title), *65*, 281
period (gene name), 176–78, 187
period, developmental, 274
permissive mechanism, 19, 107, 117, 193, 206–10, 305
permissive signal. *See* permissive mechanism
phenotypic penetrance. *See* penetrance
phenotypic variability. *See* variability
pheromone, 155–56
phototaxis, 175, 183
phylogeny, 246
Pitts, Walter, 47–48. *See also* McCulloch-Pitts neuron
place cells, 284
plasticity: developmental, 12, 40–44, 54, 199; of network function, 57, 119, 274, 290
pluripotent tissue, 234
polygenic genetics, *229*–30
pool of variation, 14, 114, *123*, *137*, 151, 170
positive feedback loop, *125*
post-specification, synaptic, 15, 130
Poulson, Donald, 178
precision: based on noise, 114, *122*–*26*, *129*; of development, 10, *132*, 302; versus flexibility, 34, 42; of network wiring, 88, *129*, *132*; as a secondary consequence, 252 (*see also* spandrel)

predisposition, genetic, 8, 176, 215–16, 227
pre-specification, synaptic, 15, 130–34
Principia Mathematica, 47, 53–54, *62*, 295
probabilistic biasing, 130, 134,
probability theory, *49*, 54
probability: algorithmic, 54; in biased random walk, 121–22; of phenotypic outcome, 85, 138, 176, 227–29 (*see also* penetrance)
programmed cell death, 145, 272
prokaryote, 120
promiscuous synapse formation, 14–15, 126–28, 130–34
protein design, 187
protein interaction network, 226
proteome, *105*
protocadherins (gene name), 225
protocognition, 120
pruning, of synapses of neuronal branches, 119, 130. *See also* synaptotropic growth
Purkinje cell, *100–101*, *152*

random architecture, of neural networks, 12, 118. *See also* random connectivity
random connectivity, of neural networks, 72, 118, 264, 292
random factors, in genetics, 136
random process: in activity-dependent fine tuning (*see* activity-dependent refinement); during algorithmic growth, 15, 57–59, 115, 134, 186–87; in branching, *123*; in evolution and selection, 14–15, 84, *123*; and flexibility, 82; indistinguishability from unpredictable deterministic growth process, 149; as the opposite of precision, 15; and precise outcomes, 59, 128–29; randomness, definition of: 98; and robustness, 20, 84, 128–29; and variability, 57–59
rational choices, of humans, 294–301
receptive field, 270
receptor protein, *105*. *See also* cell surface receptor
recognition molecule, 188–91. *See also* chemoaffinity theory
recurrent neural network, 77, 270–72, 281
recursive process, 260
reductionism, 110–11
reductionist's dilemma, 223–24
redundancy: genetic, 208, 231–32; in information storage, 29; in neural networks, 218
regeneration experiments, 40, 46, 68, 188
reinforcement learning, 273, 313
relative positioning, during network development, 71–73, 188–91, 194–97, 222
relevance: in evolutionary selection, *136*; of growth for neural networks, *45*, *126*; of information (*see* relevant information)
relevant information, 108
representation: internal, 75, 297; invariant (*see* invariant representation)
representativeness, 296. *See also* generality
repulsive signal, 71, *123–24*, 150, 196, *204*, 225
Resonance Theory, 40–42. *See also* Weiss, Paul Alfred
reticular theory, 38, 83
retina, *129*, 246–48
retinal activity waves, 59, 119, *129*
retinal ganglion axon. *See* retinal ganglion cells
retinal ganglion cells, 130, 193–94, *248*
retinotectal projections, *69*, 72, 127, 195. *See also* eye-brain mapping
robo2 (gene name), 254
robot, 26–34, 141–42, 187, 312–13. *See also* developmental robotics
robustness: of neural network development, 15, 148–49, 154; of neural network function, 55, 60, 257
roof plate, of spinal cord, *204–5*
Rosbash, Michael, 176
Rosenblatt, Frank, 63–67. *See also* Perceptron
Rothenfluh, Adrian, 185
Rubin, Gerald M., 217
rule, definition of, 235

rule 110, 5–7, 94–97, 109–10
rule of thumb, 295. *See also* heuristics

Sanes, Joshua, 128
Schank, Roger, 74–75, 298–99
schizophrenia, 138
Schmucker, Dietmar, 134
Schrödinger, Erwin, *173*, 224
scientific method, 300
scruffy versus neat dichotomy, 75–77, 130, 134, 170, 187
secondary rationalization, 249
second law of thermodynamics, 99
selection, evolutionary. *See* evolutionary selection
self-amplification, 271
self-assembling brain, 109, 181–82, 222–24, 282; and aging, 275
self-assembly, 9, *198*, 222–24; and autonomous agents, 146, 148; definition in this book, 273; and simple rules, 110
self-avoidance: of branching neurons, 116–19, *123*, 150, 225; gene for, 150, 225–26
self-modifying process, 155. *See also* dynamic normality; feedback; transcription factor cascade
self-organization, 155–56, 164–65, 267–69; in neural networks, 270–75
sensitized background, in genetics, 232–33
sensory cortex, 253, 307
sensory system, 120, *248*
sequence: and algorithmic information, 98; in developmental growth, 275; in memory storage (*see* sequential memory)
sequential memory, 244, 256, 259–61, 275, 299
sexual behavior, 178, 227–28
Shannon, Claude, *49–53*, *62*, 99
Shannon entropy. *See* information entropy
Sharma, Sansar, 68–69, 71, 127, 130, *198*
Shatz, Carla, 57–59, 128–29, 236
shortcut: to creating intelligent systems, 3, 236, 272, 275, 301–6; to predicting outcomes of growth processes, 94–98, 106, 227
short-range attraction, *205*
signaling molecule, 120, *125*, 185, 272
Simon, Herbert, *62*, 283, 294–95
simple rules, 110, 235
single cell sequencing, 104
single nucleotide polymorphism (SNP), 227
sliding scale model, 70, 191
Smalley, Richard, 302
Smith, Stephen, *152*
SNARC, first ANN machine, 52–53, 197. *See also* Minsky, Marvin
SNP. *See* single nucleotide polymorphism
soccer, 146–49
Solomonoff, Ray, *50–54*, *62*, *97*, 242
somatosensory cortex, 253, 290, 307
sonic hedgehog (gene name), 206, 234
spandrel, 250–51
spatiotemporal specificity, 15, 206, 212
speech recognition, 283
Sperry, Roger Wolcott, 42–47, 54, 57, 68–77, 83, 174, 188–92; 193–98. *See also* chemoaffinity theory
Sperry molecule, 79, 106, 189. *See also* chemoaffinity theory; guidance cue
spinal cord, 38–39, 90, *204–6*
spinal motor neurons, 151
splicing: alternative, 134, *137*; error rate, *137*
spontaneous activity, *129*. *See also* activity-dependent refinement
squirrel monkey, 253
states, neuronal during algorithmic growth, 274
stigmergy, 156
stimulation (neuronal), 128, 291
stochastic dynamics, *125*, *137*, *152*, 306
stochastic process, *125*, 134–37, 150–52, *229*
strict chemoaffinity, 70, 196–98. *See also* Sperry, Roger Wolcott
Sturtevant, Alfred, 174
Südhof, Thomas C., 183

superintelligence, 304, 312
supervised learning, 27, 263, 269, 281
surprise, and information, 256, 293–94, 298–301
swarm behavior, 140
symbol-processing logic, 3, 9, 63, 153
symmetry-breaking mechanism, *125*
synaptic matchmaking, 117–18, 208
synaptic modulator. *See* neuromodulator
synaptic promiscuity, 126–28, 133
synaptic specification, 14, 126–28, 202, 208
synaptic specificity paradox, 14
synaptic strength, 33, 242, 260
synaptic weight: in ANN, 22, 258, 266, 277, 284–86; in the brain, 110, 280
synaptogenic competency, 15. *See also* composite instruction
synaptotagmin (gene name), 183–84, 187
synaptotropic growth, 151–52, 215–16
syntax, in language and linguistics, 47, 75
systems matching, 68–70, 73, 188, 191, *198*. *See also* Gaze, Raymond Michael

target code, *69–71*, 188
target genes, of a transcription factor, *103*, 179
target specificity. *See* axon pathfinding; synaptic specification
tectum, 68–70
termite behavior, 155–56
Tessier-Lavigne, Marc, *204–6*
tetraplegia, *308–9*
tetrodotoxin, *129*
theory of computation, 53
thermodynamics, second law, *49*, 99, 112. *See also* entropy
time and energy: in algorithmic growth, 2–7, *50*, 92, 98, 109, 148, 181; in ANN development, 76, 271; in self-assembly, 273
timeless (gene name), 177
top-down technology, 142, 267, 273
Topobiology (book title), 211
topographic mapping, 194. *See also* neural superposition
topographic regulation hypothesis, 70
trait: genetic encoding of, 8, 215, 227–30. *See also* behavioral trait
transcription factor, *102–4*, 178–79, 228, 234
transcription factor cascade, *102–4*, 180
transcriptome, 104–5
transplantation experiments, 40–42, 130
Turing, Alan, 51
Turing-complete system, 6, 13, 94, 182
Turing Test, 278
Turing universality. *See* Turing-complete system
Tversky, Amos, 295–97

Ulam, Stanislaw, 92
uncertainty: in decision making, 296; in information theory, *50* (*see also* undecidability; unpredictability)
uncoordinated (unc) mutants, *204*
undecidability, 86, 94, *96*
unfolding of information, 214, 244, 255. *See also* algorithmic growth; rule 110
universal Turing machine, 6, 94–96. *See also* Turing-complete system
universe: and determinism, 14, 95, 182, 304; and increase of entropy, 112–13, 181, 313–14
unpredictability: of developmental outcome, 148, 170, 182; of information unfolding, 186, 255 (*see also* rule 110)
unreliability, of components in biological and artificial systems, 53, 55
unsupervised learning, 30, 271, 281

Varela, Francisco, 163
variability: in the context of nature vs nurture, 42, 54, 57, 136–37; of developmental outcomes (*see* phenotypic *in this entry*); of developmental processes, 124–25, 147; phenotypic, 15, 34, 134–36, 230, 252; pool of (*see* pool of variation)

Vaughn, James, 151–52. *See also* synaptotropic growth
visual cortex, 253, 270, 290
visual perception, 270
von der Malsburg, Christoph, 72–73, 109
von Foerster, Heinz, 156, 271
von Gerlach, Joseph, 38–39
von Neumann, John, 51, 55, 61, 91–92
von Neumann computer architecture, 51–53, 91, 255–58
von Waldeyer-Hartz, Gottfried Heinrich Wilhelm, 37–38

Watson, James, *173*, 199, 224
Weiner, Jonathan, 174
Weiss, Paul Alfred, 40–43
What Is Life? (book title), *173*, 224
White, John, 128, 202
whole brain emulation, 304–6
Wiener, Norbert, *49–50*, 63, 130, 268
Wiesel, Torsten, 74, 253
Wilkins, Maurice, *173*
Willshaw, D. J., 72, 109
Winograd, Shmuel, 55
wiring diagram: algorithmic growth of, 184; biological, 79, 88–89, 106, 197–202, 252, 280; of *c. elegans*, 128, *200–202*; electrical, 88–89, 252; eye-brain, 130; genetic encoding of, 106; information content of, 197–202, 240, 280
Wolfram, Steven, 94–95, 181–82, 186

XOR, perceptron limitation, 281

Young, Michael, 176

zebrafish, 254
zero sum game, 147
Zipursky, S. Lawrence, *132*

A NOTE ON THE TYPE

This book has been composed in Arno, an Old-style serif typeface in the classic Venetian tradition, designed by Robert Slimbach at Adobe.

CPSIA information can be obtained
at www.ICGtesting.com
Printed in the USA
JSHW082350091022
31410JS00002B/2

9 780691 241692